NATIONAL ACADEMIES Sciences Engineering Medicine

NATIONAL ACADEMIES PRESS
Washington, DC

Advancing Chemistry and Quantum Information Science

An Assessment of Research Opportunities at the Interface of Chemistry and Quantum Information Science in the United States

Committee on Identifying Opportunities at the Interface of Chemistry and Quantum Information Science

Board on Chemical Sciences and Technology

Board on Life Sciences

Division on Earth and Life Studies

Consensus Study Report

NATIONAL ACADEMIES PRESS 500 Fifth Street, NW Washington, DC 20001

This activity was supported by contracts 10005349 and 10005332 between the National Academy of Sciences and the Department of Energy Office of Sciences and the National Science Foundation. Any opinions, findings, conclusions, or recommendations expressed in this publication do not necessarily reflect the views of any organization or agency that provided support for the project.

International Standard Book Number-13: 978-0-309-69809-2
International Standard Book Number-10: 0-309-69809-X
Digital Object Identifier: https://doi.org/10.17226/26850
Library of Congress Control Number: 2023944701

This publication is available from the National Academies Press, 500 Fifth Street, NW, Keck 360, Washington, DC 20001; (800) 624-6242 or (202) 334-3313; http://www.nap.edu.

Printed in the United States of America.

Suggested citation: National Academies of Sciences, Engineering, and Medicine. 2023. *Advancing Chemistry and Quantum Information Science: An Assessment of Research Opportunities at the Interface of Chemistry and Quantum Information Science in the United States*. Washington, DC: The National Academies Press. https://doi.org/10.17226/26850.

The **National Academy of Sciences** was established in 1863 by an Act of Congress, signed by President Lincoln, as a private, nongovernmental institution to advise the nation on issues related to science and technology. Members are elected by their peers for outstanding contributions to research. Dr. Marcia McNutt is president.

The **National Academy of Engineering** was established in 1964 under the charter of the National Academy of Sciences to bring the practices of engineering to advising the nation. Members are elected by their peers for extraordinary contributions to engineering. Dr. John L. Anderson is president.

The **National Academy of Medicine** (formerly the Institute of Medicine) was established in 1970 under the charter of the National Academy of Sciences to advise the nation on medical and health issues. Members are elected by their peers for distinguished contributions to medicine and health. Dr. Victor J. Dzau is president.

The three Academies work together as the **National Academies of Sciences, Engineering, and Medicine** to provide independent, objective analysis and advice to the nation and conduct other activities to solve complex problems and inform public policy decisions. The National Academies also encourage education and research, recognize outstanding contributions to knowledge, and increase public understanding in matters of science, engineering, and medicine.

Learn more about the National Academies of Sciences, Engineering, and Medicine at **www.nationalacademies.org**.

COMMITTEE ON IDENTIFYING OPPORTUNITIES AT THE INTERFACE OF CHEMISTRY AND QUANTUM INFORMATION SCIENCE

THEODORE G. GOODSON, III (*Chair*), University of Michigan
DAVID D. AWSCHALOM, University of Chicago
RYAN J. BABBUSH, Google, LLC
LAWRENCE W. CHEUK, Princeton University
SCOTT K. CUSHING, California Institute of Technology
NATIA L. FRANK, University of Nevada, Reno
DANNA E. FREEDMAN, Massachusetts Institute of Technology
SINÉAD M. GRIFFIN, Lawrence Berkeley National Laboratory
STEPHEN O. HILL, Florida State University
HONGBIN LIU, Microsoft Quantum
MARILU PEREZ GARCIA, Ames Laboratory
BRENDA M. RUBENSTEIN, Brown University
ERIC J. SCHELTER, University of Pennsylvania
MICHAEL R. WASIELEWSKI, Northwestern University
DAMIAN WATKINS, Aperio Global

Staff

LINDA NHON, Study Director
ANDREW BREMER, Program Officer
AYANNA LYNCH, Research Assistant
KAYANNA WYMBS, Program Assistant

Staff

KAVITA BERGER, Board Director
NANCY CONNELL, Senior Scientist
STEVEN MOSS, Senior Program Officer
ANDREW BREMER, Program Officer
LYLY LUHACHACK, Program Officer
TRISHA TUCHOLSKI, Associate Program Officer
JESSICA DE MOUY, Research Associate
SABINA VADNAIS, Research Associate
AUDREY THEVENON, Senior Program Officer
DASIA MCKOY, Program Assistant
CHRISTL SAUNDERS, Program Coordinator
NAM VU, Program Assistant
CYNTHIA GETNER, Senior Financial Business Partner

Reviewers

This Consensus Study Report was reviewed in draft form by individuals chosen for their diverse perspectives and technical expertise. The purpose of this independent review is to provide candid and critical comments that will assist the National Academies of Sciences, Engineering, and Medicine in making each published report as sound as possible and to ensure that it meets the institutional standards for quality, objectivity, evidence, and responsiveness to the study charge. The review comments and draft manuscript remain confidential to protect the integrity of the deliberative process.

We thank the following individuals for their review of this report:

KWABENA BEDIAKO, University of California, Berkeley
FLEMING CRIM, University of Wisconsin
GREGORY HARTLAND, University of Notre Dame
RIGOBERTO HERNANDEZ, John Hopkins University
VINCENZO LORDI, Lawrence Livermore National Laboratory
KANG-KUEN NI, Harvard University
GEORGE SCHATZ, Northwestern University
FRANK SCHLAWIN, University of Hamburg
RICHARD WINPENNY, The University of Manchester

Although the reviewers listed above provided many constructive comments and suggestions, they were not asked to endorse the conclusions or recommendations of this report nor did they see the final draft before its release. The review of this report was overseen by **LOUIS BRUS (NAS)**, Columbia University, and **DAVID M. CEPERLEY**, University of Illinois at Urbana-Champaign. They were responsible for making certain that an independent examination of this report was carried out in accordance with the standards of the National Academies and that all review comments were carefully considered. Responsibility for the final content rests entirely with the authoring committee and the National Academies.

Contents

Summary

INTRODUCTION

Quantum information science (QIS) investigates how to exploit quantum behavior and its ability to encode, sense, process, and transmit information. This emerging field has developed ideas that may one day revolutionize such technological areas as communications, sensing, computing, navigation, and measurement. More recently, scientists and engineers have initiated new ideas that will significantly impact chemistry and biology. Quantum technologies follow a divergent set of operational rules beyond classical physics, which implies that these devices may surpass conventional capabilities. For example, developments in QIS could generate a faster and more secure internet, install financial systems backed by unique cryptographic codes and hardware, equip submarines with state-of-the-art surveillance systems, upgrade medical imaging machines, and generate other technologies to unparalleled levels compared to their classical counterparts. Overall, such advancements will impact U.S. economic prosperity, national security, medicine, and global research, development, and innovation competitiveness. This point becomes increasingly clear when one includes the involvement of chemistry approaches to QIS. The chemical industry as well as the development of new methods and materials could undergo a major shift in direction resulting in fruitful achievements if further developments of chemistry approaches to QIS were to be pursued. To achieve this long-term vision and sustain U.S. leadership in science and innovation, a robust QIS research enterprise needs to be established. While the physics community has enabled the field to move closer to creating QIS technologies by shedding light on the fundamental behavior of quantum properties, more details are necessary to bring these concepts into practical applications. The field of QIS is now at an inflection point, where the need for developing and measuring quantum molecular materials that are operational, practical, and efficient is paramount. Because chemistry is the study of manipulating properties and behaviors across different length scales, from subatomic to macromolecular levels, this discipline will certainly play a central role in guiding QIS toward future designs and measurements.

PROMISE OF CHEMISTRY AND QIS FOR TECHNOLOGICAL ADVANCEMENTS

To strengthen the knowledge that will require investigations conducted at the subatomic, atomic, molecular, and macromolecular scales, the Department of Energy (DOE) and National Science Foundation (NSF) requested that the National Academies of Sciences, Engineering, and Medicine convene an ad hoc committee of experts to examine the opportunities at the interface of chemistry and QIS and to provide recommendations on research and other needs to facilitate progress. Box S-1 presents the major themes and points made by the committee that illustrate the important contributions chemistry will continue to make to QIS. These contributions range from laying the fundamental blueprint for designing and characterizing quantum molecular designs and their impact on future technologies to demonstrating the significant role the chemistry enterprise has on QIS workforce development.

BOX S-1
Key Themes of the Report

Theme 1. Chemistry Provides Versatile Approaches for Solving Fundamental Scientific Challenges in QIS (Chapters 2, 3, and 4)—Chemists are developing new approaches to tackle the fundamental scientific limitations of realizing QIS ideas that can be applied to developing novel technologies. For example, creating well-designed molecular quantum bits, or qubits, that can approach the quantum limits will open a new class of memory storage applications that will redefine how information (data) such as large calculations, documents, music, and other forms of data will be collected, saved, and secured. Future molecular-based qubit systems may be designed for use in QIS applications without the requirement of a very low-temperature environment. These molecular systems also have an immense opportunity to confront the fundamental limitation of the quantum property of coherence, decoherence, and transmission of information from classical inputs to quantum systems. Achieving such parameters will alter the fabric of modern life from simply browsing the internet to sensing pollutants in the environment with quantum-enhanced sensitivity.

Theme 2. Chemists Develop Novel Instrumentations and Approaches to Drive QIS Discoveries (Chapters 2, 3, and 4)—Chemists have made great strides in utilizing a quantum concept (e.g., quantum tunneling) to inspire a medical diagnostic tool that can sense magnetic waves generated by the human brain. Through a collaboration between physicians and scientists, this approach was used in the development of a technology known as a superconducting quantum interference device (SQUID), which measures very small magnetic fields. In applications, a SQUID is used in the detection of neural disease for early diagnosis in pediatric patients. This example illustrates that a well-studied quantum concept can be translated into technologies that have a significant impact outside the realm of basic science research. Currently, chemists are seeking to take advantage of other quantum concepts, in particular quantum entanglement and superposition. These concepts are very challenging to translate into larger systems like molecules in comparison to implementing them into smaller systems like atoms. Finding ways to manipulate and control these quantum properties will give rise to unforeseen technological opportunities similar to and possibly more impactful than SQUIDs. Further improvement is needed in both the design and synthesis of novel molecular systems and in the optimization of instrumentation and measurement approaches in QIS at the interface of chemistry. Quantum computing approaches that extend beyond the noisy error-corrected approaches are also being developed by chemists and require improvement in the future. These computing hardware advancements will transform the manufacturing industry, change the speed of drug discovery, improve weather tracking, enhance aviation surveillance, and advance other technologies.

Theme 3. Chemistry Builds and Strengthens the QIS Workforce (Chapters 3 and 5)—Chemists have skills that can be capitalized for a variety of QIS projects and potential employment in companies and laboratories conducting quantum research and development without the need for a Ph.D. Interdisciplinary research in the design and characterization of novel organic and inorganic molecular systems coupled with key instrumentation and measurement tools will increase participation and lower barriers to entry in QIS research across various disciplines, especially in chemistry. Furthermore, drawing from the CHIPS* and Science Act and other initiatives with the chemist's interest in education and training in QIS is a strong avenue to increase participation in the QIS workforce. Presently, the chemical community is considerably diverse, and chemists have an opportunity to leverage this diversity in the future for broader participation.

*CHIPS stands for "Creating Helpful Incentives to Produce Semiconductors."

FUNDAMENTAL RESEARCH AREAS AND PRIORITIES AT THE INTERFACE OF QIS AND CHEMISTRY

Similar to chemistry's impact on QIS, QIS also has the potential to enable extraordinary discoveries in chemistry. The ability to model quantum properties and behaviors of chemical systems on a practical timescale could, for example, facilitate new drug and material design and sustainable energy production. Recognizing the possibility that QIS could be a disruptive technology—with the potential to create groundbreaking products and new industries—federal agencies across the U.S. government have made a coordinated effort to accelerate quantum research and development. Under the National Quantum Initiative Act (NQIA; H.R. 6227, 115th Congress (2017-2018)), a key recommendation emphasized in the National Strategic Overview for Quantum Information Science (Subcommittee on Quantum Information Science of the National Science and Technology Council 2018) was for the government to maintain research thrusts that stimulate transformative and fundamental scientific discoveries—an approach that puts science first. This language implies that before a technology can be engineered, a sound fundamental understanding of the underlying science behind the device needs to be established.

Since 2019, actions to strengthen federally funded core research programs from small grants to centers and consortia have been implemented across various agencies such as DOE, the National Institute of Standards and Technology (NIST), NSF, the Department of Defense (DOD), the National Aeronautics and Space Administration, the National Security Agency, and the Intelligence Advanced Research Projects Activity. In Fiscal Year 2022, approximately $840 million have been allocated through NQIA toward building major QIS program components. Approximately 90 percent of the budget is devoted to supporting fundamental science (~$250 million), quantum computing ($250 million), and quantum sensing (~$200 million), with more than half of the budget managed by DOE, NSF, and to a lesser extent NIST. However, a majority of the scientific programs have focused on physics and engineering (Subcommittee on Quantum Information Science and Committee on Science of the National Science and Technology Council 2021).

Considering the "science first" sentiment of the National Quantum Initiative, fundamental scientific research in chemistry and its implications on future QIS applications and vice versa is needed. The committee also addressed needs for future instrumentation and capabilities, infrastructure, database accessibility, and standards for evaluating quantum computing usage.

Recommendation 1-1. The Department of Energy and the National Science Foundation should support investigations that examine fundamental research topics to advance the field of quantum information science using chemistry-based approaches, which include experimental and theoretical studies. These research priorities are aimed at developing new approaches to scalability or addressability as well as enhanced detection of molecular systems, which may ultimately have the potential to transform this area of science and technology. Funding agencies should prioritize the following fundamental research areas:

- **Design and synthesis of molecular qubit systems,**
- **Measurement and control of molecular quantum systems, and**
- **Experimental and computational approaches for scaling qubit design and function.**

Within each research area, the committee has identified key research priorities that should be supported to advance the field of QIS and chemistry within the next decade.

Research Area 1. Design and Synthesis of Molecular Qubit Systems (Chapter 2)

Key Problem: Current Knowledge of Designing Molecular Qubits Is Limited and Will Need to Be Enhanced to Drive New Developments for QIS Applications

Qubits are the counterpart of the binary digit or bit of classical computing. Atomic control promises a new class of designs capable of functioning as sensors tuned for specific environments or analytes, as nodes that emit at desired frequencies for quantum optical networking, and as innovative new topologies for quantum computing. Synthetic molecular chemistry offers a unique tool kit of unparalleled control over structure, scalability, and quantum-scale interactions, and is poised to accelerate the development of bottom-up quantum technologies.

Within this framework, it is possible to design molecules geared for different applications. This molecular approach spans every component of synthetic chemistry, from organic systems, to inorganic molecules, to extended solids comprised of molecules. The ability to design a molecule, position atoms, tune properties, and subsequently create arrays or integrated systems is a truly unique molecular concept. The research priorities identified here are key topics to be pursued within the next decade to advance the design and synthesis of molecular qubit systems.

Research Priorities

- **Identify and tailor molecular qubit properties for specific near-term applications in quantum sensing and communications, and more long-term opportunities in quantum computing.**
- **Develop an understanding of structure–property relationships for**
  i. **increasing coherence times (T_2) in molecular qubits and quantum memories,**
  ii. **creating optically addressable molecular qubits (e.g., transition metal complexes, lanthanides, organic-based multispin qubits, and optical cycling centers), and**
  iii. **exploiting entanglement and quantum transduction.**
- **Investigate the interactions of molecular qubits with their environments.**
- **Design molecular structures with integrated chirality-induced spin selectivity effects.**
- **Target functionalization of molecular qubits for sensing and systems integration.**
- **Develop molecular quantum interconnects over broad length scales including molecule-based quantum repeaters.**
- **Fabricate scalable quantum architectures based on molecular qubits.**

Research Area 2. Measurement and Control of Molecular Quantum Systems (Chapter 3)

Key Problem: New Measurement Approaches and Techniques Are Needed for Deep Study of Chemical Systems

To further understanding of QIS, chemists are playing a large role in the development of novel measurements—for example, employing entangled photons to prepare entangled electron spins that can be probed using magnetic resonance measurements. Specifically, techniques using electron paramagnetic resonance (EPR), time-resolved EPR, and microscopy measurements provide key insights for designing new materials for qubit applications of molecules and for measuring biological systems at extremely low levels of light. Chemists are also developing novel approaches for the use of spin and spin transduction, which could be a fruitful avenue for QIS imaging and sensing applications. Utilizing both magnetic and optical approaches, chemists are advancing QIS ideas with real materials and molecules beyond the atomic systems in the gas phase used in most physics applications. This trajectory of moving beyond the gas phase means that the quantum and chemistry research community is inching closer to understanding the science behind technologies capable of being deployed in the operational regime. The research priorities identified below and discussed in Chapter 3 are directions the committee believes should be pursued in the near future to advance measurement and control of molecular quantum systems.

Research Priorities

- **Develop new approaches and techniques for addressing and controlling multiple electron and nuclear spins and optical cycling centers in molecular systems.**
- **Develop techniques to probe molecular qubits at complex interfaces to inform their systematic control.**
- **Enhance spectroscopic and microscopic techniques by creating entangled photon sources with higher yield and better spectral coverage and high-finesse cavities and nanophotonics for molecular qubit systems.**
- **Develop and exploit alternative approaches to spin polarization and coherence control (e.g., chirality-induced spin selectivity and electric field effects).**

- **Use molecular systems to teleport quantum information over distances greater than 1 μm with high fidelity.**
- **Develop molecular quantum transduction schemes that take advantage of entangled photons as well as entangled electrons and nuclear spins.**
- **Advance quantum sensing techniques to further understand biological systems.**
- **Use bio-inspired quantum processes to develop new quantum technologies.**

Research Area 3. Experimental and Computational Approaches for Scaling Qubit Design and Function (Chapter 4)

Key Problem: Scaling Up Robust Qubit Architectures Is a Major Challenge in Developing Next-Generation Quantum Systems That Requires Advances in Experiments and Classical and Quantum Computation

The primary obstacle limiting the accelerated development of quantum computers and quantum algorithms is having access to robust qubit architectures capable of being scaled up and retaining their function. To address the scalability challenge, computational theorists will need to simulate electronic structures of proposed qubit designs reliably to a high degree of accuracy. This level of prediction will expedite the qubit discovery phase and provide further insight into mechanisms limiting scalability and function. Approaches using classical and quantum computations are currently being undertaken to resolve the electronic structure problem as well as to model and predict new chemical designs of qubits. In parallel to theoretical work, experimental designs are also needed to characterize novel qubits, such as hybrid quantum architectures. Lastly, as theory and experimental research in this space advance, quantum computers will also improve, allowing for more complex chemistry problems to be solved beyond the classical regime. The following research priorities were identified by the committee as those that should be pursued in the next 5–10 years to expedite progress for scaling qubit design and function and creating the foundation for applying enhanced quantum computing capabilities.

Research Priorities

- **Develop techniques for synthesizing molecular qubits that retain their desirable quantum properties in different host chemical environments.**
- **Design hybrid quantum architectures that mutually enhance each other's quantum properties.**
- **Exploit the advantages of bottom-up chemical synthesis for constructing quantum architectures.**
- **Investigate and control the interactions among qubits and between qubits and their environments.**
- **Develop noise models and quantum error-mitigation techniques for individual qubits, systems of qubits, and quantum architectures that can be experimentally validated.**
- **Understand and advance the limits of classical electronic structure algorithms and modeling approaches that can guide the design of molecular or solid-state qubits and scalable quantum architectures.**
- **Leverage and develop machine learning, chemical informatics, chemical databases, and molecular simulations to inform and facilitate qubit design.**
- **Identify important open chemistry problems including those with applications to QIS that are unresolved due to classically intractable electronic structure.**
- **Develop more efficient methods of encoding chemical systems on quantum computers (e.g., better basis sets, quantization, fermion mappings, and embedding theories).**
- **Develop more efficient quantum algorithms for fault-tolerant quantum computers to simulate molecular systems, including those with QIS relevance.**
- **Study how quantum computing algorithms for dynamics can be used to accelerate chemistry and spectroscopy.**
- **Explore how quantum machine learning can be used to accelerate chemical research by processing quantum data from entangled sensor arrays or quantum simulations of chemistry.**

In addition to recommending research areas and research priorities, the committee was also tasked with discussing collaboration needs across different scientific disciplines, as well as identifying opportunities for building infrastructure and supporting instrumentation and tool needs at various scales (i.e., laboratory to large scale). The recommendations from these discussions were made by the committee because they are believed to have the greatest potential to advance QIS through chemistry and maximize the impact of QIS on chemistry. Furthermore, the committee assessed barriers to entry that may be limiting the size and breadth of the chemistry research community working in QIS. From these assessments, the committee identified needs and challenges for the development of a diverse, quantum-capable workforce. The recommendations on these topics are highlighted briefly in the following sections.

COLLABORATION NEEDS TO SUPPORT SCIENTIFIC PROGRESS AT THE INTERSECTION OF QIS AND CHEMISTRY

QIS and chemistry are fields that greatly benefit from scientific collaboration. In chemistry, interdisciplinary activities between researchers from various subdisciplines can lead to the development of novel solutions to chemistry challenges with applications in other industries (e.g., medicine, energy, cosmetics, agriculture, and others). Collaborations in QIS are also essential because the field involves diverse topics ranging from quantum mechanics to information processing. Historically, collaboration in QIS has involved physicists, engineers, and computer scientists. These efforts encouraged researchers to share knowledge, techniques, and technologies, which ultimately led to more efficient use of resources and accelerated scientific progress. Increasing collaborations at the interface of QIS and chemistry will also expedite discoveries and development in this emerging field.

Recommendation 2-1. The Department of Energy and the National Science Foundation should support cross-disciplinary activities that couple measurement, control, and characterization techniques traditionally employed by the physics and engineering communities with molecular systems designed by the chemistry community. Support also should be given to investigations that combine theory with experiment to take full advantage of the relationship between chemistry and quantum information science. Increasing these collaborations will be essential for scientific progress at these intersections.

IMPORTANCE OF ACCESS TO FACILITIES, CENTERS, AND INSTRUMENTATION

For the United States to maintain global competitiveness in the growing research areas at the interface between chemistry and QIS, national user facilities must remain at the cutting edge, which includes continual renewal of instrumentation. This applies not only to the major infrastructure (e.g., magnets and beamlines) but perhaps more importantly to the user end-stations (e.g., mid-scale instruments), where rapid technological advances can create a situation in which a newly developed capability can become uncompetitive (or even obsolete) within just a few years. A perfect example is microwave amplifiers and sources where EPR experiments performed today simply were not possible just 10 years ago.

Recommendation 3-1. The Department of Energy and the National Science Foundation should support the development of new instrumentation and techniques for the unique needs at the interface of chemistry and quantum information science. Broader access to laboratory-scale and mid-scale instrumentation is needed for the field to progress. For example, investments should be made in time-resolved magnetic resonance and optical spectroscopy. Support is required for professional staff to train users in the operation and utilization of these instruments, as well as to address new technique development and maintenance needs.

IMPORTANCE OF ACCESS TO DATA AND DATABASE DEVELOPMENT

Equally valuable to QIS and chemistry research are the products not only of software but of experiments—that is, data. In the case of QIS applications, both theoretical and experimental data, including calculations of decoherence times and entanglement, spectra, and structures, are invaluable not only for validating one another but for facilitating mutual method development. Some of these QIS-relevant data can be harvested from other existing databases, and such efforts to aggregate data need to be undertaken and supported. However, the amount of QIS-specific data is only expected to grow, and the community will need to establish a well-structured database with clear guidelines in the near future. FAIR—Findability, Accessibility, Interoperability, and Reusability—standards that emphasize properly labeling and determining the reusability of data would serve as baselines to be applied wherever possible to ensure professional data management and stewardship (GO FAIR 2017). As QIS-specific databases begin to emerge, ensuring that they remain open for widespread exploration will be important.

Recommendation 4-1. The Department of Energy and the National Science Foundation should establish open-access, centralized databases that include quantum information science (QIS)–relevant data to enhance predictions and expedite new discoveries. These databases should contain (1) structure–property relationships, (2) results of electronic structure calculations, (3) spectroscopic data, (4) experimental characterization of quantum devices, and (5) other data to inform QIS investigations. These data should be obtained from QIS studies contributed by scientists and engineers across industry, academia, and government. These agencies should also create a centralized database to house a body of experimental work that demonstrates discrete quantum use cases for chemistry.

NEEDS TO EVALUATE AND STANDARDIZE QUANTUM ADVANTAGE FOR CHEMISTRY

As mentioned earlier, one of the challenges faced at the intersection of QIS and chemistry is identifying chemistry problems that could be uniquely solved with a quantum computer. Currently, these types of problems are elusive; therefore, the committee has called upon DOE, NSF, and other funding agencies, both public and private, to work with the research community to continue assessing this issue as quantum computing capabilities continue to mature.

Recommendation 4-2. The Department of Energy, the National Science Foundation, and other funding agencies, both public and private, should develop initiatives to support multidisciplinary research in quantum information science to address how quantum-accelerated calculations could solve chemistry problems. In connection with these initiatives, the research community should establish a set of standards for how to evaluate quantum advantage in specific chemistry use cases.

OPPORTUNITIES TO IMPROVE EDUCATION AND BROADEN WORKFORCE DEVELOPMENT AT THE INTERSECTION OF CHEMISTRY AND QIS

Like any major modern scientific pursuit, the journey to new discoveries and developments is undoubtedly a powerful human experience that is often shared in a collaborative setting—hence DOE, NSF, and DOD's efforts to establish multidisciplinary research centers. To achieve broader access and inspire the next generation of quantum information scientists, chemistry education and outreach initiatives can be improved across various levels of workforce training and development by expanding to include nontraditional technical candidates across the country.

RECOMMENDATION 5-1. Achieving the goal of a diverse and inclusive workforce will require participation from various members across the quantum information science (QIS) and chemistry enterprise. The Department of Energy, the National Science Foundation, and other U.S. federal agencies should

support efforts to create a more diverse and inclusive chemical QIS workforce. Private and public stakeholders such as educators at various levels, nonprofit organizations, human resource personnel, and professional societies should also foster talent development and recruitment and increase public awareness related to QIS and chemistry activities. These efforts should aim to strengthen QIS in K–12, two-year degree-granting institutions, and beyond. The efforts should also lower barriers to entry for all scientists in QIS and develop the necessary skills in participants at multiple levels of education. Agencies and relevant stakeholders should prioritize actions to address the following topics:

1. **QIS and chemistry education development;**
2. **Barriers to entry at the intersection of QIS and chemistry; and**
3. **Development of a diverse, quantum-capable workforce.**

The committee identified specific opportunities that would benefit from further development within these topics.

QIS AND CHEMISTRY EDUCATION DEVELOPMENT (CHAPTER 5)

K–12 educators have many responsibilities, from academic instruction, to oversight, to fostering student social development and interaction. To task them with the additional expectation to create specialized curricula and concepts (e.g., quantum chemistry, quantum mechanics, and quantum algorithms) is a significant obstacle. Initiatives around the nation aimed at creating resources for educators will alleviate this extra burden. At the same time, these efforts will help sustain workforce development for emerging fields such as QIS. Some universities in the United States are offering a minor degree, master's degree, or certificate in quantum information science and engineering (QISE). Although chemistry is often associated as a subject deeply integrated into the multidisciplinary area of QISE, it is absent from either the course descriptions or the prerequisite lists of the QISE programs. These observations have led the committee to recommend the following actions to improve chemistry education with consideration for QIS.

Recommendation 5-2. Efforts to enhance curriculum resources and opportunities for students to gain exposure to concepts and skills at the intersection of quantum information science (QIS) and chemistry should be made. These efforts will support more learners in traditional educational and academic environments interested in pursuing research and careers at the intersection of QIS and chemistry.

- **Education development initiatives and curriculum developers should prepare curricular resources that include chemistry concepts guided by QIS principles for K–12 and undergraduate levels.**
- **Educators, human resource personnel, program managers, and communication teams should engage in outreach activities to increase exposure to QIS chemical technical concepts at varying levels of education.**

Barriers to Entry into QIS and Chemistry (Chapter 5)

Understanding the motivation behind "why" an individual would want to join the field will further the industry's understanding of how to reach a broader base. Examples of such motivations include salary, career track, company brand, and work culture. The limitation of understanding the motivation makes it difficult for the chemist to consider working in QIS as a viable career option. The committee recommends the following actions to help lower the barriers to entry in both classroom and industry to broaden the pool of diverse researchers and candidates in the QIS and chemistry workforce.

Recommendation 5-3. Efforts should be made to lower the current barriers to entry that limit members of the chemistry research community from entering quantum information science (QIS)–related research and careers. Efforts should also be made to lower barriers to entry for nontraditional participants to

provide equitable pathways to careers at the intersection of QIS and chemistry and to expand access to broader, more diverse groups of talent.

- **Industry consortiums, education organizations, federal agencies, and other relevant entities should foster cross-disciplinary and cross-sector collaborations that explore projects related to the intersection of chemistry and QIS.**
- **Program managers and administrators should create internal programmatic strategies to remove implicit and unconscious bias during the review process of grant applications and other peer-reviewed applications (e.g., Small Business Innovation Research [SBIR]).**
- **Academic institutions, SBIR programs, and other relevant stakeholders should provide support to establish incubator spaces dedicated to those pursuing QIS and chemistry innovation research and development (e.g., academic institutions, SBIR/Small Business Technology Transfer programs).**

Development of a Diverse, Quantum-Capable Workforce (Chapter 5)

Workforce demands for QIS in chemistry are rapidly expanding. At the same time, industrial corporations recognize that the field lacks diversity, both in the current workforce and new entrants. When the committee examined this challenge, it considered the value of offering internship opportunities, retraining current staff, and increasing permanent technical positions. Because the research activities at the interface of chemistry and QIS are just beginning to blossom, a unique opportunity exists in this emerging field to create a more diverse, equitable, and inclusive workforce. The committee identified the following actions as those that could be undertaken by various entities involved in programmatic development at colleges and universities; in job advertising; and in professional development for employees at national laboratories, academic institutions, and industry.

Recommendation 5-4. Increasing broader participation and diversity remains a challenge in recruiting and retaining talent in the field of quantum information science (QIS) and chemistry. Dedicated and focused efforts should be made to foster a diverse, quantum-capable chemical sciences workforce.

- **Program coordinators, researchers, educators, and other relevant personnel should recruit and support students transferring to four-year academic institutions from two-year colleges, students from minority-serving institutions, and students from historically underrepresented groups to continue in the fields of chemistry and QIS.**
- **Federal agencies and professional development coordinators should provide retraining opportunities for the academic, industrial, and national laboratory workforce of potential QIS participants with requisite professional skills that are useful for employment in a QIS field.**
- **Human resource personnel and hiring managers should provide detailed descriptions of the technical skill sets beyond doctoral prerequisites needed for jobs at the intersection of QIS and chemistry.**
- **National laboratories, industry, federal agencies, and academic institutions should increase support for hiring more permanent, professional, and diverse (in terms of demographics) technical staff at varying education and experience levels.**
- **Academic institution leaders should encourage institutions to be transparent and public about the current and aspirational demographic makeup of their leadership and workforce.**

CONCLUSION

Overall, this report is intended to guide federal agencies—specifically DOE and NSF—public and private funders, and the QIS and chemistry research community on how to navigate and support this emerging field effectively. If fulfilled, the committee believes that these recommendations will create immense opportunities to expand knowledge in this area. Such progress could place the United States in a position of technological advantage, safeguarding national security and the economy. In the end, however, genuine scientific curiosity remains the strongest driver for studying at the intersection of QIS and chemistry. In this vein, researchers will deepen their understanding of nature at a fundamental level. By diving into this frontier, scientists could uncover new knowledge and build technological treasures that may have a profound impact on society and transform the future human experience.

REFERENCES

GO FAIR. 2017. "FAIR Principles." https://www.go-fair.org/fair-principles/.

Subcommittee on Quantum Information Science of the National Science and Technology Council. 2018. "National Strategic Overview for Quantum Information Science." https://www.quantum.gov/wp-content/uploads/2020/10/2018_NSTC_National_Strategic_Overview_QIS.pdf.

Subcommittee on Quantum Information Science and Committee on Science of the National Science and Technology Council. 2021. "National Quantum Initiative Supplement to the President's FY 2022 Budget." https://www.quantum.gov/wp-content/uploads/2021/12/NQI-Annual-Report-FY2022.pdf.

1

Introduction

Quantum information science (QIS), the study of processing, storing, manipulating, measuring, and transmitting information using the principles and laws of quantum mechanics, has witnessed a dramatic rise in scientific research activities over the past decade. This enthusiasm can be observed in the numerous publications generated by the physics and computer science research communities, as reflected in recent National Academies of Sciences, Engineering, and Medicine (National Academies) consensus study reports (NASEM 2019, 2020) that focus largely on the technological capabilities of quantum computing, communications, and sensing. Many of these efforts have been supported by the National Science Foundation (NSF) and the Department of Energy (DOE) as well as many other government agencies interested in the discovery of new phenomena at the basic science level.

These two agencies have released key reports outlining the types of research (both basic and applied) required to achieve the goals of QIS and bring in the new quantum era (Bauer et al. 2020; Committee on Quantum Engineering Infrastructure 2021; U.S. Department of Energy 2017a, 2017b). The reports were instrumental in the development of the National Quantum Initiative Act (NQIA), which provided a general roadmap for the goals of quantum computing, networking, and sensing in the United States. As a result of the guidance from these efforts, the following technologies have experienced significant advancement: atomic clocks, atom interferometers, optical magnetometers, atomic electric field sensors, and quantum optical effects. The impressive accomplishments in these areas over the past two decades have inspired more scientists and engineers to enter the field of QIS.

This report acknowledges the tremendous contributions from scientists working in these areas and their role in driving the field forward. However, QIS is now at a formative point in its progression. As illustrated in Figure 1-1, new quantum systems and materials are needed to further scientific advancements. This report concludes that chemistry may present new opportunities that can broaden the scope of research and development in QIS and provide new opportunities for workforce development. The chemistry community may present new approaches to making novel QIS-related molecules, test these systems with new QIS-related experimental methods with instrumentation, and finally provide strong theoretical foundations for models and proposals. To address the challenges of developing new molecular approaches to creating and characterizing systems for QIS applications, this report takes a unique look at QIS research and development through the lens of chemical research.

A chemistry approach may expand our understanding of QIS by demonstrating how to exploit the quantum properties of molecular systems effectively. QIS also benefits from the investigations conducted using chemistry approaches and quantum tools—such as computation, simulation, or sensing, which can improve the understanding of chemical processes. In this manner, QIS and chemistry share a bidirectional relationship. A major goal of

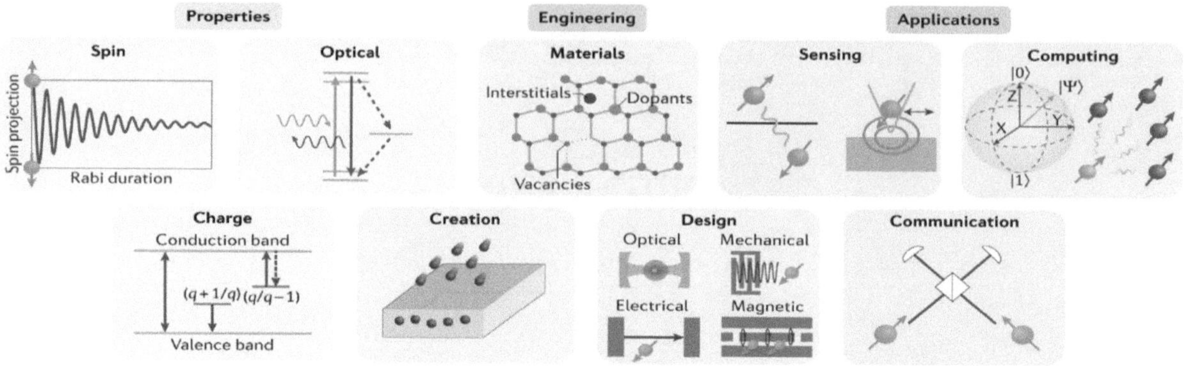

FIGURE 1-1 Illustration of key concepts in the QIS field, highlighting the three major defect properties (spin, optical, and charge) in the blue panels, engineering considerations (materials, creation, and design) in the grey panels, and quantum applications (sensing, communication, and computing) in the red panels. For the properties, the panel on spin shows the spin projection during a Rabi oscillation; the panel on optical shows pumping (green) and photoluminescence (red) in a defect with an intersystem crossing; and the panel on charge shows transition energy levels for charge conversion. For engineering considerations, the panel on materials displays various defect types and lattice sites; the panel on creation illustrates defect creation by implantation or irradiation; and the panel on design summarizes optical, mechanical, electrical, and magnetic devices for interfacing with spin defects. Regarding the main quantum applications, the panel on sensing displays the use of the electron spin as a sensor for other local spins or magnetic fields; the panel on computing shows the interaction of an electron spin with many nuclear spin registers; and the panel on communication shows entanglement between two spins via two-photon interference spectroscopies that exploit quantum phenomena and can impact our understanding of chemical systems and our ability to image them.
SOURCE: Wolfowicz et al. 2021.

this report is to assess recent and ongoing research in QIS and advances in quantum information processing and technology that have the potential to transform various aspects of chemistry and vice versa.

1.1 STUDY ORIGIN AND STATEMENT OF TASK

Recognizing the possibility that advancements in QIS could lead to the creation of disruptive technologies—with the potential to create groundbreaking products and new industries—federal agencies across the U.S. government have made a coordinated effort to accelerate quantum research and development through the NQIA (H.R. 6227, 115th Congress (2017-2018)). Under the NQIA, the Subcommittee on Quantum Information Science of the National Science and Technology Council was established under the auspices of the Committee on Science. Since 2018, the Subcommittee has published numerous strategy documents outlining research directions and opportunities for partnerships within QIS (see Subcommittee on Quantum Information Science of the National Science and Technology Council 2021, 2022a, 2022b). A key recommendation emphasized in the National Strategic Overview for Quantum Information Science was for the U.S. government to maintain research thrusts that stimulate transformative and fundamental scientific discoveries, where the approaches put science first (Subcommittee on Quantum Information Science of the National Science and Technology Council 2018). As illustrated in Box 1-1, since 2019, the United States has steadily increased its research and development budget for QIS. Under the NQIA, several multidisciplinary research centers were created. In addition, the National Defense Authorization Act (NDAA) for Fiscal Year 2020 authorized the Department of Defense to create three Quantum Research Centers (see Appendix C, Table C-1). These efforts certainly have led to fruitful advancements, particularly for the technologies mentioned above. However, historically, many of the research efforts under the NQIA and NDAA, as illustrated by the program components, have been limited in their attention to a chemistry approach to QIS. The White House Office of Science and Technology Policy hosted the National Quantum Initiative Centers Summit in the winter of 2022, where the theme of choosing a science-based approach was again echoed (The White House 2022).

BOX 1-1
Federal Agency Efforts to Support QIS Research

Actions to strengthen federally funded core research programs from small grants to centers and consortia have been implemented across various agencies. The NQIA, which became Public Law No. 115-368, specifically called upon the National Science Foundation (NSF) (sec. 301) and the Department of Energy (DOE) (sec. 401 and sec. 402) to carry out basic research programs on QIS and establish multidisciplinary centers for quantum research. NSF and DOE established 10 multidisciplinary centers across the United States, and the Quantum Leap Challenge Institutes and National Quantum Information Science Centers, respectively (Appendix C). The National Institute of Standards and Technology (NIST) (sec. 201) was charged with carrying out specific QIS activities to identify future measurement, standards, cybersecurity, and other QIS needs. In Fiscal Year (FY) 2022, approximately $840 million have been allocated through the National Quantum Initiative Act (NQIA) toward building major QIS program components areas—Quantum Sensing and Metrology (QSENS), Quantum Computing (QCOMP), Quantum Networking (QNET), QIS for Advancing Fundamental Science (QADV), and Quantum Technology (QT) (Figure 1-1-1) (Subcommittee on Quantum Information Science and Committee on Science of the National Science and Technology Council 2021). Of the total investment to date, approximately 90 percent of the budget is devoted to supporting fundamental science (~$250 million), quantum computing ($250 million), and quantum sensing (~$200 million), with more than half of the budget managed by DOE, NSF, and to a lesser extent NIST. In addition to the multidisciplinary centers, NSF and DOE have also sustained several research and development (R&D) efforts connected to QIS topics (Appendix C, Table C-2) that enable a wider exploration of quantum research, predominantly in the areas of physics, engineering, and computer science.

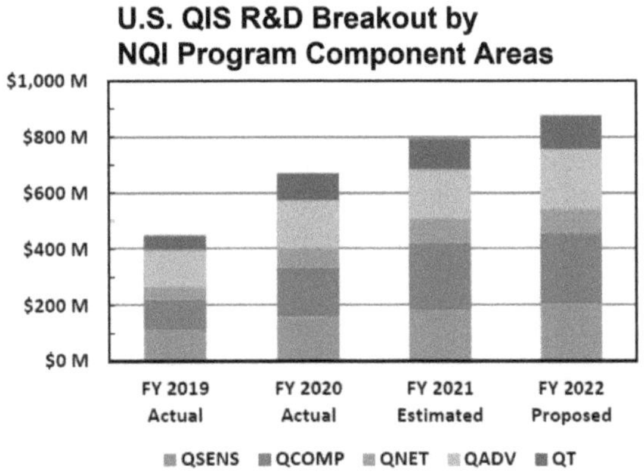

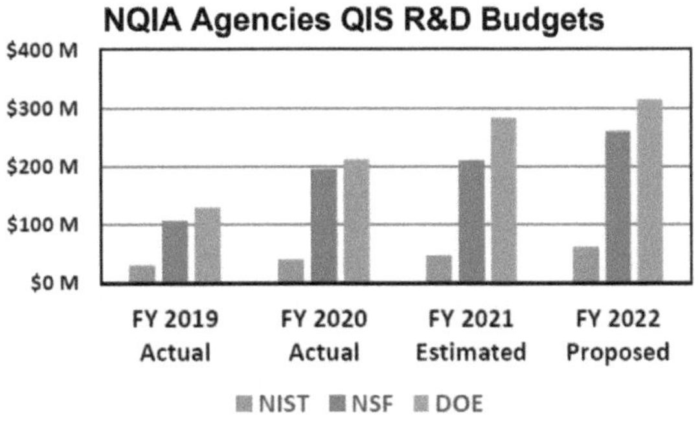

FIGURE 1-1-1 (Top) U.S. QIS R&D budget allocations by the NQI Program Component Area topic for FY 2019 and FY 2020 (actual expenditures), FY 2021 (estimated expenditures), and FY 2022 (requested budget). (Bottom) NQIA QIS R&D budgets for NIST, NSF, and DOE.
SOURCE: Subcommittee on Quantum Information Science and Committee on Science of the National Science and Technology Council 2021.

Upholding the sentiment of science first, DOE and NSF recognized the natural interplay between QIS and chemistry, as well as the timeliness for these two fields to advance together. Building on this interest, DOE and NSF asked the National Academies to examine the opportunities at the interface of chemistry and QIS and to provide recommendations on research and other needs to facilitate progress. To address this request, the consensus study Committee on Identifying Opportunities at the Interface of Chemistry and Quantum Information Science was convened by the National Academies to assess recent and ongoing research occurring at the intersection of QIS and chemistry. Box 1-2 provides the detailed statement of task for the committee; the overarching goal of the study is to provide fundamental research recommendations within chemistry (related to synthesis, measurement, tools, and theory) to advance QIS.

BOX 1-2
Statement of Task

The National Academies of Sciences, Engineering, and Medicine will convene an ad hoc committee of experts and scientific leaders to identify opportunities and research priorities at the interface of quantum information science (QIS) and chemistry. Specifically, the committee will accomplish the following tasks:

- Assess recent and ongoing research in QIS and advances in quantum information processing and technology that have the potential to transform various aspects of chemistry research. For example, the committee will consider how advances in quantum computing and quantum algorithms can impact the ability to simulate chemical systems; how quantum sensors can impact the ability to monitor and understand chemical systems and mechanisms; how quantum coherence can be used to modulate chemical reactivity; and how other quantum approaches (e.g., spectroscopies that exploit quantum phenomena including methods using entangled photons) can impact our understanding of chemical systems, including complex chemical systems such as those in the biological chemistry and nanochemical domains, and our ability to image them.
- Assess recent and ongoing research in chemistry that draws upon chemistry's unique capabilities in the synthesis, measurement, and modeling of molecular systems to advance QIS. For example, the committee will consider research efforts to understand and control quantum phenomena in molecular systems and chemical environments that could be exploited in quantum systems, such as quantum sensors and quantum computers; and the design and synthesis of novel molecular systems that manifest desired quantum behavior, including new systems with potential deployment as quantum qubits or qudits.
- Recommend fundamental research in target areas with both near- and long-term potential to advance QIS through chemistry-based approaches and comment on its potential to lead to transformational changes in science and technology.
- Identify the collaborations needed across chemistry, biochemistry, materials science, physics, engineering, and information science that will be essential for scientific progress at the intersection of chemistry and QIS, as well as assess the barriers to entry that may be limiting the size and breadth of the chemistry research community working in QIS.
- Identify needs and opportunities for infrastructure, instrumentation, and tools, ranging from laboratory scale to mid-scale to large scale (i.e., user facilities), with the potential to advance QIS through chemistry and to maximize the impact of QIS on chemistry. Identify needs and challenges for the development of a diverse, quantum-capable workforce.

The committee's report will provide guidance to the research communities in government, academia, and industry. The report's recommendations will focus on science needs and priorities rather than specific funding or organizational aspects.

The committee consisted of 15 members, who worked on a volunteer basis from November 2021 to June 2023, with expertise in a broad range of chemical sciences, physics, and engineering areas. Other experiences and backgrounds (i.e., geographic diversity, career-stage breadth, and gender balance) were also considered in the committee composition. Committee member biographies are provided in Appendix A. To address its charge, the committee held three public information-gathering meetings, where subject-matter experts across various fields—such as quantum theory, quantum computing, experimental measurements, quantum biology, and chemical synthesis—presented their perspectives on the research areas of QIS and chemistry. These experts were selected because their research involves either characterizing quantum systems or developing novel QIS applications. Public meeting agendas are listed in Appendix B. The committee also held eight closed-session meetings to develop the recommendations provided in this report.

1.2 COMMITTEE SCOPE

Some of the committee's discussion has focused on the structure–property relationships between different classes of molecules, quantum properties, and their potential use in QIS applications. These fundamental scientific discussions are also deeply aligned with the NQIA sentiment of "science first" as well as DOE and NSF's programmatic interests. At the heart of these discussions is the committee's consensus view that the research areas using chemistry-based approaches coupled with fundamental quantum physics play a central role in broadening the impact of QIS research and applications. For example, the committee examined magnetic molecules, which have a unique quantum property (e.g., spin dynamics) that makes them interesting candidates to be studied for spin-based information technology. Other molecular architectures and their potential roles as qubits (i.e., information storage units, analogous to the classical "binary bit" often discussed for quantum computers) also were explored.

Furthermore, the committee considered the relationship between characteristics of quantum properties and the molecule's potential application. For instance, spin dynamics is a critical quantum property that influences the type of role the molecule could have in QIS applications. A molecule may be considered useful as a qubit if it possesses a low spin value; in contrast, if a molecule has a high spin state, it may be more suitable as a molecular magnet than a qubit. Additionally, the chemical structure design and measurement criteria for the development of molecular qubit systems were central to the committee's investigation and are highlighted throughout the report. The use of nonclassical states of light was also emphasized during the information-gathering meetings. These discussions were devoted to the use of quantum light, which can encode information in the phase of a single photon, for example, to do spectroscopy in organic and biological molecules.

1.3 HOW TO USE THIS REPORT

Ultimately, the committee's report aims to provide guidance to the research communities in government, academia, and industry. This report targets primarily federal agencies and QIS and chemistry researchers, and it is structured to meet the needs of these two key audiences. First, the research priorities identified by the committee are labeled with bold subheadings within the chapters. These subheadings are intended to direct federal agencies and researchers to the scientific topics that interest them most. The report's recommendations focus on scientific needs and priorities rather than specific funding or organizational aspects. Second, the boxes shown throughout the report include detailed background information, such as technical concepts that educators could use as examples to expose their students to QIS and chemistry. Other boxes, especially those in Chapter 5, include policy-related information, best practices for increasing workforce development, and other pertinent background information. Third, key takeaway messages at the beginning of each chapter are useful for the broader public and policy makers who would like to have a high-level view of the content in the study. Fourth, this study provides a glossary and an acronym list (Appendix E) to support the reader who is new to this area. Finally, the research priorities and the committee's recommendations are summarized at the end of each chapter. Specific research examples are highlighted under each section and are intended to provide the reader with a brief overview of Chapters 2 through 5. The remaining section in this chapter presents the committee's first recommendation to advance research at the intersection of QIS and chemistry. Following the research recommendation are major scientific and workforce development questions framed by the committee to address the statement of task. These questions are listed in the order in which they are described in the report.

1.4 OVERALL RESEARCH RECOMMENDATION

The committee's first recommendation is intended to guide federal agencies—specifically, DOE and NSF—toward fundamental target research areas that should be supported in the near term owing to their potential to transform scientific and technological developments in QIS and chemistry. Within each research area, the committee has identified key research priorities that should be supported to advance the field of QIS and chemistry within the next decade. The research priorities will be discussed and summarized in Chapters 2, 3, and 4.

> **Recommendation 1-1. The Department of Energy and the National Science Foundation should support investigations that examine fundamental research topics to advance the field of quantum information science using chemistry-based approaches, which include experimental and theoretical studies. These research priorities are aimed at developing new approaches to scalability or addressability as well as enhanced detection of molecular systems, which may ultimately have the potential to transform this area of science and technology. Funding agencies should prioritize the following fundamental research areas:**
>
> - **Design and synthesis of molecular qubit systems,**
> - **Measurement and control of molecular quantum systems, and**
> - **Experimental and computational approaches for scaling qubit design and function.**

1.4.1 Why Should Molecules Be Studied in QIS Research?

The committee draws attention to the fact that the time has come for the chemistry community to take a larger role in QIS for several reasons. Specifically, the committee underscores that the QIS community would benefit greatly from enhanced chemistry research activities in the future. The following sections highlight some of the main messages from Chapters 2–5. These messages emphasize the importance of understanding how to manipulate and control fundamental quantum properties.

First, chemistry will provide new building blocks for QIS research by offering precise, reproducible, and possibly scalable chemical approaches toward creating qubits. Here, a qubit, or a quantum bit, is defined as a two-state quantum mechanical system capable of being placed in a state of coherent superposition (see Box 1-3 for a description of key quantum concepts curated for a broader audience). Next, a fundamental theme woven throughout the chemical design process is having the flexibility to manipulate quantum properties (e.g., zero-field splitting (ZFS)) through structural design (e.g., ligand field energies). Depending on the extent of the change, quantum properties can be controlled by changing the molecular structures in significant (e.g., coarse knob) or subtle (e.g., fine-tune knob) ways.

This level of versatility will enable chemists to address the major challenges faced when using molecules for QIS applications. One such obstacle involves the issue of decoherence, in which the information of a quantum system is affected (often negatively) by its interaction with the environment (see Box 1-3 for further details). Although decoherence is often associated with coherence, it describes the loss of coherence. Unlike in atomic systems, decoherence is a major limiting factor in larger structures like molecules and may impede their utility in QIS applications. By adopting a modular design approach, new architectures could overcome this issue of decoherence; hence, molecules with smaller decoherence rates could be designed and synthesized. In sum, the ultimate goal of the QIS and chemistry research community is to create strategies to maximize coherence times and minimize decoherence of different quantum systems. Focusing on these two parameters will unlock enhanced technological capabilities in the areas of communications, networking, computing, sensing, and more.

Chapter 2 provides a thorough assessment of chemical approaches and the measurements that are used to probe a molecule's unique quantum properties. Specifically, this chapter illustrates the opportunities for transition and lanthanide metal–containing molecular systems that offer a large energy scale of interaction to contribute to

BOX 1-3
What Are Coherence and Decoherence? What Do They Mean for Future QIS Applications?

Generally, scientists refer to coherence under the context of wave-like behavior, as illustrated by the red and blue curves in Figure 1-3-1. Coherence is a quantum property that describes the relationship between two waves that are well defined (i.e., having the same phase and amplitude). The correlations between the waves are preserved over space or time; thus, coherence can be distinguished as being either spatial or temporal, respectively. Spatial coherence refers to the correlation between the phases of a wave at different points *transverse* (perpendicular) *to the direction of propagation.* In optical measurements, this parameter is often described as the coherence area and reported as a ratio. On the other hand, temporal coherence is a measure of the correlation between the phases of a wave at different points *along the direction of propagation.* This property is usually described as coherence length and measured in unit time. The schematic in Figure 1-3-1 is focused on describing temporal coherence; although for waves to interfere, either constructively or destructively, both types of coherence need to coexist. Coherent waves can interfere constructively when they are in phase synchronization (i.e., the same oscillation behavior) (see Figure 1-3-1b, first box). The mechanisms of most chemical approaches to QIS strongly depend upon the persistence or lifetime of coherence in the molecules. Therefore, understanding and manipulating the quantum property of coherence is a fundamental challenge facing scientists who want to implement chemical systems for enhanced quantum sensing and quantum computing devices.

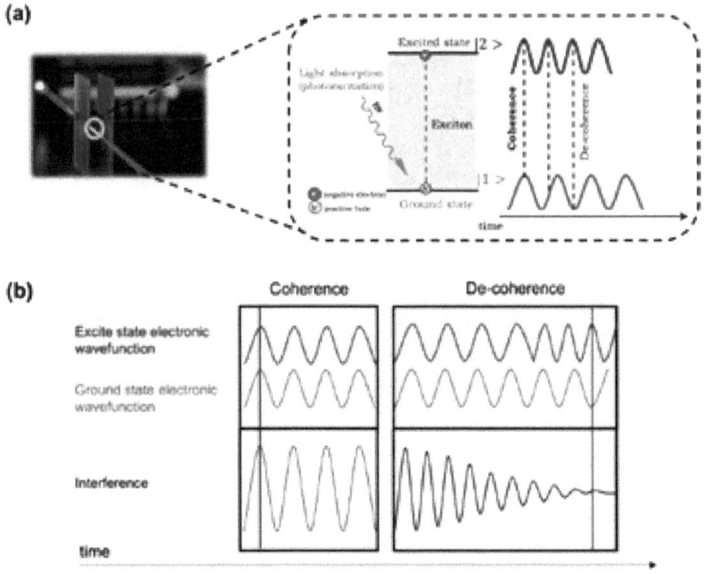

FIGURE 1-3-1 (a) (left) A photograph of a laser beam passing through a sample in a cuvette, and (right) a schematic of photoexcitation of a molecule and wave function of the excited state (red) and ground state (blue) acting in coherence and decoherence over time. (b) (left) Excited-state, ground-state, and interference wave functions are in coherence over a set time; and (right) decoherence occurs when the excited-state, ground-state, and interference wave functions are out of phase.
SOURCE: Ricci 2022.

BOX 1-3 *continued*

Figure 1-3-1 illustrates an example of how QIS applications can be studied using light. In the area of optical measurement research, scientists are often interested in the coherence or degree of synchronization between electronic-state wave functions (i.e., the function that describes all the properties of the system). As shown in Figure 1-3-1a, when a molecule is excited with a laser pulse, the photoexcitation process puts the molecule in a coherent state. The coherence time reflects the time during which the ground (blue curve) and excited (red curve) electronic states evolve with a fixed phase relationship over time (Figure 1-3-1a, right). This light–matter interaction populates higher-energy states called "excited states" in the molecule. The electronic "ground state" is defined as the lowest energy state of the molecule. When the laser light interacts with the molecule, it can cause the system to exist in a superposition state between the ground and excited states. Hence, these two states can coexist at the same time regardless of distance or location. The first box in Figure 1-3-1b shows that this coherent state possesses a fixed phase relationship between the wave functions of the ground and excited states.

Superposition of states and quantum entanglement is manipulated in QIS applications such as quantum sensors, communications, and computing because it has implications for at least doubling the resolution of images, exponentially speeding up computing time, and increasing storage space. In other words, QIS goes beyond classical physics, as shown in the following examples:

- Quantum sensors—doubling resolution at lower light intensities to image, for example, photosensitive biological samples, which impacts early and precise detections.
- Quantum computers—solving complex mathematical problems at a logarithmic computing time much less than classical computers that could require an exponential computing time, which impacts cryptographic, drug discovery, and other trend analysis.
- Quantum communications—providing secure, fast, and long-range transmittance of information, which impacts operational and organization capabilities.

Underpinning these enhanced technological capabilities is the fundamental property of coherence, which is why researchers are developing strategies and tools to maximize coherence times.

However, as illustrated in the second box of Figure 1-3-1b, another process is competing with coherence. As time evolves, the synchronized superposition of states, initialized by the laser, starts to decay. The phases (amplitude and frequency) between the ground- and excited-state wave functions will begin to shift and are no longer constant. This shift is caused by interference that primarily results from the molecule's interaction with the external environment. The process of losing coherence is known as decoherence. In general, quantum technologies require slower or smaller decoherence rates to be considered useful.

Thus, finding ways to overcome the limitations of large decoherence rates is a major goal in the design, synthesis, and assembly of molecular systems for QIS applications. This report demonstrates how chemists are using creative and productive approaches to confront this challenge and provide new strategies for creating novel materials for QIS. In sum, several different disciplinary approaches and investigations are under way to achieve longer coherence and shorter decoherence times.

QIS. These classes of molecules demonstrate how chemical approaches can be leveraged to control both ZFS and ligand field energies by creating coarse and fine-tune knobs. Figure 1-2 is an example of an echo decay experiment performed by Atzori and colleagues (2016) to characterize the quantum coherence properties and phase memory time, T_m, of a transition metal system, vanadyl phthalocyanine, as a function of temperature. Indeed, with the creation of designer quantum units, the chemical approach is poised to disrupt quantum sensing and communications in the near term and computing in the longer term. This chapter also illustrates new opportunities in QIS for chemical systems with properties for optical cycling, chiral-induced spin selectivity, electron–spin state

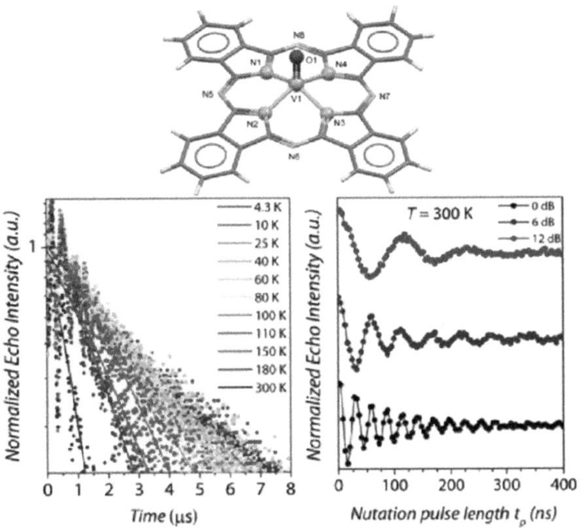

FIGURE 1-2 (Top) Molecular structure of vanadyl phthalocyanine with principal atoms labeling scheme and coordination geometry of the V^{IV} ion highlighted. (Bottom) Echo decay traces for the molecule shown at different temperature, *T* (see legend). Solid lines are the best fits. Rabi oscillations were recorded for this molecule at 300 K for different microwave attenuations. Both experiments were performed at 345 mT.
SOURCE: Atzori et al. 2016.

teleportation abilities, or quantum-entangled transduction properties. Investigating this diverse class of materials could provide new and fruitful avenues for the QIS field in the future.

The chemistry community is considering not only how one might alter the structure of the basic molecular unit of the qubit in the chemistry approach to QIS but also how to use strategies to provide multiple molecules in a supramolecular approach to provide the desired QIS effect. This form of "systems engineering" of the molecule's function has been exploited for select systems already and seems to be a promising avenue of research that is unique to the chemistry approach. For example, using supramolecular chemistry in the United Kingdom, Whitehead and colleagues (2016) have shown that a single simple module of molecular $\{Cr_7Ni\}$ octametallic rings acting as the qubits can be assembled into structures suitable for different types of quantum logic gates, such as the popular controlled NOT gate or another type of gate. These caged complexes can be curated to the gate application by switching the chemical linker in the molecular structure. Indeed, additional approaches based on this modular design motif could be useful in mitigating decoherence and illustrating the use of molecules for the operation of the quantum gate function.

1.4.2 What Tools Should Be Used to Study Chemical Systems for QIS Applications?

One of the chemical research areas that has evolved over the past decade is the development and use of new experimental tools to probe properties in organic and inorganic molecules with potential applications in QIS. The recommendations and research priorities in this area offer a way to improve the characterization of molecules useful for QIS research by seeking opportunities for less expensive advanced spectroscopy techniques, including electron paramagnetic resonance (EPR), laser sources, and other laboratory-scale instrumentation. Many of the molecules described in Chapter 2 can be investigated for their spin-related quantum properties, and these measurements heavily involve the use of EPR approaches. Additionally, optical measurements have developed substantially in the chemistry community to probe interesting and enhanced properties of organic and biological systems. As shown in Chapter 3, these techniques (both spin and optical based) allow chemists and their interdisciplinary collaborators to provide detailed insights on the structure–property relationships in particular molecules for QIS applications. Both steady-state and time-resolved (transient) EPR are powerful tools for investigating QIS properties in spin-based molecular qubits. The committee surveyed research activities (within the United States and internationally) that used transient EPR measurements to probe key quantum properties of synthesized molecules for QIS effects. The current situation in the United States suggests that a select group of laboratories is capable of making these time-resolved EPR measurements. Purchasing cost-effective instruments will increase access for more chemistry departments to provide opportunities for scientists interested in analyzing QIS-type molecules to conduct this type of research.

As shown in Figure 1-3, the publications indicate that most of the advancements in measurements for molecular systems used in QIS are being performed outside of the United States. While the references listed in the figure are not comprehensive, they serve as a representation of the international community's research efforts in this space. A deeper discussion is presented in Chapter 3. This observation elicits questions about whether the United States is missing opportunities to study and leverage chemistry research that is necessary for QIS. One of the factors contributing to this imbalance in publications is that many of the measurements in the United States are conducted at only a few laboratories. Adding to this challenge, the United States currently has a shortage of trained scientists for conducting measurements using time-resolved EPR and other available instruments such as magnetic beams. To work around this shortage, researchers engage in teams; however, the current collaboration models for synthetic chemists to work with those who have instruments need to be optimized for more fruitful interactions. These factors impede the progress of research at the interface between QIS and chemistry. Chapter 3 includes a discussion about strategies to overcome challenges related to instrumentation by exploring the possibility of having dedicated magnets and beamlines specifically for QIS researchers using chemistry-driven approaches for their studies. Overcoming challenges in infrastructure and workforce would benefit the QIS field greatly by increasing dedicated U.S.-based facilities, equipment, and knowledgeable and skilled professionals at national laboratories and university research laboratories.

Optical measurements developed in the chemical community can be used for future applications in QIS research. Time-resolved, two-dimensional techniques have been developed to enable in-depth studies of coherence. Through its information-gathering activities, the committee learned about the use of ultrafast, time-resolved, and nonlinear optical methods to probe coherent processes in organic and biological systems. This field of spectroscopy research has developed substantially over the years. Researchers now are at the point of searching for signs of clear, long-lived coherent processes in biological systems. In some respects, the field is shifting its focus from simply confirming the existence of coherence to now diving deeper and exploring the connection between coherence and function in biological systems. Chapter 3 details the state of the art in time-resolved, two-dimensional spectroscopy and its potential goals for the future in this area of research.

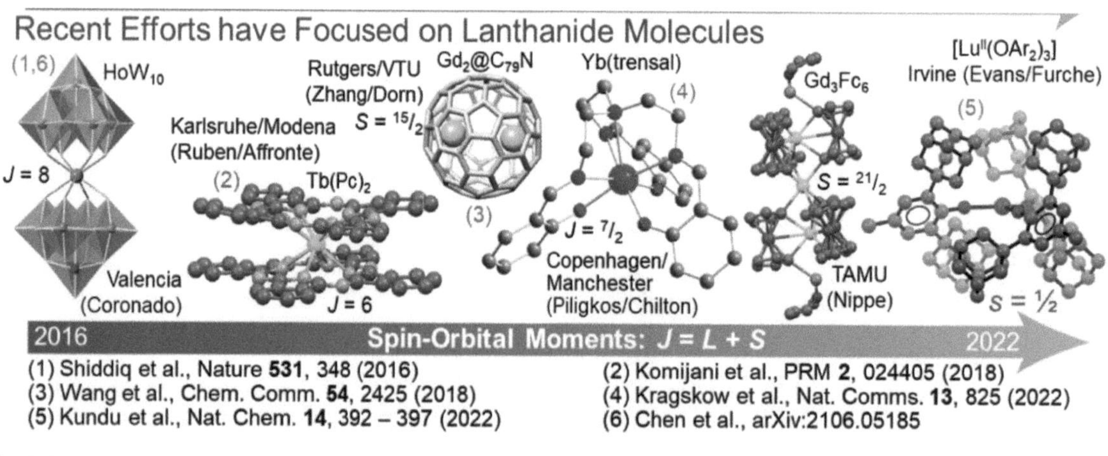

FIGURE 1-3 Molecular spins for QIS research. State-of-the-art electron paramagnetic resonance measurements have been made on transition and lanthanide molecular systems. The listed publications demonstrate that most of these research efforts are conducted outside of the United States.
SOURCE: Shiddiq et al. 2016; Komijani et al. 2018; Wang et al. 2018; Kragskow et al. 2022; Kundu et al. 2023.

The optical measurement research community has also provided the chemistry community with tools and strategies to manipulate quantum phenomena, such as nonclassical states of light, to understand fundamental coherent processes and QIS properties. The interest in nonclassical states of light, such as entangled photons, stems from the potential advantages they provide over classical laser excitation. By using these tools, scientists can take advantage of the high degree of temporal and spatial correlations in quantum light resulting from entangled pairs of photons. The use of these correlations can provide a sensing tool for chemists to detect potential molecular systems at very low levels of light. This capability can enable studies to uncover deeper insights (e.g., through advanced imaging) into the morphology and function of chemical and biological systems. Entangled pairs of photons have shown unique and lower noise characteristics when compared to classical photon sources, enabling greater detection capabilities. This feature of entangled light also can be used in new linear and nonlinear spectroscopic approaches, allowing the possibility for measurements with sensitivity beyond the standard classical limit, and can be performed at low excitation fluxes for exciting photosensitive materials.

1.4.3 What Are Some Challenges Facing Scalability and Molecular Design Approaches in QIS and Chemistry Research?

Despite some early successes using chemical systems for QIS applications, significant fundamental research is needed to provide detailed models for the design of molecules and in the development of approaches, especially in the areas of qubit designs for reduced decoherence and unwanted interactions between a qubit and its environment in molecular systems. Chapter 4 provides clear limitations about existing knowledge of the fundamental structure–property relationships for molecules in QIS. Specifically, the committee has identified a major question for the research community to address as a means of progress toward developing molecular qubit platforms: what is the origin of decoherence in molecular systems? Answering this question will allow for more accurate synthetic designs and other chemical and theoretical approaches for creating and eventually scaling molecular qubits. Additionally, for realistic large-scale, reproducible synthesis of current molecular qubits (e.g., molecular color centers), barriers to the realization of QIS platforms still exist. Chapter 4 explores the advantages of exploiting bottom-up chemical synthesis for constructing quantum architectures, leveraging novel syntheses of molecular qubits that retain their desirable quantum properties in different host chemical environments, and designing hybrid quantum architectures (i.e., molecular systems engineering) that mutually enhance the molecular and architectural quantum properties. This information allows scientists to investigate and control the interactions among qubits, and between qubits and their environments.

1.4.4 What Role Will Quantum Computers Play in Guiding the Understanding of Chemical Phenomena and Processes within QIS?

The committee took a chemistry-centric approach toward the use of quantum computing owing to the volume of research already published on quantum computing in the United States. Chapter 4 outlines how fault-tolerant quantum computers look very promising for solving problems (e.g., electronic structure) in chemistry, especially in chemical dynamics and quantum chemistry. This chapter also describes other chemical problems that are difficult for classical computers but seem to be uniquely addressed by quantum computers. From its information-gathering meetings, the committee learned that noisy intermediate-scale quantum (NISQ) computers can solve some problems near the classically intractable regime (i.e., the 50-qubit barrier), but whether they will solve useful problems without error correction is unclear. Additionally, whether quantum algorithms, annealers, NISQ computers, and other quantum computers in their current states can solve critical chemistry problems remains an open question. To address these questions, the committee explored advancements in fault-tolerant quantum computing that may allow for solving difficult chemistry problems in the future. In general, developing more initiatives to support multidisciplinary research in QIS to address how quantum-accelerated calculations could solve chemistry problems (that otherwise cannot be solved using classical systems) is an opportunity. In connection with these initiatives, the research community should establish a set of standards for how to evaluate quantum advantage in specific chemistry use cases.

1.4.5 In What Ways Can Chemistry and QIS Strengthen Economic Development and Build a Diverse and Inclusive Workforce?

Chapters 2–4 primarily focus on the areas necessary to enhance chemistry research in QIS. Those chapters also discuss the multiple benefits of advancing the QIS field in general. For research progress in QIS and chemistry to be fruitful, innovation will be driven within the academic, government, and private research sectors. As illustrated by the outcome of the NSF Project Scoping Workshop held in the summer of 2022, up to 10 large-scale public–private partnerships have been proposed to encourage the development and possible commercialization of some QIS technologies (Subcommittee on Quantum Information Science and Committee on Science of the National Science and Technology Council 2021). These efforts are proposed to be done in parallel to Small Business Innovation Research and Small Business Technology Transfer programs as a means to accelerate the economic benefit from QIS. Chapter 5 describes the need for and development of a diverse, quantum-capable workforce to support the QIS-enabled industry, and the potential for transformational impacts in science and technology from research in QIS and chemistry through the lens of education and economic development. The committee then provides its assessment of the challenges related to developing this workforce, including the barriers to entry that limit the size and breadth of the chemistry research community working in QIS.

The committee's emphasis on the development and expansion of a diverse and technically skilled workforce stems from the need for continued progress and innovation in quantum-related fields. The inclusion of scientists and engineers not historically connected with QIS research and development will enable new opportunities in this field based on skills learned from other fields. Specifically, new opportunities exist for chip-scale semiconductor researchers and others to seek QIS employment. The opportunities for developing the skills needed to pursue careers in QIS largely are concentrated at the graduate level, where the percentage of people from historically marginalized communities is low. Reported efforts to expand the introduction of QIS concepts at the K–12 level have been limited. By introducing QIS in the K–12 curriculum, a broader group of people may be provided with greater opportunities for participating in chemistry and QIS research. In addition to curriculum development, the committee also discussed reskilling and upskilling of non-Ph.D. level scientists as a way to strengthen the current QIS workforce. Furthermore, the committee discussed the potential barriers to entry into the field of QIS for chemists at various education levels and career stages. These barriers can be overcome with the right mentorship and education initiatives. Chapter 5 also highlights several quantum education and curriculum development initiatives taking place across the country. To strengthen the QIS workforce, the committee emphasizes the need to include chemistry in these initiatives. By having a solid foundation of chemistry concepts in conjunction with quantum mechanics, this workforce will be capable of creating novel quantum molecular based materials and tools.

1.5 SUMMARY

Finding sophisticated ways to manipulate and control quantum properties, such as coherence and decoherence, will advance the development of QIS technologies. Chapters 2–4 are shaped by Recommendation 1-1 and point to the untapped opportunities for discoveries, inventions, and innovations. Chapter 5 illustrates ways that the chemistry community can be included at the forefront of QIS alongside physics, computer science, and engineering. Finally, aligned with the NQIA and NDAA, this report embodies the sentiment of putting science first—by taking a deep dive into the fundamental challenges embedded at the interface of chemistry and QIS.

REFERENCES

Atzori, M., L. Tesi, E. Morra, M. Chiesa, L. Sorace, and R. Sessoli. 2016. "Room-Temperature Quantum Coherence and Rabi Oscillations in Vanadyl Phthalocyanine: Toward Multifunctional Molecular Spin Qubits." *Journal of the American Chemical Society* 138(7):2154–2157. doi.org/10.1021/jacs.5b13408.

Bauer, B., S. Bravyi, M. Motta, and G. Kin-Lic Chan. 2020. "Quantum Algorithms for Quantum Chemistry and Quantum Materials Science." *Chemical Reviews* 120(22):12685–12717. doi.org/10.1021/acs.chemrev.9b00829.

Committee on Quantum Engineering Infrastructure. 2021. "Workshop on Quantum Engineering Infrastructure Final Report." https://nnci.net/sites/default/files/inline-files/WQEI_final_report_final.pdf.

Komijani, D., A. Ghirri, C. Bonizzoni, S. Klyatskaya, E. Moreno-Pineda, M. Ruben, A. Soncini, M. Affronte, and S. Hill. 2018. "Radical-Lanthanide Ferromagnetic Interaction in a Tb^{III} Bis-Phthalocyaninato Complex." *Physical Review Materials* 2(2). doi.org/10.1103/physrevmaterials.2.024405.

Kragskow, J. G. C., J. Marbey, C. D. Buch, J. Nehrkorn, M. Ozerov, S. Piligkos, S. Hill, and N. F. Chilton. 2022. "Analysis of Vibronic Coupling in a 4f Molecular Magnet with FIRMS." *Nature Communications* 13(1):825. doi.org/10.1038/s41467-022-28352-2.

Kundu, K., J. Chen, S. Hoffman, J. Marbey, D. Komijani, Y. Duan, A. Gaita-Ariño, J. Stanton, X. Zhang, H.-P. Cheng, and S. Hill. 2023. "Electron-Nuclear Decoupling at a Spin Clock Transition." *Communications Physics* 6(1):38. doi.org/10.1038/s42005-023-01152-w.

NASEM (National Academies of Sciences, Engineering, and Medicine). 2019. *Quantum Computing: Progress and Prospects*. Washington, DC: The National Academies Press. doi.org/10.17226/25196.

NASEM. 2020. *Manipulating Quantum Systems: An Assessment of Atomic, Molecular, and Optical Physics in the United States*. Washington, DC: The National Academies Press. doi.org/10.17226/25613.

Ricci, F. 2022. "Investigation of Electronic Quantum Coherence in Semiconductor Materials Using Time-Resolved Non-Linear Optical Microscopy at Nano Level." Thesis, University of Michigan. doi.org/10.7302/4579.

Shiddiq, M., D. Komijani, Y. Duan, A. Gaita-Ariño, E. Coronado, and S. Hill. 2016. "Enhancing Coherence in Molecular Spin Qubits via Atomic Clock Transitions." *Nature* 531(7594):348–351. doi.org/10.1038/nature16984.

Subcommittee on Quantum Information Science of the National Science and Technology Council. 2018. "National Strategic Overview for Quantum Information Science." www.quantum.gov/wp-content/uploads/2020/10/2018_NSTC_National_Strategic_Overview_QIS.pdf.

Subcommittee on Quantum Information Science of the National Science and Technology Council. 2021. "A Coordinated Approach to Quantum Networking Research." www.quantum.gov/wp-content/uploads/2021/01/A-Coordinated-Approach-to-Quantum-Networking.pdf.

Subcommittee on Quantum Information Science of the National Science and Technology Council. 2022a. "Bringing Quantum Sensors to Fruition." www.quantum.gov/wp-content/uploads/2022/03/BringingQuantumSensorstoFruition.pdf.

Subcommittee on Quantum Information Science of the National Science and Technology Council. 2022b. "Quantum Information Science and Technology Workforce Development National Strategic Plan." www.quantum.gov/wp-content/uploads/2022/02/QIST-Natl-Workforce-Plan.pdf.

Subcommittee on Quantum Information Science and Committee on Science of the National Science and Technology Council. 2021. "National Quantum Initiative Supplement to the President's FY 2022 Budget." www.quantum.gov/wp-content/uploads/2021/12/NQI-Annual-Report-FY2022.pdf.

U.S. Department of Energy. 2017a. "Basic Energy Sciences Roundtable: Opportunities for Basic Research for Next-Generation Quantum Systems." doi.org/10.2172/1616258.

U.S. Department of Energy. 2017b. "Basic Energy Sciences Roundtable: Opportunities for Quantum Computing in Chemical and Materials Sciences." doi.org/10.2172/1616253.

Wang, X., J. E. McKay, B. Lama, J. van Tol, T. Li, K. Kirkpatrick, Z. Gan, S. Hill, J. R. Long, and H. C. Dorn. 2018. "Gadolinium Based Endohedral Metallofullerene $Gd_2@C_{79}N$ as a Relaxation Boosting Agent for Dissolution DNP at High Fields." *Chemical Communications* 54(19):2425–2428. doi.org/10.1039/C7CC09765D.

The White House. 2022. "Readout: National Quantum Initiative Centers Summit." www.whitehouse.gov/ostp/news-updates/2022/12/05/readout-national-quantum-initiative-centers-summit/?utm_source=link.

Whitehead, G. F. S., J. Ferrando-Soria, L. Carthy, R. G. Pritchard, S. J. Teat, G. A. Timco, and R. E. P. Winpenny. 2016. "Synthesis and Reactions of N-Heterocycle Functionalized Variants of Heterometallic $\{Cr_7Ni\}$ Rings." *Dalton Transactions* 45(4):1638–1647. doi.org/10.1039/C5DT04062K.

Wolfowicz, G., F. J. Heremans, C. P. Anderson, S. Kanai, H. Seo, A. Gali, G. Galli, and D. D. Awschalom. 2021. "Quantum Guidelines for Solid-State Spin Defects." *Nature Reviews Materials* 6(10):906–925. doi.org/10.1038/s41578-021-00306-y.

2

Design and Synthesis of Molecular Qubit Systems

Key Takeaways

- Chemistry enables the creation of designer quantum units poised to disrupt quantum sensing and communications in the near term and computing in the longer term.
- Chemistry demonstrates exquisite control over constituent materials at an atomic level central to understanding and controlling coherence and scaling in any qubit system.
- Despite recent progress in many qubit platforms, there is still a lack of knowledge of both the origins of decoherence and/or how these can be controlled through chemical design.
- Transition metals offer a large energy scale of interaction, whereby both zero-field splitting and ligand field energies can be tuned, creating coarse and fine-tune knobs.
- Molecular qubits can function like optically addressable defect centers yet have the advantages of designer properties and atomically precise spatial positioning.
- Lanthanides and actinides offer unique dual core–like and chemically accessible characteristics that make them useful for quantum information applications.
- Organic multispin molecules offer long coherence times and several demonstrated strategies for using photons to prepare well-defined spin quantum states.
- Optical cycling of molecules in the gas phase has seen much progress in identifying and predicting molecules with minimal vibrational branching, which could lead to new and improved molecular qubits.
- Photons have been used to generate four entangled electron spins in molecular crystals, which can be extended to use entangled photons for quantum transduction.
- Clock transitions in molecular qubits have been shown to be a promising approach to minimizing decoherence by the qubit environment.
- The chirality-induced spin selectivity effect can be used to strongly polarize electron spins and manipulate them even at room temperature.
- While progress has been made in understanding how two-qubit systems interact with each other and their environment, much work on the extension to multispin systems remains.

- Electron–spin state teleportation has been demonstrated in a molecular qubit system, which is an essential step toward extending this approach to nano- and microscale quantum interconnects.
- Heterogeneous systems and the presence of interfaces provide both new loss mechanisms and opportunities for quantum information transduction and coupling between dissimilar systems.
- Scaling up systems to create arrays of molecular qubits requires controlled synthesis with controllable inter-qubit distances. Examples of these types of controllable systems are metal–organic frameworks with structures tailored to particular applications.

2.1 IDENTIFYING AND TAILORING MOLECULAR QUBIT PROPERTIES FOR QIS APPLICATIONS

We are in the second quantum revolution, a time in which we are harnessing quantum properties to impact areas ranging from near-term applications, such as quantum sensing and communications, to more long-term applications, such as quantum computing. Within this broad space, unique requirements exist for each of these QIS application areas. Molecules offer three key attributes useful for QIS applications: atomic precision, reproducibility, and tunability (Aromí et al. 2012; Atzori and Sessoli 2019; Gaita-Ariño et al. 2019; Graham, Zadrozny, et al. 2017; Harvey and Wasielewski 2021; Troiani and Affronte 2011; Wasielewski et al. 2020; Yu et al. 2021). Specifically, atomic control promises a new class of designer qubits capable of functioning as sensors tuned for specific environments or analytes, as nodes that emit photons at desired frequencies for quantum optical networking, and as innovative new approaches to quantum computing. Chemical synthesis offers a unique toolkit of unparalleled atomic control over structure, scalability, and quantum properties, and is poised to accelerate the development of bottom-up quantum technologies.

Within this framework, design molecules can be synthesized to target different applications. This molecular approach spans every component of synthetic chemistry, encompassing both organic and inorganic molecules and extending to molecular solids. The ability to design a molecule, position atoms, tune various properties, and subsequently create arrays or integrated systems is unique to molecular systems. Molecules enable the creation of designer states across chemical platforms—from control over radical coupling and spin transport in organic molecules or peptides, to minute control over ligand fields in transition metal complexes, and fine-tuning over the spin ground-state manifold (M_J) in lanthanides. In the sections that follow, illustrative examples of designing molecules from the ground up to meet quantum metrics, environmental compatibility, and tunability within each of these areas are provided. For each application within QIS, chemists can use the molecular toolkit to design and synthesize the right system to achieve those goals. For example, coherent information transfer over a long distance may be mediated by an organic radical, or single-site spin-photon transduction could be achieved through manipulation of a lanthanide ground state. Transition metals reside between these extremes and may enable integration into targeted environments for sensing.

2.2 DEVELOPING AN UNDERSTANDING OF MOLECULAR STRUCTURE–PROPERTY RELATIONSHIPS NEEDED FOR QIS APPLICATIONS

2.2.1 Increasing Coherence Times in Molecular Qubits and Quantum Memories

Developing molecular systems for QIS applications depends critically on establishing and maintaining quantum-state coherence through the fundamental properties of superposition and entanglement. As described in Box 2-1, an electron spin in a superposition state is very sensitive to its environment (Wasielewski et al. 2020). Since QIS applications depend on maintaining quantum superposition and entanglement, long coherence times are essential. Spin states relax thermally, which is characterized by their longitudinal spin relaxation time T_1, while the loss of spin coherence is characterized by the transverse spin relaxation time T_2. In general $T_1 > T_2$; thus, T_1 places an upper bound on T_2. Designing molecules that feature quantum properties inherently involves synthetic control of coherence (Du et al. 1992; Eaton et al. 2001). The dominant contributions to decoherence are spin–spin coupling, either through electron spin–nuclear spin coupling or hyperfine interactions. Significant work has focused on understanding both T_1 and T_2, where increasing T_2 can be achieved by depleting the local nuclear spin environment (Canarie, Jahn, and Stoll 2020; Graham, Yu, et al. 2017; Krzyaniak et al. 2015; Morton et al. 2007; Wedge et al. 2012). Probing the impact of ligand nuclear spins on coherence properties requires the careful placement of nuclear spins within a system (Figure 2-1). For example, systematically varying the ligands decorating the periphery of a circular $\{Cr_7Ni\}$ ring was used to identify the sources of decoherence. What these

BOX 2-1
Essential Properties of Qubits: Superposition and Entanglement

"Quantum information science (QIS) exploits the intrinsic quantum nature of matter and photons to develop new approaches to computing, communications and sensing. The two basic quantum properties that QIS takes advantage of are superposition and entanglement of quantum states (Figure 2-1-1). The simplest examples of these phenomena involve two-state systems, such as electron spins and photons. For example, if we place an electron spin in a magnetic field, the two relevant quantum states are 'spin up' or 'spin down,' based on the orientation of the magnetic moment of the electron parallel or anti-parallel to the field, respectively. An analogous two-state quantum property of photons is polarization. If we focus on electron spin, observing or measuring the spin will result in spin up $|0\rangle$ or spin down $|1\rangle$ relative to the external magnetic field (top left panel). However, in the absence of a measurement, the electron spin exists in a superposition state (bottom left panel) in which its wavefunction can be described by $\psi = a|0\rangle + b|1\rangle$, where a and b are two coefficients. If two or more electron spins interact with one another through magnetic exchange or a dipole–dipole interaction, they can become entangled. The idea of entanglement is at the foundational heart of quantum mechanics. In 1935, Einstein, Podolsky and Rosen discussed the issue of how a measurement carried out on one electron spin of an interacting pair affects the other (Einstein, Podolsky, and Rosen 1935). A definitive answer to this problem waited until 1964 when John Bell published his paper proving that quantum mechanics is intrinsically 'non-local', that is, a measurement carried out on one electron would immediately result in the other particle having a definitive result, if it too was subsequently measured, no matter how far away from one another the two particles were (right panel) (Bell 1964).

These two fundamental ideas are essential to implementing QIS applications. For example, unlike classical computers that rely on bits with only two values, 0 and 1, a quantum computer takes advantage of quantum superposition, making it possible to use any combination of 0 and 1, which in principle, can greatly increase computational speed. Likewise, if we consider quantum communications, if two sites possess one photon of an entangled photon pair each, information can be transmitted between the two sites by having a third photon interact with one photon of the entangled pair. This strategy, known as quantum teleportation, was first proposed by Bennett to overcome the no-cloning theorem of quantum mechanics (Bennett et al. 1993). Any attempt to eavesdrop on the communication will result in a quantum measurement being made on the entangled photon pair, thus breaking the entanglement, which indicates that the security of the information transmission has been compromised. In the field of sensing, an electron spin that is in a superposition state is very sensitive to its surrounding environment. Even a weak interaction of an electron spin in a chemical species, such as the influence of a nearby nuclear spin, will elicit a quantum measurement resulting in sensing of that molecule. The exquisite sensitivity of the superposition state to its surroundings makes it possible to detect single molecules and to even conduct single molecule magnetic resonance spectroscopy for structural determination."

BOX 2-1 continued

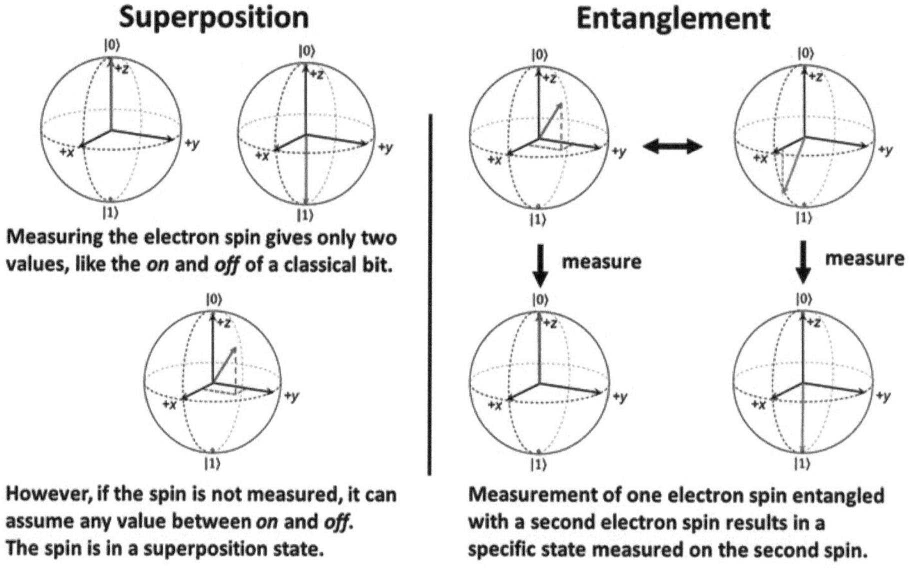

FIGURE 2-1-1 Schematic illustrating the difference between superposition and entanglement.
SOURCE: Wasielewski et al. 2020.

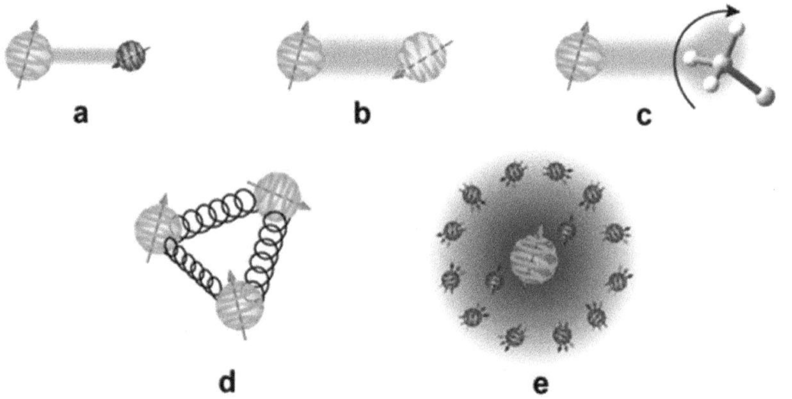

FIGURE 2-1 Factors that affect T_2: (a) nuclear spin diffusion, (b) coupling to nearby electronic spins, (c) methyl group rotation, (d) spin–lattice relaxation (T_1), and (e) spin diffusion barrier.
SOURCE: Graham, Zadrozny, et al. 2017.

studies have found is that by depleting the nuclear spins in the environment, but not those most proximal to the spin-bearing center, it is possible to increase coherence time appreciably. These experimental findings are in concert with theoretical studies on coherence. Notably, work on solid-state materials may provide insight into more complex molecular systems (Kanai et al. 2022) and other systems. Additional studies have probed extrinsic effects on coherence time—for example, protons in a solvent or lattice environment. By studying this phenomenon it is possible to evaluate the extent to which the overall environment impacts nuclear spin-based relaxation processes (Zadrozny et al. 2015). The aggregate of these recent studies demonstrates that precise positioning of nuclear spins enables control and mitigation of important contributions to decoherence (Atzori, Tesi, et al. 2016; Jackson et al. 2019, 2020; Yu et al. 2016).

Control over T_1 is another crucial component of molecular design. Phonons are the quantum unit of a crystal lattice vibration and are inherently temperature dependent because different phonon modes are accessed at different energy scales. To access higher temperature applications (>77 K), an understanding of which phonon modes are most likely responsible for spin relaxation and decoherence is needed. Elegant work from several researchers has demonstrated that certain modes are most important in decoherence processes (see also Amdur et al. 2022; Atzori et al. 2017; Atzori, Morra, et al. 2016; Atzori, Tesi, et al. 2016; Fataftah et al. 2019; Kazmierczak and Hadt 2022; Kazmierczak, Mirzoyan, and Hadt 2021; Lunghi and Sanvito 2020; Tesi et al. 2016). Due to an improved understanding of the contributions of phonon processes in molecules, there are now a handful of examples of molecules that can be controlled at room temperature (Atzori, Tesi, et al. 2016; Bader et al. 2014; Kazmierczak and Hadt 2022; Warner et al. 2013). As spin–phonon coupling affects both spin relaxation and decoherence, and T_1 provides an upper limit on T_2, developing robust models for spectroscopic identification of the unique vibrational modes that contribute to T_1 and T_2 is an area ripe for research (Kazmierczak and Hadt 2022).

While most of these molecules do not have coherence times comparable to defect centers in semiconductors at room temperature, they nevertheless are sufficiently long for certain applications—for example, quantum sensing or potentially communications. Specifically, the utility of a qubit will be related to the number of operations that one needs to execute within the coherence time. There are numerous sensing applications, such as measuring magnetic field or temperature, that have been previously demonstrated with defect centers that require under five pulses, thereby making them compatible with relatively short coherence times (~10 μs). A clear opportunity exists to develop new materials that target specific applications. Future work on phonon engineering, a key research area, may enable room-temperature operation of molecular QIS systems, while increased coherence times at lower temperatures may be valuable for quantum interconnects essential to communications applications.

2.2.2 Creating Optically Addressable Molecular Qubits

The history of harnessing inorganic chemistry to create quantum systems is extensive and has its origins in the field of molecular magnetism. In the early 1990s, researchers discovered that molecules could display magnetic bistability mimicking classical magnets (Gatteschi, Sessoli, and Villain 2006; Sessoli, Gatteschi, et al. 1993; Sessoli, Tsai, et al. 1993). To understand the properties that enabled this unusual behavior, there was a drive to translate ideas of slow magnetic relaxation into information about T_1. One of the key features of molecular magnetism that arose to have critical importance in QIS is control over magnetic parameters of axial and rhombic zero-field splitting (ZFS), D and E, respectively, and spin–orbit coupling. This knowledge has enabled scientists researching QIS to control spin relaxation systems with some degree of understanding of the contributions of ZFS and spin–orbit coupling to spin dynamics. Notably, ZFS exists on the ~1–50 cm^{-1} energy scale, spin–orbit coupling exists on the 10–500 cm^{-1} energy scale, and ligand field interactions tune energy levels on the 10,000 cm^{-1} energy scale. Bringing together these fine and coarse energy-tuning knobs precisely positions molecules to address challenges within QIS. While single-molecule magnetism is a vital field for several applications and comprises the foundation for a molecular approach to QIS, its core goals of achieving, accessing, and manipulating high spin states and magnetic bistability are orthogonal to the aims of QIS (Fataftah et al. 2014; Gaita-Ariño et al. 2019).

The power of synthetic chemistry to create new QIS systems can be envisioned through the illustrative example of developing the molecular analog of color centers (i.e., defect sites) in solid-state materials. Numerous candidates for defects in semiconductors feature optical readout of spin information. These color centers are leading candidates for spin-based quantum information. One approach to creating molecular qubits is to develop molecular analogs of these systems. One of the most promising solid-state qubits is the anionic nitrogen-vacancy (NV) pair defect in diamond (Awschalom et al. 2018; Doherty et al. 2013). This system features optical readout of spin information, a powerful feature that coupled with its optical initialization makes it a core quantum technology. Indeed, initialization, which is essential and challenging for molecules, is a key feature. Initialization fundamentally means placing the system into a specific quantum state, which DiVincenzo (2000) highlighted as a fundamental requirement for a good qubit (see Box 2-2). The DiVincenzo criteria are difficult to fulfill using the Boltzmann populations of electron spin states in molecular systems because of the small energy gap between states. For example, achieving >95 percent spin polarization requires millikelvin temperatures. Even if strategies are used to increase the energy gap (e.g., manipulating Zeeman interactions), magnetic field strengths of ~10 T still require temperatures on the order of 4 K (Harvey and Wasielewski 2021).

Developing similar optical approaches for spin-based molecules is very powerful and enables molecules to feature initialization and optical readout. The combination of these properties enables molecules to be treated as quantum objects and to open up approaches to single-molecule readout, analogous to that of defect-based systems. The quantum properties of defects in semiconductors, such as the NV center, are exemplary, but their spatial properties are sub-optimal. Defects are generally synthesized through ion bombardment, which leads potentially to nanometer-scale control, not the sub-angstrom control desired for most quantum applications. Coupling the optical interface of diamond with a system featuring tunability and atomistic precision would enable the seamless integration of molecules with state-of-the-art readout technology. To access such a system, one has to probe the

BOX 2-2
What Constitutes a Good Qubit? The DiVincenzo Criteria

"In 2000, DiVincenzo enumerated a set of now well-known criteria for designing a good qubit. These include:

i) The ability to initialize a qubit in a well-defined quantum state. For example, this often means preparing a typical two-state system so that all the population is in one of the two states.
ii) The qubit must exhibit long coherence times, although systems with shorter coherence times may also be of interest. Electronic coherences most often decay in less than < 1 ps, vibronic and vibrational coherences last several ps, whereas electron and nuclear spin coherences are much longer-lived, extending from μ–ms and ms–s, respectively. Tailoring molecular structures and their surrounding environment to maximize coherence times is currently a major challenge.
iii) The system must provide a set of universal quantum logic gates that operate on one or two entangled molecular qubits. Quantum gates can be generated by physical and chemical perturbations of the system, such as applied fields and chemical reactions.
iv) The outcomes of qubit operations should be measured in a specific manner, following their unique spectroscopic signatures, where these measurements are frequently performed using time-resolved spectroscopies. Stationary and propagating (moveable) qubits used for QIS protocols should be easy to produce and control. Furthermore, faithful transmission of these propagating qubits between specific sites needs to be assured. Photons are most often employed as moveable qubits, so that photophysical and photochemical knowledge of molecular systems can be exploited in this regard.

From a chemistry perspective, ease of synthesis, ability to prepare complex architectures, scalability and low cost should be added to the above criteria."

SOURCE: Wasielewski et al. 2020.

electronic structure of an NV center in diamond, which enables the center to access optical readout and polarization of spin information.

2.2.2a Creating Optically Addressable Molecular Qubits: Transition Metals

NV centers feature a polarization scheme that relies upon a ground-state triplet, an excited-state triplet, and an intervening singlet (Dobrovitski et al. 2013). Critically, this polarization scheme circumvents the need for millikelvin temperatures for polarization—instead of polarizing a ~1 K temperature difference thermally, spins can be initialized by light. To design the ground state, which is a triplet with a small axial ZFS, the electronic structure that can support this approach needs to be considered. With transition metal complexes, for example, highly symmetric d^2, d^6, and d^8 systems in tetrahedral, trigonal bipyramidal, or octahedral geometries will support a triplet, $S = 1$ ground state (Fataftah and Freedman 2018). By holding the symmetry rigid in a tetrahedral d^2 system, for example, the ZFS will be relatively low, which is crucial for microwave manipulation of the quantum unit. The excited-state design needs to consider the dynamics of both the excited and ground spin states to enable photodriven spin polarization and ensure that the singlet state is higher in energy than the triplet state. The triplet state necessarily arises from the promotion of an electron from the lower energy level to the higher one, meaning it will be extremely sensitive to the ligand field strength. The singlet state, however, will be relatively insensitive to the ligand field strength because it is a spin-flip transition. Therefore, a strong ligand field is required for the correct alignment of the excited-state manifold. By mimicking this structure, it is possible to initialize the system first through a selective polarization scheme. Once the system is initialized, the system can be manipulated with microwaves and optical information can be read out. By developing a series of organometallic Cr^{4+} molecules (Figure 2-2), an interdisciplinary team of chemists and physicists created and manipulated such a system (Bayliss, Laorenza, et al. 2020; Laorenza et al. 2021). This proof of concept demonstrates the power of molecular chemistry to create new systems by design. Furthermore, optical readout of spin information in the ground state integrates seamlessly with quantum technology developed for reading out and manipulating defects in semiconductors.

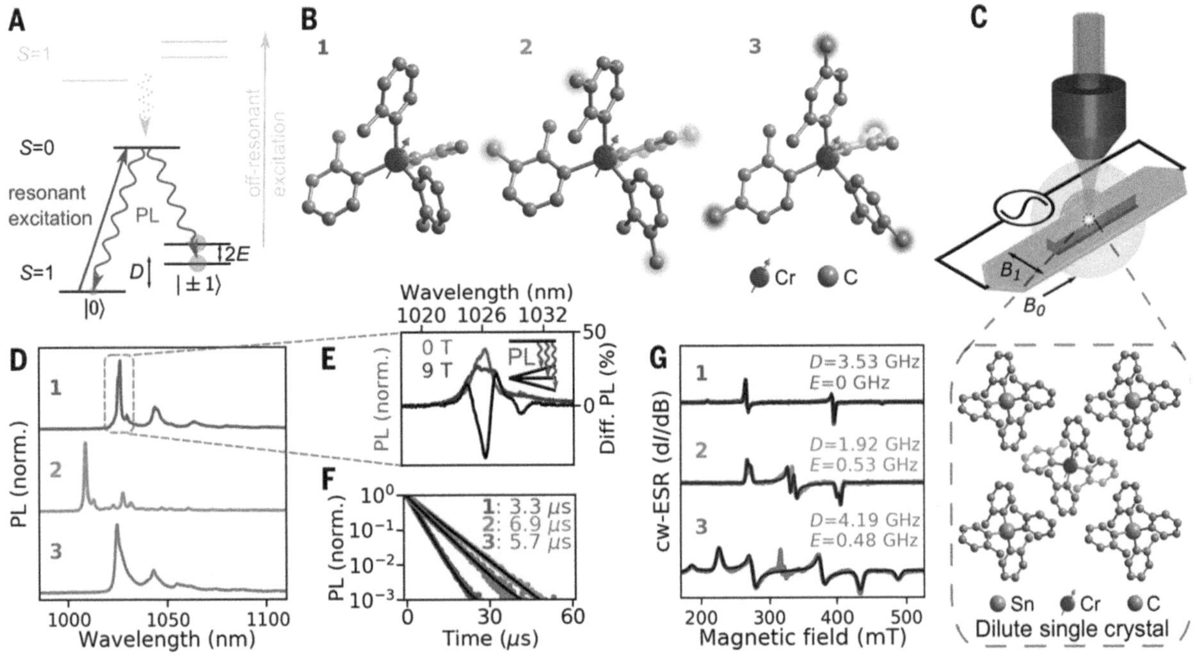

FIGURE 2-2 Molecular color centers featuring optical readout of spin information.
SOURCE: Bayliss, Laorenza, et al. 2020.

With this design, we can move forward and consider the periphery of the system. Careful molecular design can use tethering groups, such as siloxanes, to bind to SiO_2 or imbue the molecule with water solubility. The ability to design a molecule from the ground up with the desired properties, couple it to other molecular centers, and tune its properties offers a tremendous opportunity for chemistry to target QIS applications.

It is important to note that this approach is broadly generalizable; by identifying the key quantum attributes that are desired, chemists can build from the ground up. While the example is based on transition metals, lanthanides and organic molecules offer similar versatility as illustrated below. Lanthanides offer narrow emission lines—for example, erbium emission lies in the telecom range—while organic molecules feature radicals with long coherence times that can be readily initialized using photophysical processes. With all of these systems, individual units can be connected via a building block approach into larger conjugates. Metal–organic frameworks, covalent organic frameworks, polymers, self-assembly chemistry, DNA origami, and other chemical strategies are well established for creating large architectures that prove useful for assembling multiqubit arrays.

2.2.2b Creating Optically Addressable Molecular Qubits: Lanthanides and Actinides

The study of quantum phenomena and the development of quantum information processing systems have benefited from understanding the chemistry and physics of f-electron compounds and materials. f-electron compounds include those formed from lanthanide or actinide elements. In this part of the periodic table, the f-electron subshells are filled from f^0 to f^{14} electronic configurations for the tri-positive cations from La–Lu in the lanthanides and Ac–Lr in the actinides, respectively. Ions and materials with occupied (or partially occupied) f orbitals have provided key test beds for fundamental studies of a range of quantum phenomena including superconductivity (H. Wang et al. 2019), quantum tunneling of magnetization (Goodwin et al. 2017; Guo et al. 2018), quantum critical points (Kaluarachchi et al. 2018), time crystals (Zhang et al. 2017), teleportation (Olmschenk et al. 2009), optical cycling (Siyushev et al. 2014), clock transitions and others. Open-shell f-electron compounds provide unique electronic and magnetic properties derived from their quantum mechanical characteristics. Coupling of their spin and unquenched orbital angular momenta results in total angular momentum J states that are split by crystal field levels into M_j sublevels with large magneto-crystalline anisotropies.

Quantum information can be controlled in f-electron spins because of their inherently *almost* core-like, quantum mechanical characteristics (Cheisson and Schelter 2019). The 4f shells of lanthanides and the 5f shells of actinides are in the "Goldilocks zone" between valence and core (Jochen Autschbach, private communication, April 2019). The valence-like properties allow for f-shells to be tuned using chemical modifications in the first coordination sphere of the metal cations for QIS applications. Conversely, the core-like nature of f-electrons provides inherent protection against decoherence for local and emergent phenomena. f-element materials that display notable quantum phenomena include ceramics and intermetallics.

In an orthogonal area of research, molecular complexes of f-elements have also been explored extensively for quantum effects in dilute spin systems, such as single-molecule magnets. While single-molecule magnets operate on the opposite principles as qubits, the deep knowledge that has been developed about the electronic structure of such species can be repurposed toward QIS, in particular, based on understanding magnetic relaxation (Reta, Kragskow, and Chilton 2021). In these and related compounds, blocked magnetization is relaxed through quantum tunneling processes and one- (Orbach) or two-phonon (Raman) phonon scattering processes. Understanding how the factors that influence electronic spin and molecular vibration can benefit both fields (Chilton 2022). Additional understanding of electronic structure for quantum information can be gleaned from work on new modes of bonding in f-elements that can contribute to new types of strong spin polarization. Gould and colleagues (2022) have made progress in this regard with the isolation of a compound showing a 0.5-order metal–metal bond between two lanthanide centers. This new family of compounds—$(Cp^{i}Pr_5)_2Ln_2I_3$—shows magnetic blocking below 65 K for Ln = Tb and 72 K for Ln = Dy, as well as unprecedented strong magnetic anisotropy owing to very strong magnetic exchange. Building off this foundation of knowledge to create molecules that are not magnetically bistable, such as single-molecule magnets, but can be manipulated in a superposition will benefit the emerging area of molecular QIS.

Furthermore, lanthanides have core electrons that are essentially protected from the environment; thus, this attribute can be used to increase coherence or maintain properties across a range of external environments.

Certain lanthanides have the capability to interconvert light and spin information (i.e., to perform quantum transduction). Laorenza and Freedman (2022) demonstrated in one example the potential to interface molecular spins with telecommunication photons by designing systems with Er^{3+} emitters. These molecules have a natural transition at ~1540 nm and, when placed in the appropriate crystal field environment, function as an effective $S = ½$ ground state. This attractive optical transition has sparked impressive work with Er^{3+} dopants in yttrium orthosilicate, including single-spin control, quantum nondemolition measurements, and coherent control of multiple spin centers at the same optical spot. Molecular approaches provide a complementary approach to designing Er^{3+} emitters. As the natural 4f-4f transition at ~1540 nm is largely insensitive to the surrounding ligand environment, bottom-up design may be used to place nuclear spin memories on the surrounding ligands while developing the principles to achieve long spin coherence in systems that interface with telecom fibers.

A second example is the potential for molecular lanthanides to serve as optical memories. According to Laorenza and Freedman (2022), "These dopants in solid-state hosts (e.g., yttrium aluminum garnet, yttrium orthosilicate, yttrium orthovanadate, lithium niobate) have emerged as a valuable platform for quantum optical networking with single-spin optical readout of Ce^{3+}, Pr^{3+}, Nd^{3+}, Yb^{3+}, and Er^{3+}, as well as long-lived quantum optical memories for ensembles of rare-earth ions. The trivalent ions Pr^{3+}, Eu^{3+}, and Tm^{3+} have demonstrated an optically addressable ground-state *nuclear* spin. Furthermore, the $^{151}Eu^{3+}$ spins have been shown to exhibit coherence times as long as six hours. These dopants offer 4f-4f transitions that are highly shielded from their environment. This level of shielding gives rise to a narrow homogenous linewidth, Γ_{hom}, for spin initialization, long optical coherence time, $T_{2,\,opt}$, and high quantum yields. Translating these features into molecular Eu^{3+} spins, recent work from Kumar and colleagues (2021) demonstrated optical initialization and readout of the ground-state nuclear spin in dinuclear Eu^{3+} molecules, $[Eu_2Cl_6(4\text{-picoline N-oxide})_4(\mu_2\text{-4-picoline N-oxide})_2]\cdot 2H_2O$, denoted ($Eu_2$) (Figure 2-3a). The narrow Γ_{hom} of 22 MHz enabled all optical initialization of the ground state. Using a mononuclear Eu^{3+} complex, [piperidin-1-ium][Eu(benzoylacetonate)$_4$], Serrano and colleagues (2022) demonstrated a three-order-of-magnitude improvement with a recorded Γ_{hom} value of 13 kHz (Figure 2-3b)." The energy diagram shown in Figure 2-3c illustrates the relevant transitions that are occurring in the dopant to allow for selective optical pumping; > 95% of the spin population is was initialized into the $F_0M_I = \pm ½$ sublevels.

Complexes of actinide elements, where the 5f- and/or 6d-valence electron shells are partially filled, also present unique opportunities related to the manipulation of electron spins for QIS applications. One possible advantage of actinides is that the larger radial extent of the 5f principal quantum shell and indirect relativistic effects that are operative for these heavy atoms render the 5f electrons more accessible for covalent bonding interactions and modifications through structural chemistry, as compared to the 4f electrons. Thus, actinide elements, especially early actinides, have been considered as conferring advantages that derive from a chemical behavior intermediate between lanthanide and d-block transition metal systems. While thus far the work on understanding the electronic

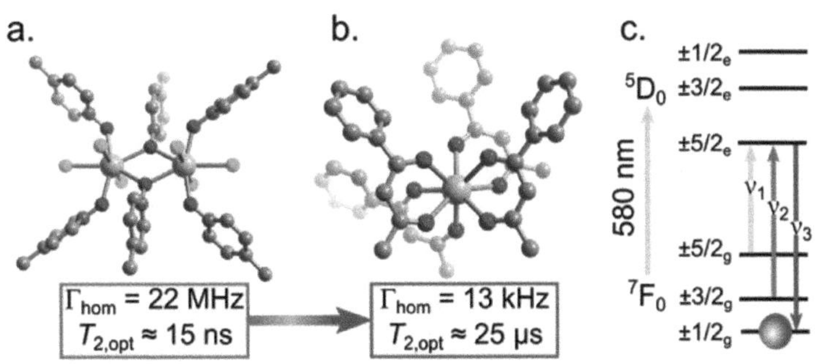

FIGURE 2-3 Optically addressable quantum memories. Molecular structures determined from single crystal X-ray diffraction for (Eu_2), where Eu is represented as the pink ball (a) and [piperidin-1-ium][Eu(benzoylacetonate)$_4$] (b), with carbon, nitrogen, oxygen, chlorine, and europium shown. (c) Energy level diagram.
SOURCE: Laorenza and Freedman 2022.

structure of these systems focused on uranium, significant space exists beyond that element. All isotopes of the actinides are radioactive, but hazards associated with the radioactivity of uranium, primarily ^{238}U, and thorium, primarily ^{232}Th, are relatively minor such that those isotopes can be handled in a conventional laboratory setting. Rinehart and Long (2009) reported the first uranium single-molecule magnet with a spin relaxation barrier of 20 cm^{-1}: $U(Ph_2BPz_2)_3$, $[Ph_2BPz_2]^-$ = diphenylbis(pyrazolyl)borate. Evidently, the complex valence electronic structures of uranium complexes comprising strongly mixed electronic wave functions and a tendency for those to couple strongly with molecular vibrations undermines better performance. Deconvoluting these competing interactions represents an opportunity for creating a first generation of molecular qubits based off of these ideas that were developed for single-molecule magnets (Escalera-Moreno et al. 2019).

For actinide elements heavier than uranium, so-called transuranic elements, there are notable opportunities but also practical challenges due to radiotoxicity and requirements for the safe handling of these isotopes. The key isotopes of interest are ^{237}Np and ^{239}Pu, whose non-integer nuclear spins could couple with the valence electron spin. To advance the study of transuranic QIS systems, it is necessary to have access to rare isotopes and to the infrastructure to be able to synthesize and study them safely. Engineering or administrative requirements for sample handling of transuranics typically demand only small amounts of material (e.g., <5 mg of the isotope). This situation has resulted in the reporting of only a small number of Np and Pu complexes. Within the United States, much of the work on transuranic isotopes is performed in the national laboratory system, largely due to both safety and security concerns; a smaller fraction is performed in certain academic laboratories where the necessary licensing and infrastructure are in place.

2.2.2c Creating Optically Addressable Molecular Qubits: Organic Multispin Qubits

Molecular qubit design principles based on fully organic systems utilize the premise that decoherence in metal-based molecular spin-qubits has significant contributions from spin–orbit coupling and ZFS. In organic systems, spin–orbit coupling is generally very weak, leading to the prediction of longer decoherence times in organic spin-based molecular qubit architectures. Contributions to spin relaxation T_1 are dominated by the direct process, spin–phonon coupling, and the Raman process, while contributions to T_2 decoherence times are dominated by hyperfine coupling to spin ½ hydrogen atoms within organic molecules (Canarie, Jahn, and Stoll 2020).

Based on the DiVincenzo criteria (Box 2-2), a critical requirement for a physical qubit is the preparation of a pure initial state. In addition, the preparation of two-qubit entangled states is necessary to execute fundamental quantum gate operations. The primary challenge is generating well-defined initial quantum states of the system that maintain their spin coherence for times long enough to permit a useful number of spin manipulations to carry out quantum gate operations. As was discussed earlier in Section 2.2.2, these criteria are difficult to fulfill using the Boltzmann populations of the electron spin states due to the small energy gap between the states (Harvey and Wasielewski 2021).

Electron spins are good qubits because their two spin states constitute the quintessential two-level quantum system, in which the two states can exist in a superposition. In addition, coupling two or more spins via the spin-spin exchange (J) and/or dipolar (D) interactions results in rich spin physics that allow for quantum entanglement as well as implementation of two-qubit gates essential for quantum gate operations. Sub-nanosecond photodriven electron transfer (ET) from a molecular electron donor (D) to an acceptor (A) can generate two spatially separated, entangled electron spins that function as a spin qubit pair (SQP) in a well-defined pure initial singlet quantum state even at ambient temperature (Figure 2-4a) (Closs, Forbes, and Norris 1987; Thurnauer and Norris 1980).

Generally, the two spins of the SQP experience different magnetic environments as a consequence of differing electron-nuclear hyperfine interactions in $D^{\bullet+}$ and $A^{\bullet-}$, as well as differing spin–orbit interactions in each radical leading to different electronic g-factors. This results in coherent spin evolution from the singlet to the triplet SQP state (Closs, Forbes, and Norris 1987; Hore et al. 1987). In the absence of SQP spin relaxation or recombination, the spin coherence between the singlet and triplet SQP states can persist indefinitely.

However, spin decoherence and charge recombination to either the ground state via the singlet channel or a neutral molecular triplet via the triplet channel can occur. Upon application of a magnetic field that is much larger than J, D, and the electron-nuclear hyperfine interactions, only the S and T_0 states mix (Figure 2-4b), which results

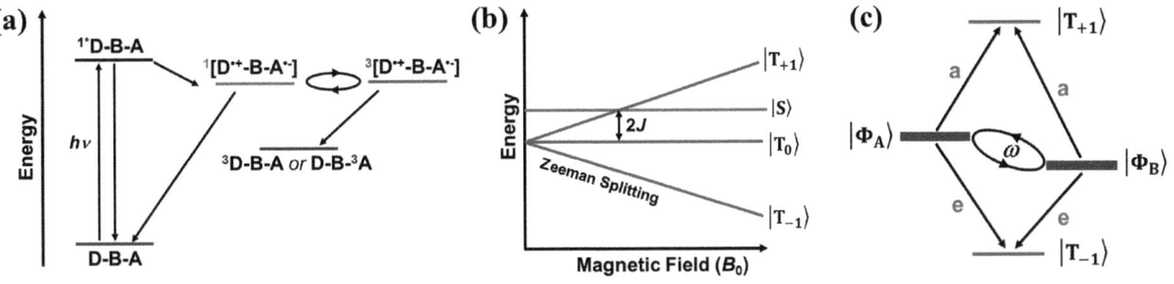

FIGURE 2-4 (a) Electron transfer and intersystem crossing pathways in a donor-acceptor (D-A) system. (b) Radical ion pair energy levels as a function of magnetic field for $J > 0$, $D = 0$. (c) Radical pair energy levels in the high magnetic field limit showing the result of mixing $|S\rangle$ and $|T_0\rangle$ states, where ω is the mixing frequency.
SOURCE: Harvey and Wasielewski 2021.

in their overpopulation and produces strong spin-polarization that can be observed readily using time-resolved electron paramagnetic resonance (EPR) or optically detected magnetic resonance (ODMR) spectroscopies (Figure 2-4c) to yield detailed data on the spin dynamics of the system. For example, using time-resolved EPR techniques, researchers have shown that photogenerated SQPs can polarize a third spin (Colvin et al. 2013; Horwitz et al. 2016, 2017; Mi et al. 2006), engage in quantum–spin state teleportation (Rugg et al. 2019), and serve as a controlled NOT (CNOT) gate (Nelson et al. 2020). In addition, using g-factor engineering, individual spin qubit addressability can be achieved within an SQP system (Fernandez et al. 2015; Nakazawa et al. 2012; Olshansky et al. 2020).

A second promising approach to photoinitialized molecular qubits having optical pumping and addressability properties similar to those of NV centers uses photoexcited, covalently linked chromophore-stable radical (C-R•) systems. Three-spin systems that produce photogenerated molecular quartet states were first observed using porphyrin and fullerene chromophores connected to stable nitroxide radicals via covalent or coordination bonds (Corvaja et al. 1995; Fujisawa et al. 2001; Ishii et al. 1996, 1998; Mizuochi, Ohba, and Yamauchi 1997). Metalloporphyrins with paramagnetic metals also exhibit photogenerated quartet spin states (Gouterman 1970; Kandrashkin, Asano, and van der Est 2006a, 2006b; Poddutoori et al. 2019). In addition, more recent work has demonstrated that robust perylenediimide chromophores linked to nitroxide radicals can produce quartet states that subsequently spin-polarize the nitroxide doublet ground state (Giacobbe et al. 2009; Maylaender et al. 2021).

Figure 2-5 depicts a typical photophysical pathway for a C-R• molecule. Upon photoexcitation, the chromophore of the doublet ground state (D_0) is optically pumped to its first excited state (D_1), followed by enhanced

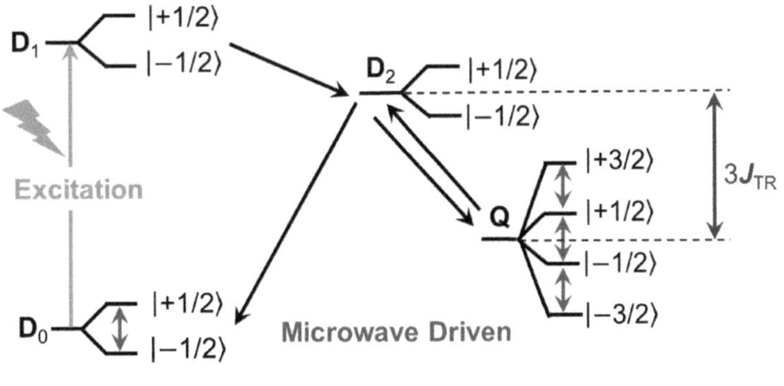

FIGURE 2-5 Representation of the photophysical and spin dynamics following photoexcitation of a chromophore-radical system. Resonant microwave pulses (blue arrows) induce the change of population in different spin sublevels, which can be detected using optically detected magnetic resonance.
SOURCE: Qiu et al. 2022.

intersystem crossing driven by the exchange interaction between the two spin-paired electrons on $^{1*}C$ with the unpaired electron on R$^{•}$ to generate $^{3*}C$-$^{2}R^{•}$. The resulting three-spin system is best described at high magnetic fields as an excited doublet state (D_2) and a quartet state (Q). The D_2 and Q states are separated by an energy difference of $3J_{TR}$, and Q is typically lower in energy than D_2. Since D_1 and D_2 have the same spin multiplicity, the transition from D_1 to D_2 is more rapid than to Q, which is populated by intersystem crossing from the D_2 state driven by the ZFS (Kandrashkin and van der Est 2003, 2004; Teki 2020).

Furthermore, the decay of Q to D_0 is spin forbidden, allowing a sufficiently long lifetime to probe and manipulate Q using resonant microwave pulses. The three electronic spins in the photogenerated quartet state Q form a multilevel qubit, or qudit, which can reduce the number of entangled gates required in quantum algorithms, thus improving algorithmic efficiency (Wang et al. 2020). C-R$^{•}$ molecules also offer the possibility of optical readout of spin information using ODMR because the chromophore excited state is usually photoluminescent or has readily observable excited-state absorptions.

2.2.2d Creating Optically Addressable Molecular Qubits: Optical Cycling Centers

Optically active molecules could provide high-fidelity quantum-state initialization through optical pumping and high-fidelity qubit readout through the detection of laser-induced fluorescence. Both purposes require the ability to scatter optical photons repeatedly (i.e., optically cycle). To achieve optical cycling, it is necessary to find an optically closed molecular system, in which a molecule excited by resonant laser light returns to its initial quantum state with high probability and continues to interact with the laser light. Although optical cycling is routinely achieved in neutral atoms and atomic ions, it is challenging to realize in molecular systems owing to the large number of quantum states. A successful approach has been to engineer atom-like electronic structures—for example, using the NV centers or designer organometallic qubits discussed above. A complementary approach has been to identify molecules in the gas phase that permit optical cycling even in isolation. These molecules, known as optical cycling centers (OCCs), have electronic transitions with nearly diagonal Franck–Condon factors that enable the optical cycling of large numbers of photons.

The topic of OCCs lies at the intersection of atomic physics and chemistry. Initial work with small OCCs started when the first candidates were identified in the early 2000s (Di Rosa 2004; Stuhl et al. 2008). Early work had been pursued by the atomic physics community, whose motivation was to extend established optical techniques of atomic control such as optical pumping and laser-cooling to molecules. This effort has been successful. To date, many OCCs have been identified; experimentally, a variety of optical control techniques in atoms have been extended to OCCs. Notably, diatomic OCCs have emerged as promising molecular qubit candidates. As first proposed by DeMille (2002), trapped polar molecules can be a promising platform for quantum computation. A qubit can be encoded in two rotational states of a polar molecule to provide long-lived and highly coherent quantum memory. By relying on the long-ranged electric dipolar interaction that couples rotational states of a molecule, two-qubit gates sufficient for universal quantum computation can be naturally implemented (Ni, Rosenband, and Grimes 2018). On this frontier, recent work with laser-cooled OCCs trapped in programmable optical tweezer arrays has demonstrated the DiVincenzo criteria (see Box 2-2) required for quantum computing (Bao et al. 2022; Holland, Lu, and Cheuk 2022). In particular, the abilities to prepare and detect defect-free arrays of single molecules with high fidelity were demonstrated; coherent electric dipolar interactions and on-demand creation of entangled pairs of molecules were also shown. These results establish optical tweezer arrays of molecular OCCs as a promising new platform for quantum science. Notably, such molecular tweezer arrays are scalable to hundreds of qubits in the near term and naturally offer the ability for single-qubit resolved addressing (Figure 2-6).

We note in passing that another successful approach to creating ultracold and trapped molecules is coherently assembling them from samples of ultracold atoms, which can be routinely created and controlled to a high precision in the laboratory. These molecules, however, do not permit the high degree of optical cycling necessary for direct fluorescent detection, quantum-state preparation, or laser cooling. Nevertheless, assembled molecules are also being pursued successfully for quantum science applications. In fact, coherent electric dipolar interactions between polar molecules were first observed using assembled $^{40}K^{87}Rb$ molecules trapped in an optical lattice

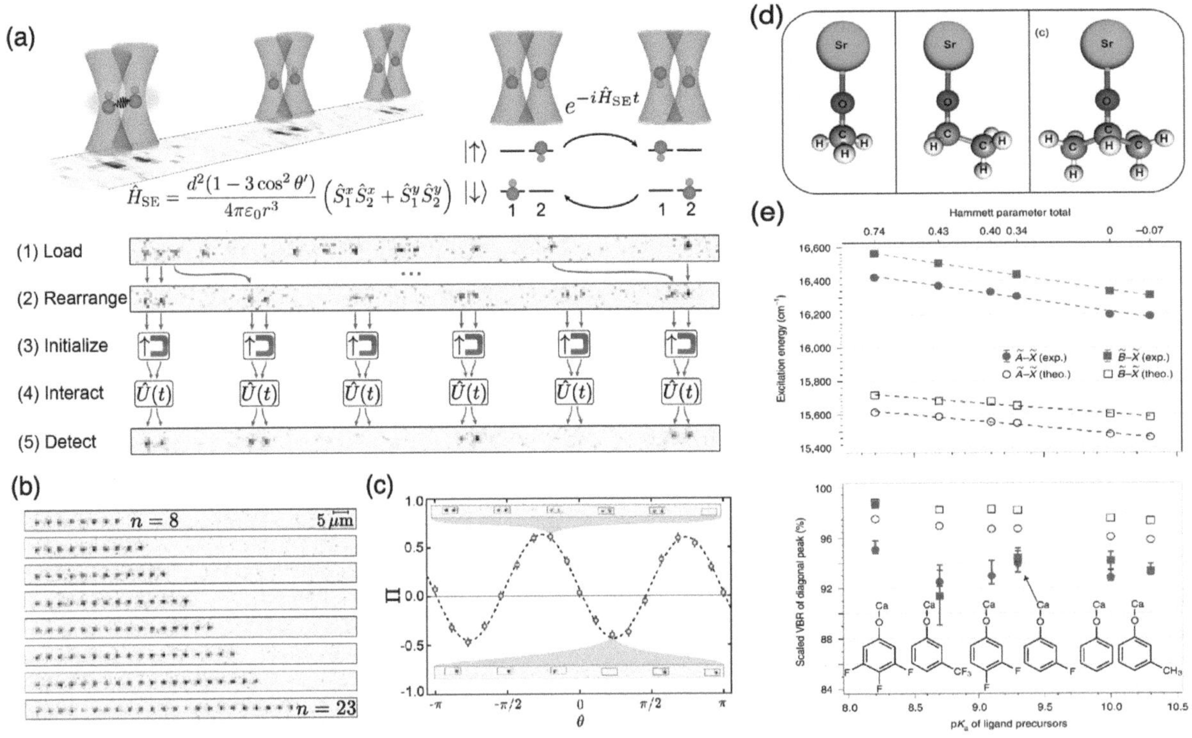

FIGURE 2-6 (a) Diatomic optical cycling centers (OCCs) trapped in programmable optical tweezer arrays as a new molecular qubit platform. The electric dipolar interaction between OCCs allows two-qubit gates to be implemented. (b) Images of defect-free arrays of single diatomic OCCs. (c) Parity oscillations establish the creation of maximally entangled Bell pairs of OCCs. (d) M-O-R motif for optical cycling. (e) Correlations of excitation energies and vibrational branching ratios with Hammett total and ligand pK_a.
SOURCES: (a, b, c) Holland, Lu, and Cheuk 2022; (d) Kozyryev, Baum, Matusda, and Doyle 2016; (e) Zhu et al. 2022.

(Yan et al. 2013). Recently, single $^{23}Na^{133}Cs$ molecules have also been successfully created and fully controlled in optical tweezer traps (Liu et al. 2018).

Parallel to the successful developments with diatomic OCCs, efforts to find larger OCCs have significantly increased. Larger OCCs could open up new areas in quantum science and chemistry, including new schemes for processing quantum information (Yu et al. 2019), enhanced sensitivity to new fundamental physics (Augenbraun et al. 2020; Hutzler 2020; Kozyryev, Lasner, and Doyle 2021), and new methods to control and witness molecular dynamics (Zhu et al. 2022). Recent proposals have also envisioned OCCs that are chemically bonded to surfaces, which could give rise to practical quantum devices such as sensors and quantum interconnects (Guo et al. 2021).

The chemistry community is increasingly interested in the topic of larger OCCs. Since the general structural and chemical principles that determine optical cycling properties are not fully known, identifying larger molecular candidates for OCCs remains challenging. Nevertheless, progress has been rapid, and varied organic molecules functionalized with OCCs have been discovered. Notably, recent collaborations between chemists and physicists have identified new chemical principles that determine optical properties in certain classes of organic molecules (Dickerson, Guo, Shin, et al. 2021; Lao et al. 2022; Zhu et al. 2022). These chemical principles, and those yet to be discovered, could guide the identification of candidate OCCs. Below, we describe in more detail some OCCs that have been explored to date.

Diatomic OCCs are well explored. Early work identified optical cyclable diatomic OCCs and an optical cycling scheme with only a few lasers (Di Rosa 2004; Stuhl et al. 2008). Initial experimental work with diatomic OCCs has focused on ^2S radicals such as SrF (Barry et al. 2014), CaF (Anderegg et al. 2017; Truppe et al. 2017), and YO

(Collopy et al. 2018). With successful optical cycling, many atomic techniques such as sub-Doppler laser cooling and conservative trapping have been extended to diatomic OCCs to bring them into the ultracold quantum regime (Anderegg et al. 2018; Cheuk et al. 2018; Williams et al. 2018). Notably, single OCCs have been trapped and detected with high fidelity in optical tweezer arrays (Anderegg et al. 2019). Recent work in this platform has created additional defect-free arrays of OCCs, achieved programmable control over individual OCCs, and demonstrated an entangling two-qubit gate sufficient for universal quantum computation (Holland, Lu, and Cheuk 2022). These results establish OCCs as a viable platform for quantum simulation, quantum computing, and quantum-enhanced sensing, and further open the door to exploring chemistry with entangled matter.

In addition to ^{2}S OCCs, a new class of ^{1}S molecules such as aluminum monofluoride (Hofsäss et al. 2021) and aluminum monochloride (Daniel et al. 2021) has garnered increased attention over the past few years. These molecules have favorable properties, such as magnetic insensitivity and narrow optical transitions, useful for highly coherent qubits and new types of molecular clocks but that come at the expense of ultraviolet transitions that are technically difficult to work with. Whether ^{1}S OCCs with convenient optical transitions exist remains an open question.

Motivated by the alkaline-earth fluorine motif in diatomic OCCs of CaF and SrF, initial work on polyatomic OCCs extended to alkaline earth alkoxide, M-O-R, molecules, such as SrOH, CaOH, and $CaOCH_3$ (Kozyryev et al. 2019; Kozyryev, Baum, Matsuda, et al. 2016; Kozyryev, Baum, Matsuda, and Doyle 2016). These efforts have been successful, and optical control over CaOH and $CaOCH_3$ is rapidly approaching that in diatomic OCCs (Hallas et al. 2023; Mitra et al. 2020; Vilas et al. 2022). Recent work has extended this idea to a large class of M-O-R molecules (Dickerson et al. 2022; Dickerson, Guo, Shin, et al. 2021; Dickerson, Guo, Zhu, et al. 2021; Mitra et al. 2022; Zhu et al. 2022). Notably, in M-O-R aromatic compounds functionalized with OCCs, the transition energies and degree of optical cycling have been found to correlate well with the electron-withdrawing strength of the R-ligand (Dickerson, Guo, Shin, et al. 2021; Zhu et al. 2022), providing a new chemical design principle to tune Franck–Condon factors. While the M-O-R appears to be a very successful design principle, whether other motifs for creating larger OCCs exist remains an open question. In a closely related area, recent work has also explored small molecules with multiple OCCs, where the OCCs could be separately optimized for different purposes such as sensitivity to new physics and favorable laser-cooling and optical trapping properties (Ivanov, Gulania, and Krylov 2020; Kłos and Kotochigova 2020; Yu et al. 2022).

2.2.2e Exploiting Entanglement and Quantum Transduction

The coherent coupling of molecules with photons allows long-range entanglement of qubits, opening up the possibility of quantum networks over long distances. Separately, photons can be used as a universal bus to transduce quantum information to other physical platforms such as superconducting qubits. In addition, visible to near-infrared (NIR) radiation provides a wealth of information through its frequency, phase, and polarization; it also provides spatial information down to the length scale of the optical wavelength or below using near-field techniques. Optical detection ensures that molecular qubits can be addressed on a single-molecule level. In particular, the strong interaction of molecular chromophores with light (visible to NIR radiation) provides a facile means of optically addressing spins to initialize and read out qubit states (Awschalom et al. 2018). Strategies for optically gating spin states on the single-molecule level have been demonstrated in transition metal-semiquinone complexes via photoisomerization-induced spin-charge excited states (Paquette et al. 2018) and single-photon-induced spin polarization strategies (Kirk, Shultz, Reddy Marri, et al. 2022).

According to Mao (2023), a promising route for realizing multiple-spin entangled molecular qubits in organic systems is the photogeneration of high spin states in organic semiconductors using singlet fission (SF). In SF, the absorption of a single photon generates a triplet-pair multiexciton, $^1(T_1T_1)$, in an initial singlet state ($S = 0$). (Smith and Michl 2010). If the $^1(T_1T_1)$ lifetime is sufficiently long, spin evolution can occur to produce the quintet ($S = 2$) state, $^5(T_1T_1)$, before the spins decohere (Figure 2-7) (Basel et al. 2017; Kumarasamy et al. 2017; Matsuda, Oyama, and Kobori 2020; Sakai et al. 2018; Tayebjee et al. 2017; Weiss et al. 2017). The $^5(T_1T_1)$ state comprises four entangled spins that can be initialized into a pure, well-defined quantum state by optical pumping. In addition, the spins of the $^5(T_1T_1)$ state can be addressed and manipulated using pulsed microwaves to execute quantum gate

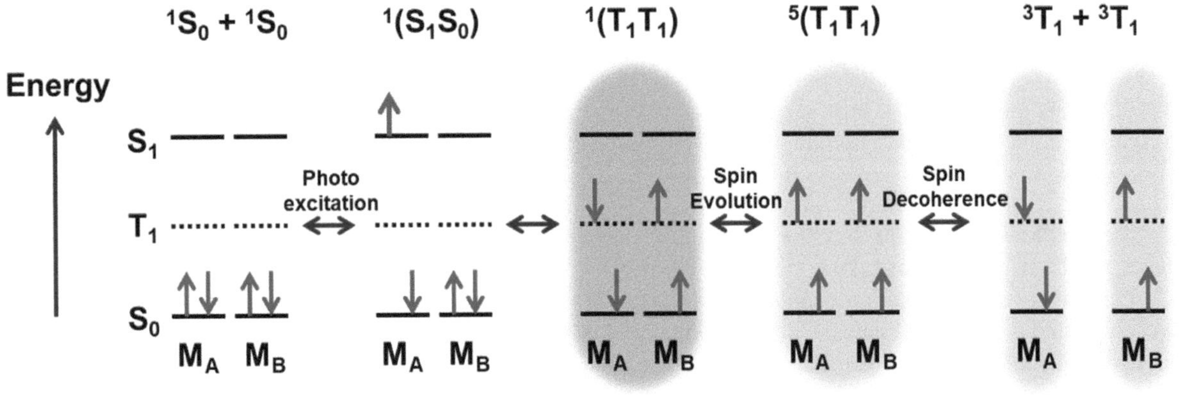

FIGURE 2-7 Schematic representation of the photophysical and spin dynamics of singlet fission.
SOURCE: Mao et al. 2023.

operations, the results of which can be read out using either pulse-EPR or ODMR spectroscopies (Smyser and Eaves 2020; Tayebjee et al. 2017; Weiss et al. 2017; Yunusova et al. 2020). The latter potentially offers single-spin sensitivity and is possible because the decay of the $^5(T_1T_1)$ state by triplet-triplet annihilation produces delayed fluorescence. Using entangled photons to produce the $^5(T_1T_1)$ state may also enable transduction between photons and spins that serve as propagating and stationary qubits, respectively.

Several other qualities of the $^5(T_1T_1)$ state also make it attractive for QIS applications. Due to its high spin multiplicity, $^5(T_1T_1)$ can be utilized as a five-level qudit, which may facilitate greater storage and processing of quantum information compared to systems with fewer accessible quantum levels (Moreno-Pineda et al. 2018; Wang et al. 2020; Wang et al. 2021). The $^5(T_1T_1)$ state can also be photogenerated on demand at arbitrary locations, potentially with high spatial resolution using nanophotonic architectures (Kauranen and Zayats 2012; Tame et al. 2013). Investigations of the $^5(T_1T_1)$ state for QIS applications have been limited, despite its intriguing potential properties (Bayliss, Weiss, et al. 2020; Jacobberger et al. 2022; Weiss et al. 2017). Consequently, the mechanisms of its decoherence are not well understood, and strategies for extending coherence lifetimes have not been fully realized. While SF has been observed in molecular dimers or higher aggregates of more than 200 chromophores in both solution and the solid state (Casillas et al. 2020), $^5(T_1T_1)$ has been relatively elusive in solid-state materials (Bae et al. 2020; Lubert-Perquel et al. 2018; Matsuda, Oyama, and Kobori 2020; Pace et al. 2020; Weiss et al. 2017). To realize a long-lived $^5(T_1T_1)$ state, the interchromophore electronic coupling must be sufficiently strong to allow efficient SF and prevent dissociation of the triplet-pair state, yet sufficiently weak to minimize triplet-triplet annihilation and triplet diffusion. Striking this delicate balance is a major challenge. In the solid state, electronic coupling is dictated by the crystal structure, which is difficult to predict or rationally control due to the various weak intermolecular forces that govern molecular packing (Corpinot and Bucar 2019; Day and Cooper 2018). Recently, the advantages of engineering the crystal morphology of tetracene (Bayliss, Weiss, et al. 2020; Jacobberger et al. 2022; Weiss et al. 2017) and related polyacene (Rugg et al. 2022) single crystals were demonstrated. In the most favorable case reported, the lifetime of $^5(T_1T_1)$ is 130 μs, and the spin coherence lifetime is 3 μs at 5 K with $^5(T_1T_1)$ readily observable even at room temperature (Jacobberger et al. 2022). The single crystals spatially align and organize the interchromophore spacing to optimize the electronic coupling needed to achieve favorable $^5(T_1T_1)$ properties. This enables more facile identification of decoherence sources, facilitating the development of guidelines to further tailor the $^5(T_1T_1)$ state for QIS applications.

2.2.3 Beyond Qubits: Multispin Systems, Error-Mitigation and Error-Correction

While challenges in the field of molecular QIS have focused primarily on the structural factors that control decoherence rates and relaxation dynamics of qubit states, significant challenges remain in the design and control of both superposition and entangled states critical for quantum algorithms, error correction and error mitigation, communications, and sensing. Within this context, "qudits," or d-dimensional quantum systems, have significant advantages over simple two-state qubits. The entangled quantum states lead to the implementation of quantum algorithms with decreased code sizes and fewer entangling gates, quantum error correction, and robust quantum cryptography protocols. For example, the possibility of information-scrambling protocols with entangled qutrits ($d = 3$) leads to effective teleportation protocols that can be carried out even in the presence of decoherence and decreased fidelities for quantum communication (Blok et al. 2021). The physical realization of qudits and qutrits has been explored within the physics realm through entangled superconducting qubits and entangled photons (Kues et al. 2017; Sciara et al. 2021).

Early approaches to quantum control in molecular systems involved the physical realization of single-qubit, two-qubit, and qudit logical gates within rotational and vibrational states of a diatomic molecule with operations via resonant Raman transitions (Shapiro et al. 2003). The entanglement of nuclear and electron spin or electron and electron spin that is possible in molecular systems allows the discovery of molecular qudits. Unlike two-level qubits ($b = 2$), qudits are quantum systems with dimensional space ($d > 2$) (Moreno-Pineda et al. 2018). The entanglement among electron-nuclear ($S + I$), nuclear-nuclear ($I + I$), and electron-electron ($S + S$) states in molecular dimers or trimers allows the development of strategies to uniquely address error correction and mitigation critical to quantum computation, sensing, and telecommunication protocols.

Within the context of information theory, a question arises as to how one transmits information reliably over a noisy channel. Shannon (1948) developed the noisy channel coding theorem, which quantifies the upper limit of information transmitted over a noisy channel and provides an upper limit to the protection afforded by error-correcting codes. Although no analog to Shannon's noisy channel theorem exists for quantum information, the theory of quantum error-correcting codes is sufficiently developed to allow for reliable communication over noisy quantum channels. The basic theory of quantum error correction protects quantum information against noise by encoding quantum states with redundant information, followed by decoding to recover the original state. Within the theory of quantum error-correcting codes, there are those derived from classical theory of linear codes that give rise to the Calderbank–Shor–Steane codes and stabilizer codes (Nielsen and Chuang 2009). Effective quantum error correction need not assume that encoding and decoding of quantum states must be carried out without error, as would be the case for noisy quantum gates. Fault tolerance, as defined by the threshold theorem for quantum computation, allows for a certain degree of faulty gates. The theorem roughly states that if the noise in individual quantum gates is below a certain constant threshold, it is possible to efficiently perform an arbitrarily large quantum computation. In current practice, quantum error correction is a method of protecting quantum information from errors by encoding that quantum information in quantum error-correcting codes and making a series of measurements (known as syndrome measurements) in order to determine whether errors have caused the quantum information to leave the valid space of those codes. Unlike classical information, quantum information cannot be copied (i.e., the no-cloning theorem); thus, simply redundant encodings like a repetition code are not possible. Instead, quantum error-correction codes often encode logical information in global properties of quantum states, such as in topological features of the states. Critically, quantum error correction involves extraction of entropy from the quantum state through a series of syndrome measurements and a reset of qubits.

It is important to distinguish between error correction and error mitigation. Error mitigation is a loosely defined set of techniques that are usually algorithmic in nature and that aim to extract not-noisy quantities from noisy quantum computers. Usually error mitigation does not include techniques that extract entropy from the system via reset and measurement (such techniques are in the realm of error correction). As a result of this, error mitigation often amounts to making a trade-off between the number of circuit repetitions and the fidelity of the estimated quantities. For example, many methods of error mitigation aspire to the limit of perfect error detection and postselection. Although error-mitigation techniques are generally not scalable in the way that error-correction techniques are, all state-of-the-art noisy intermediate-scale quantum (NISQ) experiments leverage some form of error mitigation. However, error mitigation is often used as a catchall and includes techniques like spin echoes, which are basically pulses equating to the identity gate that are used to decorrelate errors.

A critical figure of merit for error mitigation is quantum fidelity, which is the squared overlap of some errant quantum state with the intended (error-free) quantum state. If the fidelity of a quantum system is 1, then there were no errors and the output state is perfect. If the fidelity is 0, then there was an error and the states are now orthogonal. The infidelity is just 1 – fidelity.

In the context of error correction, the gates need to have a certain fidelity in order to be below the threshold for an error-correcting code to work. When describing the fidelity of a gate, if this gate is applied to an input state, what is the output fidelity (squared overlap) of the state that is produced with respect to the state it is supposed to produce (i.e., a measure of the error rate of the gate)? If that gate has 0.99, or "two nines," fidelity, that means an error will occur roughly 1 in 100 times that we apply the gate. The threshold error rate for gates in the surface code is approximately 0.999. For error correction, an important goal is to develop a scalable system that includes a sufficient number of physical qubits with gate fidelities below the threshold (e.g., 0.99). In such a case, the probability of an error occurring at all is suppressed exponentially, leading to a scalable system in which error correction can be carried out. However, if the intention is to run a NISQ algorithm with error mitigation only and without error correction, in order to detect one correct output, the number of times the circuit must be run to see one correct output (the number of gates × gate fidelity) is very large and is not scalable.

The physical implementation toward error correction and error mitigation on the molecular level relies primarily on single-ion magnets, in which the electron and nuclear spin states are weakly anisotropic and exchange coupled, leading to multiple spin states that are entangled. For suitably engineered systems, unequal energy spacings allow addressing via microwave resonant pulses (Aguilà, Roubeau, and Aromí 2021; Chicco et al. 2021; Ferrando-Soria et al. 2016; Gimeno et al. 2021; Godfrin et al. 2017; Hussain et al. 2018; Jenkins et al. 2017; Luis et al. 2011, 2020; Moreno-Pineda et al. 2017, 2018). Molecular electron-nuclear spin-based qudits have been utilized for implementation of quantum error-mitigation codes, where the molecular systems function as NISQ systems in which the electronic structure is "protected from decoherence" (Chiesa et al. 2020; Macaluso et al. 2020). Taking advantage of nuclear spin structures that function as qudits leads to the possibility for long coherence times due to isolation of the system from the environment but, consequently, long manipulation times. Strategies to shorten the manipulation times for gate operations involve taking advantage of electron–nuclear coupling (hyperfine interactions) to perform operations on nuclear spin states at rates much shorter than the decoherence times (Castro et al. 2022; Chizzini et al. 2022; Hussain et al. 2018).

Coherent control of molecular qudit states requires magnetic dilution in order to minimize dipolar coupling, alignment of molecular axes within the crystalline environment (such as can be obtained in a diamagnetic crystalline host environment or encapsulation), and sufficient magnetic anisotropy to address each transition independently. The magnetic anisotropy, however, needs to be small enough to access transitions within experimentally accessible microwave frequencies. As mentioned in Section 2.2.2b, lanthanides with zero spin–orbit coupling (e.g., orbitally quenched f^7 configuration) and small ZFS, such as Gd (S = 7/2), have been investigated extensively (Jenkins et al. 2017). X-band pulsed EPR experiments allow independent addressing of observed transitions, with little dependence of T_1 and T_2 on each transition and the observation of Rabi oscillations. Implementation of a universal Toffoli gate (controlled-controlled NOT [CCNOT] gate) in the GdPOM system was demonstrated, as well as implementation of the Deutsch (Kiktenko et al. 2015), Grover (Godfrin et al. 2017), and quantum phase estimation algorithms (Wang et al. 2020) with shorter times and fewer operations than on qubit (S = ½) systems. The implementation of quantum error-correction gates and algorithms successfully with microwave or radiowave pulse control in molecular qudits suggests feasibility for coherence control of entangled states, making this a promising area for new breakthroughs in the molecular sciences.

Molecular electron-nuclear spin-based qudits have been utilized for implementation of quantum error-mitigation codes, in which the electronic structure is "protected from decoherence" and the molecular systems can function as NISQ systems. The dominant source of decoherence or error can be ascribed to the electron spin units' high susceptibility to interact with the nuclear spin bath (typically 10^2 spins) via hyperfine interactions, which results in a nonexponential decay behavior. Simulations of the coupled system-bath dynamics predict that while the squared fidelity of the recovered state is above 0.9 for up to ~30 μs for S = ½, the recovered fidelity is above this value for 40–300 μs for qudit spin states of S > ½ (3/2 and 9/2, respectively) (Petiziol et al. 2021). The high fidelity accompanied by long evolution times is significant for the development of quantum error-correction and error-mitigation schemes, in which the long implementation operations (encoding, detection, recovery) must fall within the evolution time. The assumption here is that the electronic qudit energy gaps are larger than the energy gaps in the nuclear spin bath or, for a nuclear qudit, than the hyperfine interaction (electron-nuclear interaction). Hussain and colleagues (2018) showed that coupling of a nuclear qudit to an S = ½ electronic spin ancillae (d = 2) offers the combination of the long decoherence times associated with nuclear degrees of freedom and the large reduction of nutation time induced by electron-nuclear (hyperfine) mixing to enable coherent control of a qudit by radiofrequency pulses. For systems in which the electron-nuclear transitions are well resolved, coherence times are longer than operation times, and coherent control of dynamics allows for implementation of simple gates; qubit–qudit systems can be exploited to implement quantum error-mitigation and quantum simulation algorithms (Chicco et al. 2021; Hussain et al. 2018).

Large energy splitting between M_J states, large *g*-factors that provide polarizability for better initialization capacities, and electron-nuclear transitions present in lanthanides make this class of molecular candidates attractive as targets for error-correction and error-mitigation codes. Quantum transitions can be driven coherently (coherent manipulations) via electromagnetic pulses, and the time that quantum coherence is maintained is the phase memory time T_m. Decoherence occurs predominantly through dipole-dipole interactions and hyperfine coupling (all spin-spin interactions) versus T_1 spin relaxation, in which spin-phonon, Orbach (one-phonon), and Raman (two-phonon) mechanisms play a role. By applying spin-echo pulse sequences in GdW_{30} systems S = 7⁄2, 99 percent fidelity can be achieved in less than 10 ns (much shorter than $T_2 \approx 2$ μs); reaching the same result with a sequence of monochromatic pulses would take more than 1 μs (Castro et al. 2022). Such strategies allow reaching a high fidelity of the outcome wave function with a single control pulse.

Challenges in implementation arise due to the limited frequencies of EPR spectrometers, giving rise to electrically gated techniques that may offer additional possibilities for pulse manipulation (Castro et al. 2022). The multilevel structure of the qudit [Yb(trensal)] (I = 5⁄2) can encode a d = 6 qudit that can be exploited for coherent control of the nuclear-spin degrees of freedom by nuclear magnetic resonance and via hyperfine coupling to electron spin (S = ½) (Hussain et al. 2018). A minimal code protecting against amplitude or phase shift errors can be implemented within a Gd-oxalate complex, suggesting a strategy for error mitigation with reasonably high fidelity. The possibility of carrying out error-correction protocols in heterobimetallic lanthanide complexes (LnLn) has been demonstrated (Aguilà, Roubeau, and Aromí 2021). As shown in Figure 2-8, a LnLnLn trimer [000]-[111]

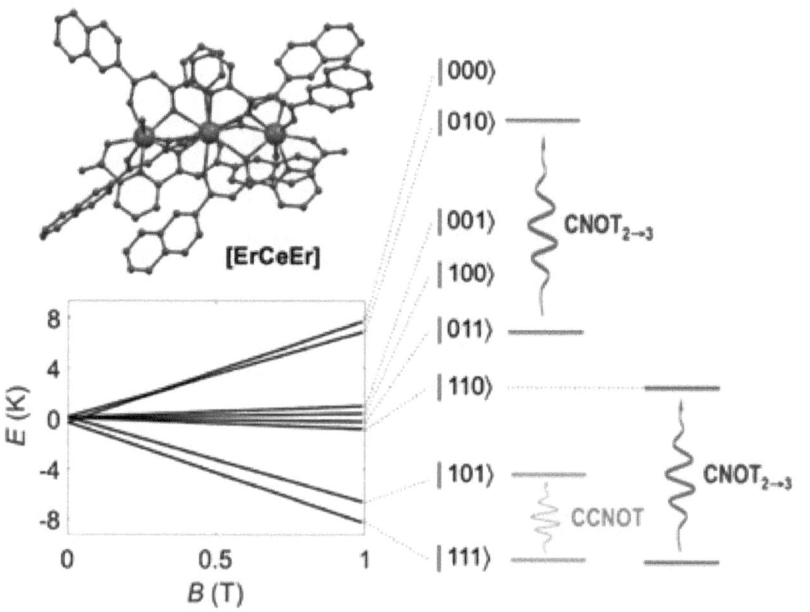

FIGURE 2-8 The crystal structure of [ErCeEr] and the energy levels as a function of the external magnetic field, *B*, applied along the *z* axis (the Er-Ce direction). The qubit states for the eight levels are depicted together with the transitions corresponding to the controlled NOT $(CNOT)_{2\rightarrow 3}$ and controlled-controlled NOT (CCNOT) quantum gates.
SOURCE: Aguilà, Roubeau, and Aromí 2021.

was investigated, and the quantum error code for a three-qubit phase-flip repetition code was successfully carried out by resonant pulses. The first step (encoding) involved two CNOT operations; followed by a $\pi/2$ pulse; followed by T_m, with a reverse step (decoding); followed by the correction step, consisting of a CCNOT operation with the central qubit as the target and the two ancillae as control qubits. Time-dependent numerical simulations were performed to assess the protocol, and the error was found to be efficiently reduced by the correction code, allowing implementation of 50–100 gates before repetition was required (Aguilà, Roubeau, and Aromí 2021).

Purely organic molecular qudits can be designed by coupling an organic stable radical with a doublet state (S = ½, qubit) to a photoexcited organic triplet state (S = 1, qutrit) to generate spin-polarized quartet and doublet states, depending on the sign and magnitude of exchange coupling between the triplet and doublet states (Qiu et al. 2022; Wang et al. 2021). As organic systems have extremely small spin–orbit coupling, the decoherence times are long (10–100 μs) and dominated by hyperfine and/or dipolar coupling to the nuclear spin bath. As optical initialization is possible in these systems, future work in this area may involve the demonstration of simple gates with high fidelities, which have so far not been implemented for organic qudit states. Alternate strategies for optical initialization of molecular organic qudit systems have been demonstrated in semiquinone radical–cobalt complexes (Kirk, Shultz, Hewitt, et al. 2022; Paquette et al. 2018). Critical to this strategy is an understanding of the sign and magnitude of exchange in the excited state, the quantum yield of excited-state (triplet) population, and the competition between rates of excited-state relaxation to the ground state versus decoherence of the resultant spin states. Ultimately, in order to implement desired quantum protocols, the excited-state relaxation must be at least an order of magnitude slower than the decoherence rate.

2.3 INVESTIGATING THE INTERACTIONS OF MOLECULAR QUBITS WITH THEIR ENVIRONMENTS

Synthetic chemistry can play a key role in the design of molecules with so-called clock transitions (Gaita-Ariño et al. 2019), which can lead to enhanced coherence by providing protection from various environmental decoherence sources. Spin clock transitions are found at avoided level crossings associated with the Zeeman splitting of qubit states in a magnetic field. Named after the principle that gives atomic clocks their exceptional temporal stability, spin clock transitions provide an optimal operating point at which the qubit resonance frequency, f, becomes insensitive to the local magnetic field (B_0) fluctuations (i.e., $df/dB_0 = 0$) (Brantley et al. 2022). In this way, a molecular clock qubit is immune to magnetic noise. There are several synthetic strategies for generating molecular clock transitions. The key requirement is a term in the spin Hamiltonian that does not commute with the Zeeman interaction. For molecules containing metal centers with integer total spin (an even number of unpaired electrons), the crystal or ligand field interaction can do the job: clock transitions can be tuned through manipulation of the coordination environment around the metal center (Giménez-Santamarina et al. 2020; Sørensen et al. 2017). In the first molecular example, a crystal-field clock transition was demonstrated for a Ho(III) ion ([Xe]$4f^{10}$ electronic configuration) encapsulated within a polyoxometalate moiety, resulting in a significant enhancement in spin coherence for a crystal rich in fluctuating electron and nuclear spins (Kundu et al. 2023; Shiddiq et al. 2016) (Figure 2-9).

An alternative strategy that works for molecules possessing half-integer spin states (i.e., an odd number of unpaired electrons) involves the electron-nuclear hyperfine interaction, which has been employed widely in trapped-ion quantum devices (Wright et al. 2019). Crucially, in the molecular case, the hyperfine interaction can again be controlled using coordination chemistry to maximize unpaired electron spin density at the relevant nuclear site. Recent examples have shown that this is possible by varying the degree of s-orbital mixing into a spin-bearing molecular d-orbital (Kundu et al. 2022; McInnes 2022; Zadrozny et al. 2017). This has the added advantage of increasing the s-orbital character, which reduces spin–orbit coupling and suppresses spin-lattice relaxation.

Additional recent work (Bayliss et al. 2022) has demonstrated that by tuning the local structural environment in the vicinity of a molecular qubit, it is possible to prepare molecules with clock-like transitions operating in practical frequency ranges that were not possible in their native structures (Figure 2-10). The potential to access these transitions lies not only within the molecular design but within instrumental flexibility. Using cavity-based EPR spectrometers, one is often limited by the field and frequency range that can be accessed (see Section 3.4 of this report). Moving to custom broadband microwave resonators analogous to the types that are used to probe

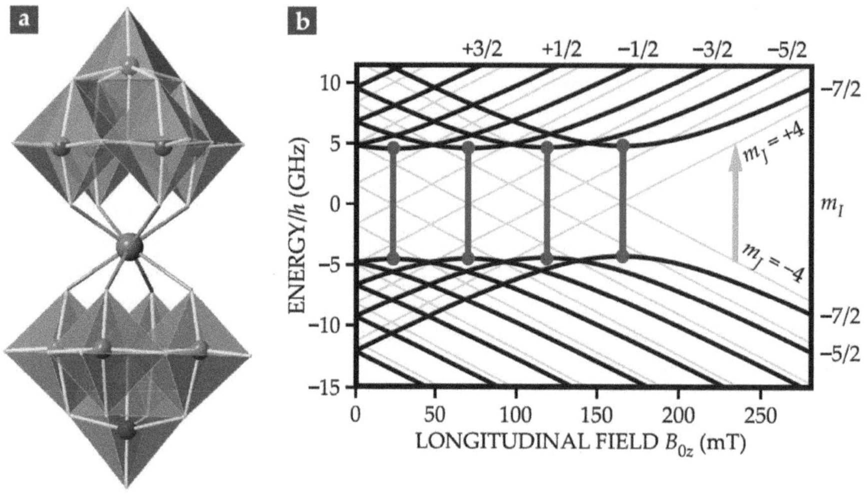

FIGURE 2-9 Energy level diagram for a holmium molecular nanomagnet. (a) The magnetic holmium ion consists of a holmium ion (purple) sandwiched between two complexes of tungsten atoms (blue) and oxygen atoms (at the vertices of the polyhedra). (b) The hyperfine energy levels in a fully symmetric molecule (light grey lines) are shifted and bent back (black lines) because the molecule has a slight axial asymmetry. Transitions that connect extrema of the curves (red lines) are the clock transitions. The frequency of such transitions is insensitive to fluctuations in the magnetic field.
SOURCE: Levi 2016.

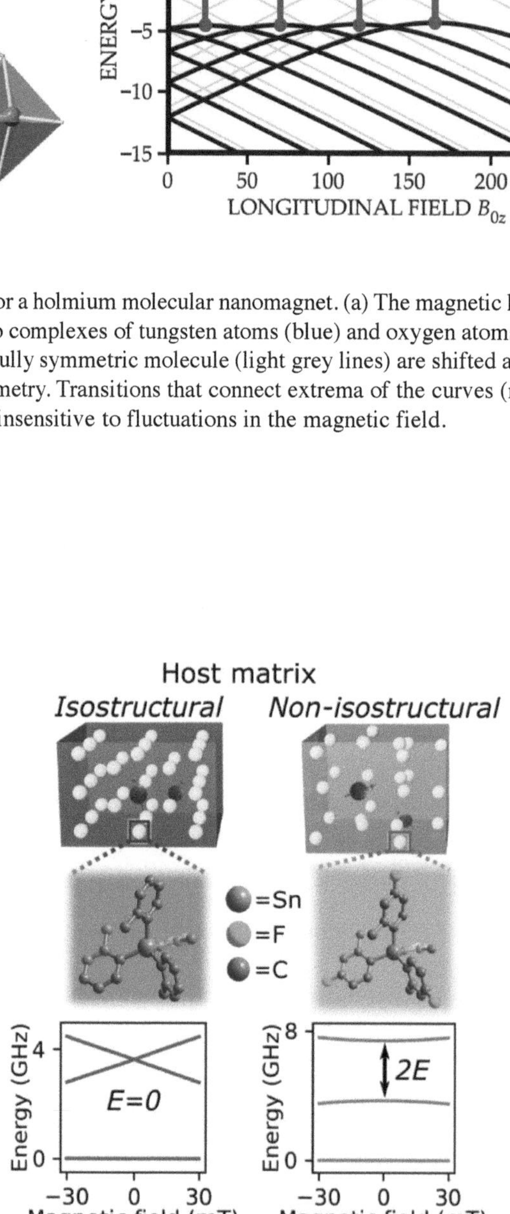

FIGURE 2-10 By modifying the host and symmetry, the ground-state magnetic parameters can be tuned to create clock-like transitions.
SOURCE: Bayliss et al. 2022.

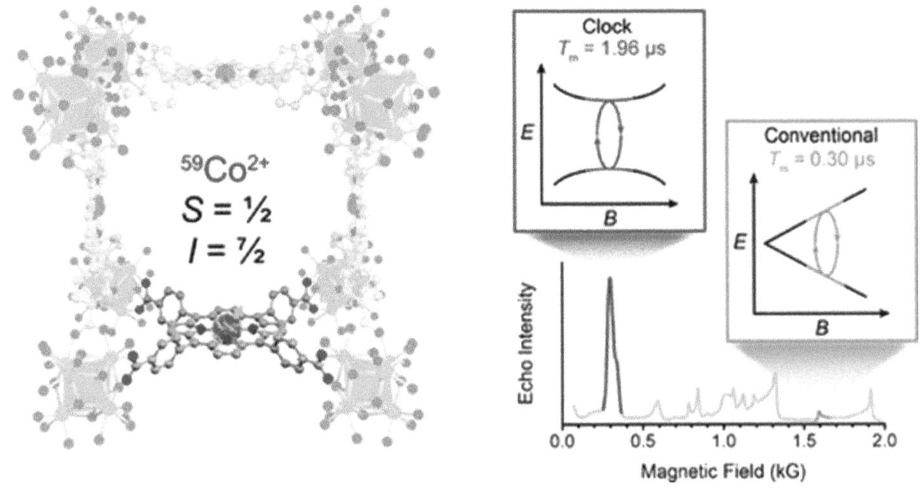

FIGURE 2-11 Schematic of a clock-like transition in a metal–organic framework.
SOURCE: Yu et al. 2021.

defects in semiconductors enables access to much wider field/frequency combinations. The potential for tuning into clock-like transitions with a more flexible operating field and frequency could enable more robust transitions to be harnessed for future devices.

Clock transitions can also be incorporated into frameworks, bringing together concepts of arrays of spins with structurally induced control over coherence time. Figure 2-11 illustrates one such example: by tuning the interaction of nuclear and electronic spins, a clock-like transition was engineered within a metal–organic framework (MOF).

With the discovery that scalable quantum applications are possible using single-photon sources, linear optical elements, and single-photon detectors, approaches based on cluster states or error encoding have made all-optical architectures promising targets for the ultimate goal of large-scale quantum devices. Challenges involve the development of high-efficiency sources of indistinguishable single photons, scalable optical circuits, high-efficiency single-photon detectors, and low-loss interfacing of these components (O'Brien 2007). Controlled coupling of organic chromophores to photonic structures has led to the development of single-photon quantum emitters, which could play a key role in molecular systems for QIS. With transition linewidths of ~10 MHz at low temperatures, organic chromophores can function as single-photon sources with long coherence times that are scalable and compatible with diverse integrated platforms. In addition, such chromophores can be used as transducers for the optical readout of electrical and/or magnetic fields and material properties for quantum sensing with single-quantum resolution (Dickerson, Guo, Zhu, et al. 2021; Toninelli et al. 2021; Wang, Kelkar, et al. 2019). Recent strategies for the generation and manipulation of organic-based photonic qubits require the generation of polarized light emission (or absorption) by utilizing the crystallographic order of the resulting environment.

Organic crystals doped with organic emitters function as single quantum emitters with a well-defined polarization relative to the crystal axes, making them amenable to alignment with optical nanostructures. The radiative lifetime and saturation intensity varies little within the crystalline environment, and a large fraction of these emitters can be excited more than 10^{12} times without photobleaching (Polisseni et al. 2016).

2.4 DESIGNING MOLECULAR STRUCTURES WITH INTEGRATED CHIRALITY-INDUCED SPIN SELECTIVITY EFFECTS

Chirality is a key property of molecules important in many chemical and nearly all biological processes. Recent observations have shown that electron transport through chiral molecules attached to solid electrodes can induce high spin polarization even at room temperature (Wasielewski 2023). The ability to produce highly spin-polarized electrons at ambient temperatures is potentially important for developing room-temperature quantum devices.

Electrons with their spin aligned parallel or antiparallel to the ET displacement vector are preferentially transmitted depending on the chirality of the molecular system, resulting in chirality-induced spin selectivity (CISS). The first evidence of the relationship between chirality and electron motion dates back to 1999 when Naaman and Waldeck (2012) observed a large asymmetry in the transmission of oppositely spin-polarized electrons by thin films of chiral molecules. The coupling of orbital angular momentum to spin angular momentum in directional ET processes can provide a method of manipulating spin polarization. However, little is known about the interplay of CISS with the spin dynamics of molecular ET processes. One way to address this question is to explore the ET dynamics of covalently linked donor–chiral bridge–acceptor (D-Bχ-A) molecules following photoexcitation. The CISS effect on the coherent spin dynamics of photogenerated radical pairs in these systems depends on competing photophysical processes, most of which operate on a sub-nanosecond timescale (Aiello et al. 2022; Harvey and Wasielewski 2021). While femtosecond and nanosecond transient absorption spectroscopies can be used to probe the ET dynamics of D-Bχ-A molecules, time-resolved EPR spectroscopy is essential to elucidating their spin dynamics.

Photoexcitation of the donor (D) or acceptor (A) in a molecular electron donor-bridge-acceptor (D-B-A) system can result in the formation of a $D^{\bullet+}$-B-$A^{\bullet-}$ quantum-entangled electron SQP initially in a pure singlet state. Coherent spin evolution of this system results in a partial triplet character, which results in strong electron spin polarization that can be observed by time-resolved EPR or ODMR spectroscopies. If charge transfer occurs through a chiral bridge, such as in D-Bχ-A, the CISS effect induces a spin polarization that depends on the chirality and the direction of the ET (Aiello et al. 2022). As a result, only one of the four two-spin states is populated in the weak-coupling limit, with vanishing entanglement between the hole and electron spin (Figure 2-12a) (Luo and Hore 2021). Moreover, CISS retards the rate of the radical pair recombination reaction, making it possible to extend the radical pair lifetime (Hafner et al. 2018).

Pulse-EPR techniques have been used to obtain detailed information about magnetic exchange (*J*) and dipolar coupling (*D*) for photogenerated spin-correlated radical pairs in a variety of systems, where *D* gives detailed distance and structural information, as well as to provide a direct probe of spin coherence in the radical pair (Aiello et al. 2022). For example, if photogeneration of the radical pair is followed by a microwave Hahn echo pulse sequence, $\pi/2$ - τ - π - τ - echo (Figure 2-12b), and the time delay τ is scanned, coherent oscillations between $|\Phi_A\rangle$ and $|\Phi_B\rangle$ that are related to both *J* and *D* result in modulation of the spin echo amplitude (Figure 2-12c) (Tang, Thurnauer, and Norris 1994; Thurnauer and Norris 1980). When this experiment is performed on spin-coherent radical pairs, the echo appears "out of phase" (i.e., in the detection channel in quadrature to the one in which it is expected) and is therefore termed out-of-phase electron-spin-echo envelope modulation (OOP-ESEEM). Recent theory has shown that based on the phase relationships of these coherent oscillations, OOP-ESEEM can be used to detect the CISS effect on the spin coherence of spin-correlated radical pairs (Chizzini, et al. 2021; Fay 2021; Fay and Limmer 2021).

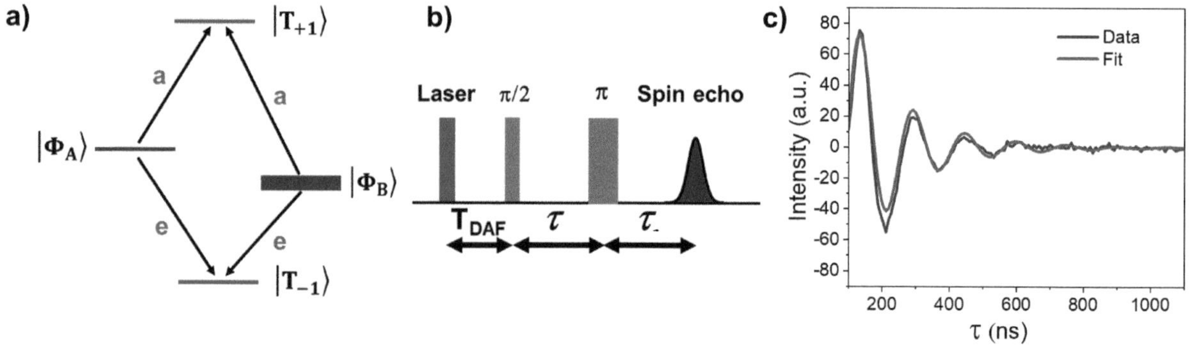

FIGURE 2-12 (a) $D^{\bullet+}$-Bχ-$A^{\bullet-}$ energy levels with chirality-induced spin selectivity in which only $|\Phi_B\rangle$ populated. Process "a" corresponds to absorption and process "e" shows emission. (b) Hahn echo pulse sequence for out-of-phase electron-spin-echo envelope modulation (OOP-ESEEM) experiments, where T_{DAF} is the delay after the laser flash. (c) Typical OOP-ESEEM behavior of a spin-correlated radical pair.
SOURCE: Michael Wasielewski.

2.5 TARGETING FUNCTIONALIZATION OF MOLECULAR QUBITS FOR SENSING AND SYSTEMS INTEGRATION

Taking the foundation of molecular spin qubits that have been developed and translating it to areas such as quantum sensing and quantum communication will require the synthesis of new molecules, new two-spin systems, and multiqubit arrays. For example, innovations that specifically target quantum sensing by preparing molecules with desirable quantum properties to serve as bioconjugates would be transformative for the nascent area of quantum biosensing. In an orthogonal area, tethering molecules to inert substrates, such as hexagonal boron nitride, would enable the measurement of properties at a spatial scale too small to resolve using other means and could open up new directions within condensed matter physics.

As described previously, the initial step in understanding and creating more complex molecular systems for QIS applications is progressing beyond individual molecular qubits to two, three, and ultimately many qubits. For example, by exploiting the atomistic control inherent to synthetic chemistry, recent work on two-qubit systems has addressed a fundamental question of how the spin–spin distance between two qubits impacts electronic spin coherence (von Kugelgen et al. 2021). To achieve this goal, a series of molecules featuring two spectrally distinct qubits—an early transition metal, Ti^{3+}, and a late transition metal, Cu^{2+}—with increasing separation between the two metals was examined (Figure 2-13a). The spectral separation between the two metals enabled each metal to be probed individually in the bimetallic species along with comparisons with the monometallic control samples (Figure 2-13b).

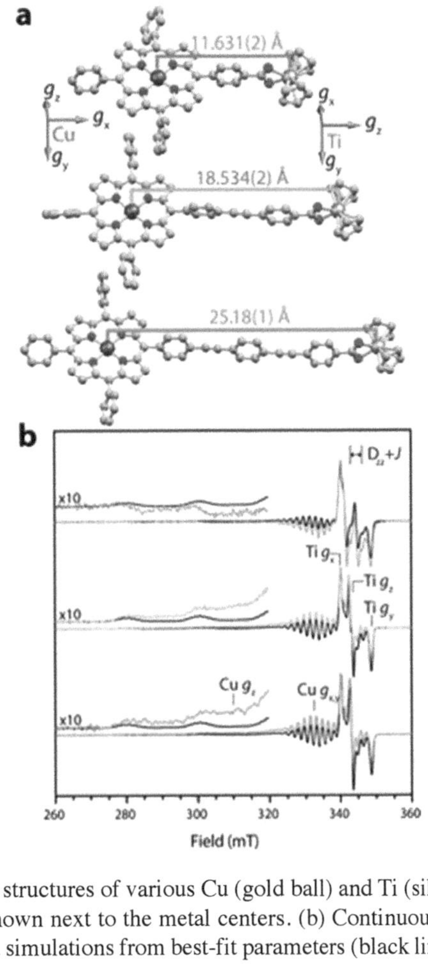

FIGURE 2-13 (a) Single-crystal X-ray structures of various Cu (gold ball) and Ti (silver ball) systems having different spacer lengths. The *g*-tensor alignments are shown next to the metal centers. (b) Continuous-wave electron paramagnetic resonance spectra (pink, blue, and green lines) and simulations from best-fit parameters (black line) for each of the bimetallic compounds. SOURCE: von Kugelgen et al. 2021.

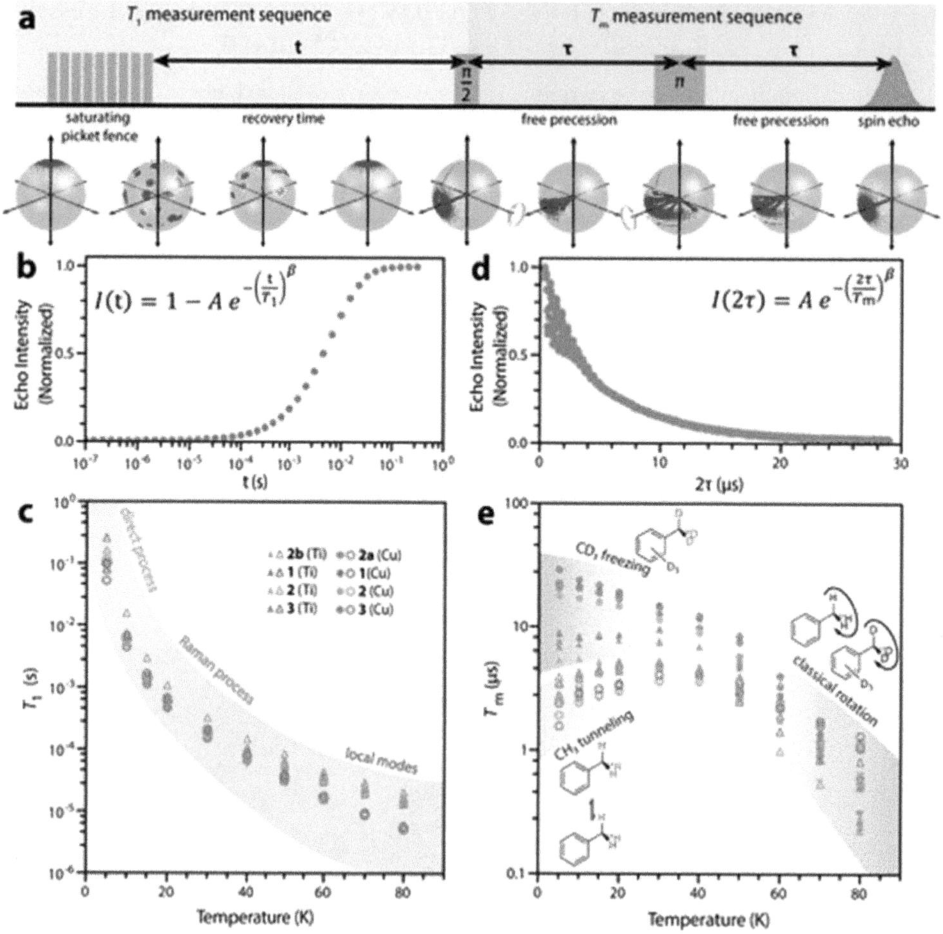

FIGURE 2-14 (a) Pulse sequence and Bloch sphere representation of pulse electron paramagnetic resonance experiments measuring spin-relaxation time, T_1, and phase-memory time, T_m. (b) Representative saturation-recovery curve, with the formula for its relationship to T_1 (inset). (c) Temperature dependence of T_1 for various Cu and Ti compounds (grey, pink, blue, green) in toluene. (d) Representative Hahn-echo decay curve for Ti^{3+} compound with the formula for its relationship to T_m (inset). (e) Temperature dependence of T_m in different solvents.
SOURCE: von Kugelgen et al. 2021.

Across a range of 1.2–2.5 nm, von Kugelgen and colleagues (2021) found that the electron spins have a negligible effect on coherence times, a finding attributed to the distinct resonance frequencies. Coherence times are governed, instead, by the distance to nuclear spins on the other qubit's ligand framework (Figure 2-14). This finding offers guidance for the design of spectrally addressable molecular qubits. This work lays the foundation for the bottom-up integration of multiple qubits with distinct functions into custom quantum systems, such as pairs of anisotropic quantum sensors for mapping vector fields in three-dimensional or multifunctional arrays featuring unique quantum elements designed for initialization, sensing, storage, and readout.

Moving beyond individual units, there has been a significant amount of work on bimetallic systems, which are powerful because they allow simple operations with the potential for frequency addressability that will enable their use as quantum gates. Some notable examples include bimetallic lanthanide systems (Aguilà et al. 2014) and systems connecting two wheel-like species featuring ground states amenable to quantum manipulation (Ardavan et al. 2015). In each of these approaches, the modularity of the linker enables

FIGURE 2-15 Modularity of bimetallic systems with switchable gate.
SOURCE: Walsh and Freedman 2016.

control over both the separation between the spins and potentially the form of gate that could be executed; for example, an electrochemical gate could be enabled by a redox switchable linker (Figure 2-15) (Walsh and Freedman 2016).

Moving from discrete systems to arrays of spins requires careful design of the local spin, linker, and interaction of molecular systems with both interfaces and the environment. There are numerous examples of MOFs featuring spins; however, there are a limited number of studies in which the coherence properties of an array of spins have been examined. Arguably, the first such example of a fully concentrated array is a two-dimensional porphyrin lattice, which was measured to have spin coherence (Urtizberea et al. 2018) and is an important achievement. A second approach to measuring coherence in a lattice of spins connects clock transitions with the design of individual moieties that form the building blocks of an MOF (Zadrozny et al. 2017). Clock transitions were described earlier in this chapter and are particularly powerful for mitigating the effects of magnetically noisy environments such as those found in MOFs.

MOFs provide many new opportunities for QIS. For example, a recent result demonstrates using an MOF for lithium-ion sensing (Figure 2-16) (Sun et al. 2022). Specifically, the porosity of MOFs can be harnessed for sensing intercalated analytes. In a different approach, one could envision using a two-dimensional MOF to position individual quantum units, such as sensors, onto an analyte with a great level of precision. Using framework chemistry, it is possible to employ a building block approach for positioning quantum units. The linking unit tunes both the distance and the strength of magnetic coupling between the quantum units, while different quantum units can be positioned within the framework. Significant work has focused on creating MOFs with designer topologies and incorporating different core units (Figure 2-17). The molecular nature of MOFs also enables tuning their phonon spectrum, which determines the interaction of qubits with the thermal energy of the environment and, therefore, T_1.

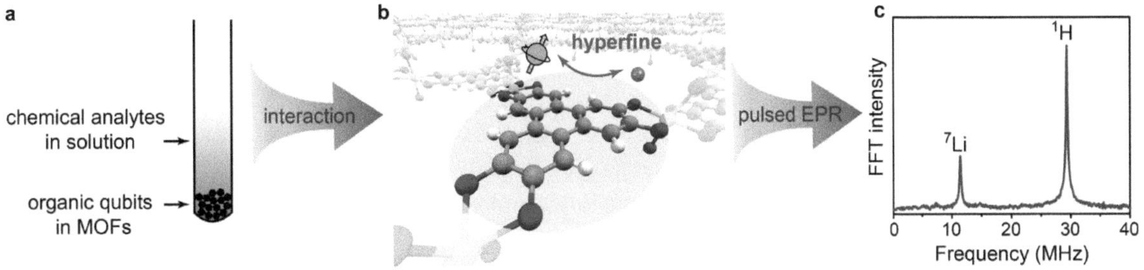

FIGURE 2-16 (a) Porosity in metal–organic frameworks (MOFs). MOFs can be used to sense chemical analytes in solution. (b) Chemical analytes are adsorbed into the MOF and interact with the embedded radicals through hyperfine coupling. (c) Hyperfine spectroscopy can be used to identify the nuclear species and possibly characterize coupling strengths, for example chemical analytes.
SOURCE: Sun et al. 2022.

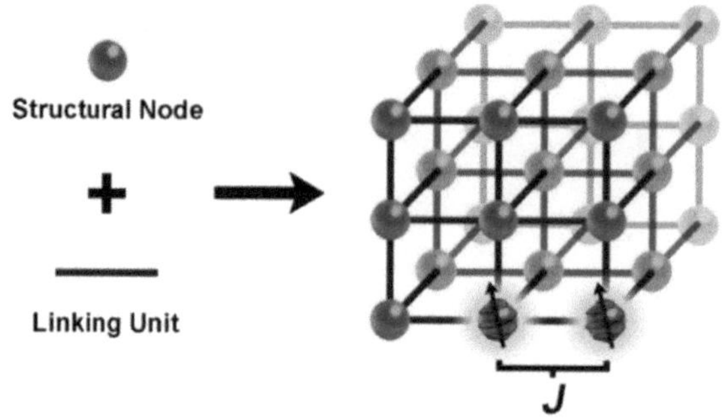

FIGURE 2-17 A framework approach enables construction of arrays of quantum units perhaps for quantum sensing. SOURCE: Graham, Zadrozny, et al. 2017.

2.6 DEVELOPING MOLECULAR QUANTUM INTERCONNECTS OVER BROAD LENGTH SCALES INCLUDING MOLECULE-BASED QUANTUM REPEATERS

A powerful opportunity for chemistry in QIS and engineering is the control of coherent wave function transfer/transduction through functional inorganic–organic interfaces. The coherent transport of quantum information across tailored interfaces between two different materials systems offers extraordinary opportunities spanning biology, chemistry, and physics, as well as impacts the development of advanced instrumentation across the disciplines. Past experiments with widely disparate solid-state materials have suggested that coherent quantum transport and transfer of angular momentum can take place with high efficiency even when charge transport may be challenging, such as two materials with markedly different bandgaps and stoichiometry (Crooker et al. 1996). Being able to move information coherently across an inorganic–organic interface allows researchers to exploit the technological advantages of integrated semiconductor electronics with the synthesis of molecular electronic structures for molecular quantum memories and targeted biological and chemical sensing, as well as to investigate the fundamental mechanisms of coherent spin exchange at the atomic and subatomic levels. Notably, in quantum communications, designing molecules that emit in the telecom region has proven challenging but could lead to the development of quantum repeaters with onboard memory elements. Creating systems with these features would lead to the integration of molecules in the quantum internet and would support a modular approach to quantum communication.

For example, ultrafast photodriven ET within an organic donor-acceptor (D-A) molecule can produce a radical pair that can function as two entangled spin qubits ($D^{\bullet+}$–$A^{\bullet-}$), giving rise to an entangled two-spin singlet or triplet state (Wasielewski et al. 2020). This strategy has been used to achieve electron–spin state teleportation; this is essential to preserve coherent quantum information transfer across an ensemble of covalent D-A-$R^{\bullet}$ molecules, in which it is possible to move the initially prepared spin state of a stable radical $R^{\bullet}$ to $D^{\bullet+}$ (Figure 2-18) (Pirandola et al. 2015; Rugg et al. 2019).

Following preparation of a specific electron spin state on $R^{\bullet}$ using a microwave pulse to rotate the spin orientation with respect to an externally applied magnetic field, photoexcitation of A results in the formation of a singlet entangled electron spin pair $D^{\bullet+}$-$A^{\bullet-}$. The spontaneous ultrafast chemical reaction $D^{\bullet+}$-$A^{\bullet-}$-$R^{\bullet} \rightarrow D^{\bullet+}$-A-$R^{-}$ constitutes the Bell state measurement step necessary to achieve spin-state teleportation. Quantum-state tomography of the initial $R^{\bullet}$ and final $D^{\bullet+}$ spin states using pulse-EPR spectroscopy shows that the spin state of $R^{\bullet}$ is teleported to $D^{\bullet+}$ with high fidelity. Extensions of this strategy may be able to transfer a spin state coherently across the 10–1,000 nm distances that are important for quantum interconnects.

2.7 FABRICATING SCALABLE MOLECULAR QUANTUM ARCHITECTURES BASED ON MOLECULAR QUBITS

The exquisite ability to manipulate inorganic materials at the nanoscale can potentially address many of the bottlenecks for all-molecular systems. Such "nanomaterials" can have a broad range of tuning parameters such as heterostructuring, patterning, and intercalation, allowing them to play a variety of roles. They can function as tunable scaffolds to produce patterned arrays (Gong et al. 2018); can act as hosts for spin-based qubits (whether defect-based

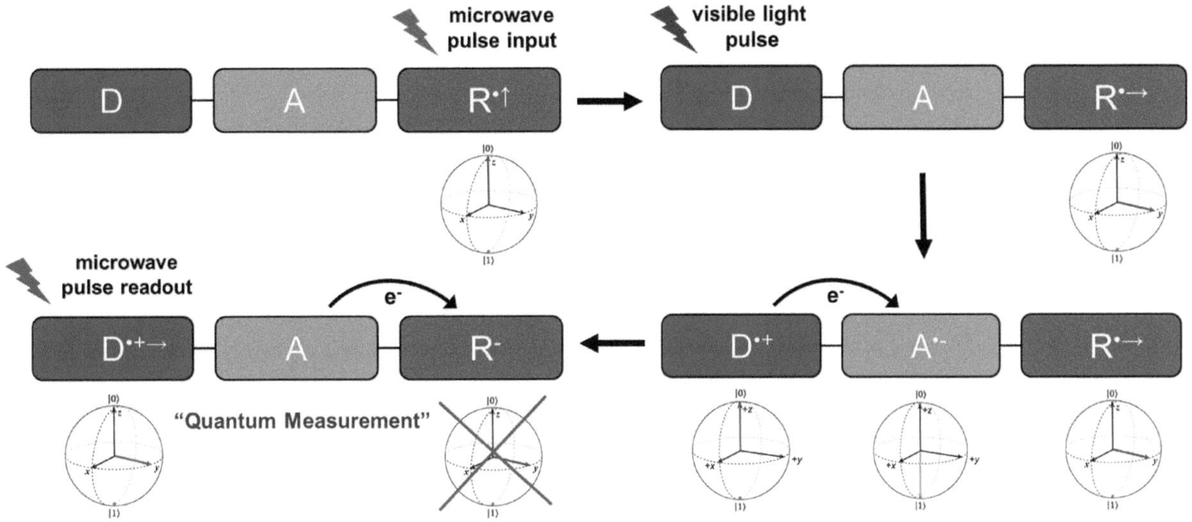

FIGURE 2-18 Quantum–spin state teleportation scheme using molecular qubits.
SOURCE: Michael Wasielewski.

or molecular); augment the existing functionality to incorporate initialization, addressability, and/or readout in hybrid architectures; and can be used to control and mitigate primary decoherence channels. The synthesis, fabrication, characterization, and understanding of nanomaterials are mature with applications ranging from microelectronics and energy storage to renewable energy—these same approaches can be used to accelerate discovery and control in molecular qubits systems. Chemical inclusion or covalent design of qubits/qudits into carbon-based materials offer promising strategies for development of multiqubit arrays. By taking advantage of delocalization in carbon frameworks, strategies for generating structurally well-defined nanographene biradicals (Lombardi et al. 2019) and porphyrin tapes (Van Raden et al. 2022) reveal long memory times with well-defined environments for inclusion into graphene-based nanostructures. Chapters 3 and 4 will discuss more general applications of inorganic materials for qubits.

One of the challenges within a variety of wide-bandgap semiconductor materials is maintaining stable charge defect states of spin qubits for quantum computing, communication, and sensing. In particular, charged vacancy and divacancy states in these systems are often subject to noise-driven ionization processes and surface potential fluctuations that rapidly decohere and ultimately destroy the quantum state, typically requiring re-initializing/resetting the system. These unstable and often random dynamics limit the stability of qubits for broad applications in QIS and engineering. There is a high-impact opportunity for chemical passivation and controlled electronic stabilization of these materials that would enable longer coherent lifetimes and a subsequent increase in quantum sensing sensitivity and readout fidelity. A combination of theoretical modeling of electronic structure and targeted chemical synthesis offers a promising solution to this problem while providing fundamentally new mechanisms of quantum-state stability and control (see below).

Atomic-scale gating and local doping would accelerate the engineering of scalable semiconductor qubits and controlled entanglement of both electronic and nuclear spin states. Moreover, this level of control would provide new routes for spin-to-charge conversion to achieve high-fidelity single-shot readout and extend both spin relaxation and coherence times in the solid state (Anderson et al. 2022). Opportunities to model, design, and synthesize electrically controllable molecular states with functional solid-state interfaces would provide a spatial level of control difficult to achieve through traditional inorganic fabrication techniques and may offer opportunities for multifunctional quantum control through optical and electronic properties. Chemically activated local gating has the potential to realize the atomistic control needed for activating quantum sensors, controlling quantum memories, and potentially gating entangled quantum registers between adjacent spin states (both electronic and nuclear). In particular, controllable atomic-scale doping through proximal molecular systems may activate coherent states as well as enable photonic processes where optical properties are dependent on the local ionization state.

2.8 SUMMARY OF RESEARCH PRIORITIES AND RECOMMENDATION

The following fundamental research priorities have been identified by the committee and discussed in Chapter 2 as those that the Department of Energy and the National Science Foundation should prioritize within the target research area of "design and synthesis of molecular qubit systems."

Research Priorities:

- **Identify and tailor molecular qubit properties for specific near-term applications in quantum sensing and communications, and more long-term opportunities in quantum computing.**
- **Develop an understanding of structure–property relationships for**
    - **i. increasing coherence times (T_2) in molecular qubits and quantum memories,**
    - **ii. creating optically addressable molecular qubits (e.g., transition metal complexes, lanthanides, organic-based multispin qubits, and optical cycling centers), and**
    - **iii. exploiting entanglement and quantum transduction.**
- **Investigate the interactions of molecular qubits with their environments.**
- **Design molecular structures with integrated chirality-induced spin selectivity effects.**
- **Target functionalization of molecular qubits for sensing and systems integration.**
- **Develop molecular quantum interconnects over broad length scales including molecule-based quantum repeaters.**
- **Fabricate scalable quantum architectures based on molecular qubits.**

Recommendation 2-1. The Department of Energy and the National Science Foundation should support cross-disciplinary activities that couple measurement, control, and characterization techniques traditionally employed by the physics and engineering communities with molecular systems designed by the chemistry community. Support also should be given to investigations that combine theory with experiment to take full advantage of the relationship between chemistry and quantum information science. Increasing these collaborations will be essential for scientific progress at these intersections.

REFERENCES

Aguilà, D., L. A. Barrios, V. Velasco, O. Roubeau, A. Repollés, P. J. Alonso, J. Sesé, S. J. Teat, F. Luis, and G. Aromí. 2014. "Heterodimetallic [LnLn′] Lanthanide Complexes: Toward a Chemical Design of Two-Qubit Molecular Spin Quantum Gates." *Journal of the American Chemical Society* 136(40):14215–14222. doi.org/10.1021/ja507809w.

Aguilà, D., O. Roubeau, and G. Aromí. 2021. "Designed Polynuclear Lanthanide Complexes for Quantum Information Processing." *Dalton Transactions* 50(35):12045–12057. doi.org/10.1039/d1dt01862k.

Aiello, C. D., J. M. Abendroth, M. Abbas, A. Afanasev, S. Agarwal, A. S. Banerjee, D. N. Beratan, et al. 2022. "A Chirality-Based Quantum Leap." *ACS Nano* 16(4):4989–5035. doi.org/10.1021/acsnano.1c01347.

Amdur, M. J., K. R. Mullin, M. J. Waters, D. Puggioni, M. K. Wojnar, M. Gu, L. Sun, P. H. Oyala, J. M. Rondinelli, and D. E. Freedman. 2022. "Chemical Control of Spin–Lattice Relaxation to Discover a Room Temperature Molecular Qubit." *Chemical Science* 13(23):7034–7045. doi.org/10.1039/D1SC06130E.

Anderegg, L., B. L. Augenbraun, Y. Bao, S. Burchesky, L. W. Cheuk, W. Ketterle, and J. M. Doyle. 2018. "Laser Cooling of Optically Trapped Molecules." *Nature Physics* 14(9):890–893. doi.org/10.1038/s41567-018-0191-z.

Anderegg, L., B. L. Augenbraun, E. Chae, B. Hemmerling, N. R. Hutzler, A. Ravi, A. Collopy, J. Ye, W. Ketterle, and J. M. Doyle. 2017. "Radio Frequency Magneto-Optical Trapping of CaF with High Density." *Physical Review Letters* 119(10):103201. doi.org/10.1103/PhysRevLett.119.103201.

Anderegg, L., L. W. Cheuk, Y. Bao, S. Burchesky, W. Ketterle, K.-K. Ni, and J. M. Doyle. 2019. "An Optical Tweezer Array of Ultracold Molecules." *Science* 365(6458):1156–1158. doi.org/10.1126/science.aax1265.

Anderson, C. P., E. O. Glen, C. Zeledon, A. Bourassa, Y. Jin, Y. Zhu, C. Vorwerk, A. L. Crook, H. Abe, J. Ul-Hassan, T. Ohshima, N. T. Son, G. Galli, and D. D. Awschalom. 2022. "Five-Second Coherence of a Single Spin with Single-Shot Readout in Silicon Carbide." *Science Advances* 8(5):eabm5912. doi.org/10.1126/sciadv.abm5912.

Ardavan, A., A. M. Bowen, A. Fernandez, A. J. Fielding, D. Kaminski, F. Moro, C. A. Muryn, M. D. Wise, A. Ruggi, E. J. L. McInnes, K. Severin, G. A. Timco, C. R. Timmel, F. Tuna, G. F. S. Whitehead, and R. E. P. Winpenny. 2015. "Engineering Coherent Interactions in Molecular Nanomagnet Dimers." *npj Quantum Information* 1(1):15012. doi.org/10.1038/npjqi.2015.12.

Aromí, G., D. Aguilà, P. Gamez, F. Luis, and O. Roubeau. 2012. "Design of Magnetic Coordination Complexes for Quantum Computing." *Chemical Society Reviews* 41(2):537–546. doi.org/10.1039/C1CS15115K.

Atzori, M., and R. Sessoli. 2019. "The Second Quantum Revolution: Role and Challenges of Molecular Chemistry." *Journal of the American Chemical Society* 141(29):11339–11352. doi.org/10.1021/jacs.9b00984.

Atzori, M., E. Morra, L. Tesi, A. Albino, M. Chiesa, L. Sorace, and R. Sessoli. 2016. "Quantum Coherence Times Enhancement in Vanadium(IV)-based Potential Molecular Qubits: The Key Role of the Vanadyl Moiety." *Journal of the American Chemical Society* 138(35):11234–11244. doi.org/10.1021/jacs.6b05574.

Atzori, M., L. Tesi, S. Benci, A. Lunghi, R. Righini, A. Taschin, R. Torre, L. Sorace, and R. Sessoli. 2017. "Spin Dynamics and Low Energy Vibrations: Insights from Vanadyl-Based Potential Molecular Qubits." *Journal of the American Chemical Society* 139(12):4338–4341. doi.org/10.1021/jacs.7b01266.

Atzori, M., L. Tesi, E. Morra, M. Chiesa, L. Sorace, and R. Sessoli. 2016. "Room-Temperature Quantum Coherence and Rabi Oscillations in Vanadyl Phthalocyanine: Toward Multifunctional Molecular Spin Qubits." *Journal of the American Chemical Society* 138(7):2154–2157. doi.org/10.1021/jacs.5b13408.

Augenbraun, B. L., Z. D. Lasner, A. Frenett, H. Sawaoka, C. Miller, T. C. Steimle, and J. M. Doyle. 2020. "Laser-Cooled Polyatomic Molecules for Improved Electron Electric Dipole Moment Searches." *New Journal of Physics* 22(2):022003. doi.org/10.1088/1367-2630/ab687b.

Awschalom, D. D., R. Hanson, J. Wrachtrup, and B. B. Zhou. 2018. "Quantum Technologies with Optically Interfaced Solid-State Spins." *Nature Photonics* 12(9):516–527. doi.org/10.1038/s41566-018-0232-2.

Bader, K., D. Dengler, S. Lenz, B. Endeward, S.-D. Jiang, P. Neugebauer, and J. van Slageren. 2014. "Room Temperature Quantum Coherence in a Potential Molecular Qubit." *Nature Communications* 5(1):5304. doi.org/10.1038/ncomms6304.

Bae, Y. J., X. Zhao, M. D. Kryzaniak, H. Nagashima, J. Strzalka, Q. Zhang, and M. R. Wasielewski. 2020. "Spin Dynamics of Quintet and Triplet States Resulting from Singlet Fission in Oriented Terrylenediimide and Quaterrylenediimide Films." *Journal of Physical Chemistry C* 124(18):9822–9833. doi.org/10.1021/acs.jpcc.0c03189.

Bao, Y., S. S. Yu, L. Anderegg, E. Chae, W. Ketterie, K-K. Ni, and J. M. Doyle. 2022. "Dipolar Spin-Exchange and Entanglement between Molecules in an Optical Tweezer Array." *ArXiv*. doi.org/10.48550/arXiv.2211.09780.

Barry, J. F., D. J. McCarron, E. B. Norrgard, M. H. Steinecker, and D. DeMille. 2014. "Magneto-Optical Trapping of a Diatomic Molecule." *Nature* 512(7514):286–289. doi.org/10.1038/nature13634.

Basel, B. S., J. Zirzlmeier, C. Hetzer, B. T. Phelan, M. D. Krzyaniak, S. R. Reddy, P. B. Coto, N. E. Horwitz, R. M. Young, F. J. White, F. Hampel, T. Clark, M. Thoss, R. R. Tykwinski, M. R. Wasielewski, and D. M. Guldi. 2017. "Unified Model for Singlet Fission within a Non-Conjugated Covalent Pentacene Dimer." *Nature Communications* 8:15171. doi.org/10.1038/ncomms15171.

Bayliss, S. L., P. Deb, D. W. Laorenza, M. Onizhuk, G. Galli, D. E. Freedman, and D. D. Awschalom. 2022. "Enhancing Spin Coherence in Optically Addressable Molecular Qubits through Host-Matrix Control." *Physical Review X* 12(3):031028. doi.org/10.1103/PhysRevX.12.031028.

Bayliss, S. L., D. W. Laorenza, P. J. Mintun, B. D. Kovos, D. E. Freedman, and D. D. Awschalom. 2020. "Optically Addressable Molecular Spins for Quantum Information Processing." *Science* 370(6522):1309–1312. doi.org/10.1126/science.abb9352.

Bayliss, S. L., L. R. Weiss, F. Kraffert, D. B. Granger, J. E. Anthony, J. Behrends, and R. Bittl. 2020. "Probing the Wave Function and Dynamics of the Quintet Multiexciton State with Coherent Control in a Singlet Fission Material." *Physical Review X* 10(2):021070. doi.org/10.1103/PhysRevX.10.021070.

Bell, J. S. 1964. "On the Einstein Podolsky Rosen Paradox." *Physics Physique Fizika* 1(3):195–200. doi.org/10.1103/PhysicsPhysiqueFizika.1.195.

Bennett, C. H., G. Brassard, C. Crépeau, R. Jozsa, A. Peres, and W. K. Wootters. 1993. "Teleporting an Unknown Quantum State via Dual Classical and Einstein-Podolsky-Rosen Channels." *Physical Review Letters* 70(13):1895–1899. doi.org/10.1103/PhysRevLett.70.1895.

Blok, M. S., V. V. Ramasesh, T. Schuster, K. O'Brien, J. M. Kreikebaum, D. Dahlen, A. Morvan, B. Yoshida, N. Y. Yao, and I. Siddiqi. 2021. "Quantum Information Scrambling on a Superconducting Qutrit Prcoessor." *Physical Review X* 11(2):021010. doi.org/10.1103/PhysRevX.11.021010.

Brantley, C., D. Lubert-Perquel, S. Hill, and G. Christou. 2022. "An Unusual Co_2 Complex a Potential Molecular Route to MultiQubit Systems." *9th North America Greece Cyprus Conference on Paramagnetic Materials*.

Canarie, E. R., S. M. Jahn, and S. Stoll. 2020. "Quantitative Structure-Based Prediction of Electron Spin Decoherence in Organic Radicals." *Journal of Physical Chemistry Letters* 11(9):3396–3400. doi.org/10.1021/acs.jpclett.0c00768.

Casillas, R., I. Papadopoulos, T. Ullrich, D. Thiel, A. Kunzmann, and D. M. Guldi. 2020. "Molecular Insights and Concepts to Engineer Singlet Fission Energy Conversion Devices." *Energy & Environmental Science* 13(9):2741–2804. doi.org/10.1039/D0EE00495B.

Castro, A., A. García Carrizo, S. Roca, D. Zueco, and F. Luis. 2022. "Optimal Control of Molecular Spin Qudits." *Physical Review Applied* 17(6). doi.org/10.1103/physrevapplied.17.064028.

Cheisson, T., and E. J. Schelter. 2019. "Rare Earth Elements: Mendeleev's Bane, Modern Marvels." *Science* 363(6426):489–493. doi.org/10.1126/science.aau7628.

Cheuk, L. W., L. Anderegg, B. L. Augenbraun, Y. Bao, S. Burchesky, W. Ketterle, and J. M. Doyle. 2018. "Lambda-Enhanced Imaging of Molecules in an Optical Trap." *Physical Review Letters* 121(8):083201. doi.org/10.1103/PhysRevLett.121.083201.

Chicco, S., A. Chiesa, G. Allodi, E. Garlatti, M. Atzori, L. Sorace, R. De Renzi, R. Sessoli, and S. Carretta. 2021. "Controlled Coherent Dynamics of [VO(TPP)], a Prototype Molecular Nuclear Qudit with an Electronic Ancilla." *Chemical Science* 12(36):12046–12055. doi.org/10.1039/d1sc01358k.

Chiesa, A., M. Chizzini, E. Garlatti, E. Salvadori, F. Tacchino, P. Santini, I. Tavernelli, R. Bittl, M. Chiesa, R. Sessoli, and S. Carretta. 2021. "Assessing the Nature of Chiral-Induced Spin Selectivity by Magnetic Resonance." *Journal of Physical Chemistry Letters* 12(27):6341–6347. doi.org/10.1021/acs.jpclett.1c01447.

Chiesa, A., E. Macaluso, F. Petiziol, S. Wimberger, P. Santini, and S. Carretta. 2020. "Molecular Nanomagnets as Qubits with Embedded Quantum-Error Correction." *Journal of Physical Chemistry Letters* 11(20):8610–8615. doi.org/10.1021/acs.jpclett.0c02213.

Chiesa, A., F. Petiziol, E. Macaluso, S. Wimberger, P. Santini, and S. Carretta. 2021. "Embedded Quantum-Error Correction and Controlled-Phase Gate for Molecular Spin Qubits." *AIP Advances* 11(2):025134. doi.org/10.1063/9.0000166.

Chilton, N. F. 2022. "Molecular Magnetism." *Annual Review of Materials Research* 52(1):79–101. doi.org/10.1146/annurev-matsci-081420-042553.

Chizzini, M., L. Crippa, L. Zaccardi, E. Macaluso, S. Carretta, A. Chiesa, and P. Santini. 2022. "Quantum Error Correction with Molecular Spin Qudits." *Physical Chemistry Chemical Physics* 24(34):20030–20039. doi.org/10.1039/d2cp01228f.

Closs, G. L., M. D. E. Forbes, and J. R. Norris. 1987. "Spin-Polarized Electron-Paramagnetic Resonance-Spectra of Radical Pairs in Micelles—Observation of Electron Spin-Spin Interactions." *Journal of Physical Chemistry* 91(13):3592–3599. doi.org/10.1021/j100297a026.

Collopy, A. L., S. Ding, Y. Wu, I. A. Finneran, L. Anderegg, B. L. Augenbraun, J. M. Doyle, and J. Ye. 2018. "3D Magneto-Optical Trap of Yttrium Monoxide." *Physical Review Letters* 121(21):213201. doi.org/10.1103/PhysRevLett.121.213201.

Colvin, M. T., R. Carmieli, T. Miura, S. Richert, D. M. Gardner, A. L. Smeigh, S. M. Dyar, S. M. Conron, M. A. Ratner, and M. R. Wasielewski. 2013. "Electron Spin Polarization Transfer from Photogenerated Spin-Correlated Radical Pairs to a Stable Radical Observer Spin." *Journal of Physical Chemistry A* 117(25):5314–5325. doi.org/10.1021/jp4045012.

Corpinot, M. K., and D. K. Bucar. 2019. "A Practical Guide to the Design of Molecular Crystals." *Crystal Growth & Design* 19(2):1426–1453. doi.org/10.1021/acs.cgd.8b00972.

Corvaja, C., M. Maggini, M. Prato, G. Scorrano, and M. Venzin. 1995. "C-60 Derivative Covalently-Linked to a Nitroxide Radical-Time-Resolved EPR Evidence of Electron-Spin Polarization by Intramolecular Radical-Triplet Pair Interaction." *Journal of the American Chemical Society* 117(34):8857–8858. doi.org/10.1021/Ja00139a022.

Crooker, S. A., J. J. Baumberg, F. Flack, N. Samarth, and D. D. Awschalom. 1996. "Terahertz Spin Precession and Coherent Transfer of Angular Momenta in Magnetic Quantum Wells." *Physical Review Letters* 77(13):2814-2817. doi.org/10.1103/PhysRevLett.77.2814.

Daniel, J. R., C. Wang, K. Rodriguez, B. Hemmerling, T. N. Lewis, C. Bardeen, A. Teplukhin, and B. K. Kendrick. 2021. "Spectroscopy on the $A^1\Pi \leftarrow X^1\Sigma^+$ Transition of Buffer-Gas-Cooled AlCl." *Physical Review A* 104(1):012801. doi.org/10.1103/PhysRevA.104.012801.

Day, G. M., and A. I. Cooper. 2018. "Energy-Structure-Function Maps: Cartography for Materials Discovery." *Advanced Materials* 30(37). doi.org/10.1002/adma.201704944.

DeMille, D. 2002. "Quantum Computation with Trapped Polar Molecules." *Physical Review Letters* 88(6):067901. doi.org/10.1103/PhysRevLett.88.067901.

Di Rosa, M. D. 2004. "Laser-cooling Molecules." *European Physical Journal D* 31(2):395–402. doi.org/10.1140/epjd/e2004-00167-2.

Dickerson, C. E., C. Chang, H. Guo, and A. N. Alexandrova. 2022. "Fully Saturated Hydrocarbons as Hosts of Optical Cycling Centers." *Journal of Physical Chemistry A* 126(51):9644–9650.

Dickerson, C. E., H. Guo, A. J. Shin, B. L. Augenbraun, J. R. Caram, W. C. Campbell, and A. N. Alexandrova. 2021. "Franck-Condon Tuning of Optical Cycling Centers by Organic Functionalization." *Physical Review Letters* 126(12):123002. doi.org/10.1103/PhysRevLett.126.123002.

Dickerson, C. E., H. Guo, G.-Z. Zhu, E. R. Hudson, J. R. Caram, W. C. Campbell, and A. N. Alexandrova. 2021. "Optical Cycling Functionalization of Arenes." *Journal of Physical Chemistry Letters* 12(16):3989–3995. doi.org/10.1021/acs.jpclett.1c00733.

DiVincenzo, D. P. 2000. "The Physical Implementation of Quantum Computation." *Fortschritte der Physik* 48(9–11):771–783. doi.org/10.1002/1521-3978(200009)48:9/11<771:AID-PROP771>3.0.CO;2-E.

Dobrovitski, V. V., G. D. Fuchs, A. L. Falk, C. Santori, and D. D. Awschalom. 2013. "Quantum Control over Single Spins in Diamond." *Annual Review of Condensed Matter Physics* 4(1):23–50. doi.org/10.1146/annurev-conmatphys-030212-184238.

Doherty, M. W., N. B. Manson, P. Delaney, F. Jelezko, J. Wrachtrup, and L. C. L. Hollenberg. 2013. "The Nitrogen-Vacancy Colour Centre in Diamond." *Physics Reports* 528(1):1–45. doi.org/10.1016/j.physrep.2013.02.001.

Du, J.-L., K. M. More, S. S. Eaton, and G. R. Eaton. 1992. "Orientation Dependence of Electron Spin Phase Memory Relaxation Times in Copper(II) and Vanadyl Complexes in Frozen Solution." *Israel Journal of Chemistry* 32(2–3):351–355. doi.org/10.1002/ijch.199200041.

Eaton, S. S., J. Harbridge, G. A. Rinard, G. R. Eaton, and R. T. Weber. 2001. "Frequency Dependence of Electron Spin Relaxation for Threes=1/2 Species Doped into Diamagnetic Solid Hosts." *Applied Magnetic Resonance* 20(1):151–157. doi.org/10.1007/BF03162316.

Einstein, A., B. Podolsky, and N. Rosen. 1935. "Can Quantum-Mechanical Description of Physical Reality Be Considered Complete?" *Physical Review* 47(10):777–780. doi.org/10.1103/PhysRev.47.777.

Escalera-Moreno, L., J. J. Baldoví, A. Gaita-Ariño, and E. Coronado. 2019. "Exploring the High-Temperature Frontier in Molecular Nanomagnets: From Lanthanides to Actinides." *Inorganic Chemistry* 58(18):11883–11892. doi.org/10.1021/acs.inorgchem.9b01610.

Fataftah, M. S., and D. E. Freedman. 2018. "Progress Towards Creating Optically Addressable Molecular Qubits." *Chemical Communications* 54(98):13773–13781. doi.org/10.1039/C8CC07939K.

Fataftah, M. S., M. D. Krzyaniak, B. Vlaisavljevich, M. R. Wasielewski, J. M. Zadrozny, and D. E. Freedman. 2019. "Metal–Ligand Covalency Enables Room Temperature Molecular Qubit Candidates." *Chemical Science* 10(27):6707–6714. doi.org/10.1039/C9SC00074G.

Fataftah, M. S., J. M. Zadrozny, D. M. Rogers, and D. E. Freedman. 2014. "A Mononuclear Transition Metal Single-Molecule Magnet in a Nuclear Spin-Free Ligand Environment." *Inorganic Chemistry* 53(19):10716–10721. doi.org/10.1021/ic501906z.

Fay, T. P. 2021. "Chirality-Induced Spin Coherence in Electron Transfer Reactions." *Journal of Physical Chemistry Letters* 12(5):1407–1412. doi.org/10.1021/acs.jpclett.1c00009.

Fay, T. P., and D. T. Limmer. 2021. "Origin of Chirality Induced Spin Selectivity in Photoinduced Electron Transfer." *Nano Letters* 21(15):6696–6702. doi.org/10.1021/acs.nanolett.1c02370.

Fernandez, A., E. Moreno Pineda, C. A. Muryn, S. Sproules, F. Moro, G. A. Timco, E. J. L. McInnes, and R. E. P. Winpenny. 2015. "g-Engineering in Hybrid Rotaxanes to Create AB and AB2 Electron Spin Systems: EPR Spectroscopic Studies of Weak Interactions between Dissimilar Electron Spin Qubits." *Angewandte Chemie* 54(37):10858–10861. doi.org/10.1002/anie.201504487.

Ferrando-Soria, J., E. Moreno Pineda, A. Chiesa, A. Fernandez, S. A. Magee, S. Carretta, P. Santini, I. J. Vitorica-Yrezabal, F. Tuna, G. A. Timco, E. J. L. McInnes, and R. E. P. Winpenny. 2016. "A Modular Design of Molecular Qubits to Implement Universal Quantum Gates." *Nature Communications* 7(1). doi.org/10.1038/ncomms11377.

Fujisawa, J., Y. Iwasaki, Y. Ohba, S. Yamauchi, N. Koga, S. Karasawa, M. Fuhs, K. Moebius, and S. Weber. 2001. "Excited Quartet and Doublet States in the Complex of Tetraphenylporphine Zinc (II) and a Nitroxide Radical in Solution. X- and W-Band Time-Resolved EPR Studies." *Applied Magnetic Resonance* 21(3–4):483–493. doi.org/10.1007/BF03162422.

Gaita-Ariño, A., F. Luis, S. Hill, and E. Coronado. 2019. "Molecular Spins for Quantum Computation." *Nature Chemistry* 11(4):301–309. doi.org/10.1038/s41557-019-0232-y.

Gatteschi, D., R. Sessoli, and J. Villain. 2006. *Molecular Nanomagnets*. Vol. 376. Oxford: Oxford University Press.

Giacobbe, E. M., Q. Mi, M. T. Colvin, B. Cohen, C. Ramanan, A. M. Scott, S. Yeganeh, T. J. Marks, M. A. Ratner, and M. R. Wasielewski. 2009. "Ultrafast Intersystem Crossing and Spin Dynamics of Photoexcited Perylene-3,4:9,10-Bis(Dicarboximide) Covalently Linked to a Nitroxide Radical at Fixed Distances." *Journal of the American Chemical Society* 131(10):3700–3712. doi.org/10.1021/ja808924f.

Giménez-Santamarina, S., S. Cardona-Serra, J. M. Clemente-Juan, A. Gaita-Ariño, and E. Coronado. 2020. "Exploiting Clock Transitions for the Chemical Design of Resilient Molecular Spin Qubits." *Chemical Science* 11(39):10718–10728. doi.org/10.1039/d0sc01187h.

Gimeno, I., A. Urtizberea, J. Román-Roche, D. Zueco, A. Camón, P. J. Alonso, O. Roubeau, and F. Luis. 2021. "Broad-Band Spectroscopy of a Vanadyl Porphyrin: A Model Electronuclear Spin Qudit." *Chemical Science* 12(15):5621–5630. doi.org/10.1039/d1sc00564b.

Godfrin, C., A. Ferhat, R. Ballou, S. Klyatskaya, M. Ruben, W. Wernsdorfer, and F. Balestro. 2017. "Operating Quantum States in Single Magnetic Molecules: Implementation of Grover's Quantum Algorithm." *Physical Review Letters* 119(18). doi.org/10.1103/physrevlett.119.187702.

Gong, C., K. Hu, X. Wang, P. Wangyang, C. Yan, J. Chu, M. Liao, L. Dai, T. Zhai, C. Wang, L. Li, and J. Xiong. 2018. "2D Nanomaterial Arrays for Electronics and Optoelectronics." *Advanced Functional Materials* 28(16):1706559. doi.org/10.1002/adfm.201706559.

Goodwin, C. A. P., F. Ortu, D. Reta, N. F. Chilton, and D. P. Mills. 2017. "Molecular Magnetic Hysteresis at 60 Kelvin in Dysprosocenium." *Nature* 548(7668):439–442. doi.org/10.1038/nature23447.

Gould, C. A., K. R. McClain, D. Reta, J. G. C. Kragskow, D. A. Marchiori, E. Lachman, E.-S. Choi, J. G. Analytis, R. D. Britt, N. F. Chilton, B. G. Harvey, and J. R. Long. 2022. "Ultrahard Magnetism from Mixed-Valence Dilanthanide Complexes with Metal-Metal Bonding." *Science* 375(6577):198–202. doi.org/10.1126/science.abl5470.

Gouterman, M. 1970. "Porphyrins. XIX. Tripdoublet and Quartet Luminescence in Cu and VO Complexes." *Journal of Chemical Physics* 52(7):3795–3802. doi.org/10.1063/1.1673560.

Graham, M. J., C.-J. Yu, M. D. Krzyaniak, M. R. Wasielewski, and D. E. Freedman. 2017. "Synthetic Approach to Determine the Effect of Nuclear Spin Distance on Electronic Spin Decoherence." *Journal of the American Chemical Society* 139(8):3196–3201. doi.org/10.1021/jacs.6b13030.

Graham, M. J., J. M. Zadrozny, M. S. Fataftah, and D. E. Freedman. 2017. "Forging Solid-State Qubit Design Principles in a Molecular Furnace." *Chemistry of Materials* 29(5):1885–1897. doi.org/10.1021/acs.chemmater.6b05433.

Guo, F.-S., B. M. Day, Y.-C. Chen, M.-L. Tong, A. Mansikkamäki, and R. A. Layfield. 2018. "Magnetic Hysteresis up to 80 Kelvin in a Dysprosium Metallocene Single-Molecule Magnet." *Science* 362(6421):1400–1403. doi.org/10.1126/science.aav0652.

Guo, H., C. E. Dickerson, A. J. Shin, C. Zhao, T. L. Atallah, J. R. Caram, W. C. Campbell, and A. N. Alexandrova. 2021. "Surface Chemical Trapping of Optical Cycling Centers." *Physical Chemistry Chemical Physics* 23(1):211–218. doi.org/10.1039/D0CP04525J.

Hafner, R. J., L. Tian, J. C. Brauer, T. Schmaltz, A. Sienkiewicz, S. Balog, V. Flauraud, J. Brugger, and H. Frauenrath. 2018. "Unusually Long-Lived Photocharges in Helical Organic Semiconductor Nanostructures." *ACS Nano* 12(9):9116–9125. doi.org/10.1021/acsnano.8b03165.

Hallas, C., N. B. Vilas, L. Anderegg, P. Robichaud, A. Winnicki, C. Zhang, L. Cheng, and J. M. Doyle. 2023. "Optical Trapping of a Polyatomic Molecule in an l-Type Parity Doublet State." *Physical Review Letters* 130:153202. doi.org/10.1103/PhysRevLett.130.153202.

Harvey, S. M., and M. R. Wasielewski. 2021. "Photogenerated Spin-Correlated Radical Pairs: From Photosynthetic Energy Transduction to Quantum Information Science." *Journal of the American Chemical Society* 143(38):15508–15529. doi.org/10.1021/jacs.1c07706.

Hofsäss, S., M. Doppelbauer, S. C. Wright, S. Kray, B. G. Sartakov, J. Pérez-Ríos, G. Meijer, and S. Truppe. 2021. "Optical Cycling of AlF Molecules." *New Journal of Physics* 23(7):075001. doi.org/10.1088/1367-2630/ac06e5.

Holland, C. M., Y. Lu, and L. W. Cheuk. 2022. "On-Demand Entanglement of Molecules in a Reconfigurable Optical Tweezer Array." *ArXiv*. doi.org/10.48550/arxiv.2210.06309.

Hore, P. J., D. A. Hunter, C. D. Mckie, and A. J. Hoff. 1987. "Electron-Paramagnetic Resonance of Spin-Correlated Radical Pairs in Photosynthetic Reactions." *Chemical Physics Letters* 137(6):495–500. doi.org/10.1016/0009-2614(87)80617-6.

Horwitz, N. E., B. T. Phelan, J. N. Nelson, M. D. Krzyaniak, and M. R. Wasielewski. 2016. "Picosecond Control of Photogenerated Radical Pair Lifetimes Using a Stable Third Radical." *Journal of Physical Chemistry A* 120(18):2841–2853. doi.org/10.1021/acs.jpca.6b02621.

Horwitz, N. E., B. T. Phelan, J. N. Nelson, C. M. Mauck, M. D. Krzyaniak, and M. R. Wasielewski. 2017. "Spin Polarization Transfer from a Photogenerated Radical Ion Pair to a Stable Radical Controlled by Charge Recombination." *Journal of Physical Chemistry A* 121(23):4455–4463. doi.org/10.1021/acs.jpca.7b03468.

Hussain, R., G. Allodi, A. Chiesa, E. Garlatti, D. Mitcov, A. Konstantatos, K. S. Pedersen, R. De Renzi, S. Piligkos, and S. Carretta. 2018. "Coherent Manipulation of a Molecular Ln-Based Nuclear Qudit Coupled to an Electron Qubit." *Journal of the American Chemical Society* 140(31): 9814–9818. doi.org/10.1021/jacs.8b05934.

Hutzler, N. R. 2020. "Polyatomic Molecules as Quantum Sensors for Fundamental Physics." *Quantum Science and Technology* 5(4):044011. doi.org/10.1088/2058-9565/abb9c5.

Ishii, K., J. Fujisawa, A. Adachi, S. Yamauchi, and N. Kobayashi. 1998. "General Simulations of Excited Quartet Spectra with Electron-Spin Polarizations: The Excited Multiplet States of (Tetraphenylporphinato)zinc(II) Coordinated by p- or m-Pyridyl Nitronyl Nitroxides." *Journal of the American Chemical Society* 120:3152–3158. doi.org/10.1021/ja973146f.

Ishii, K., J. Fujisawa, Y. Ohba, and S. Yamauchi. 1996. "A Time-Resolved Electron Paramagnetic Resonance Study on the Excited States of Tetraphenylporphinatozinc(II) Coordinated By P-Pyridyl Nitronyl Nitroxide." *Journal of the American Chemical Society* 118(51):13079–13080. doi.org/10.1021/ja961661s.

Ivanov, M. V., S. Gulania, and A. I. Krylov. 2020. "Two Cycling Centers in One Molecule: Communication by Through-Bond Interactions and Entanglement of the Unpaired Electrons." *Journal of Physical Chemistry Letters* 11(4):1297–1304. doi.org/10.1021/acs.jpclett.0c00021.

Jackson, C. E., C.-Y. Lin, S. H. Johnson, J. van Tol, and J. M. Zadrozny. 2019. "Nuclear-Spin-Pattern Control of Electron-Spin Dynamics in a Series of V(IV) Complexes." *Chemical Science* 10(36):8447–8454. doi.org/10.1039/C9SC02899D.

Jackson, C. E., C.-Y. Lin, J. van Tol, and J. M. Zadrozny. 2020. "Orientation Dependence of Phase Memory Relaxation in the V(IV) Ion at High Frequencies." *Chemical Physics Letters* 739:137034. doi.org/10.1016/j.cplett.2019.137034.

Jacobberger, R. M., Y. Qiu, M. L. Williams, M. D. Krzyaniak, and M. R. Wasielewski. 2022. "Using Molecular Design to Enhance the Coherence Time of Quintet Multiexcitons Generated by Singlet Fission in Single Crystals." *Journal of the American Chemical Society* 144:2276–2283. doi.org/10.1021/jacs.1c12414.

Jenkins, M. D., Y. Duan, B. Diosdado, J. J. García-Ripoll, A. Gaita-Ariño, C. Giménez-Saiz, P. J. Alonso, E. Coronado, and F. Luis. 2017. "Coherent Manipulation of Three-Qubit States in a Molecular Single-Ion Magnet." *Physical Review B* 95(6). doi.org/10.1103/physrevb.95.064423.

Kaluarachchi, U. S., V. Taufour, S. L. Bud'Ko, and P. C. Canfield. 2018. "Quantum Tricritical Point in the Temperature-Pressure-Magnetic Field Phase Diagram of $CeTiGe_3$." *Physical Review B* 97(4). doi.org/10.1103/physrevb.97.045139.

Kanai, S., F. J. Heremans, H. Seo, G. Wolfowicz, C. P. Anderson, S. E. Sullivan, M. Onizhuk, G. Galli, D. D. Awschalom, and H. Ohno. 2022. "Generalized Scaling of Spin Qubit Coherence in Over 12,000 Host Materials." *Proceedings of the National Academy of Sciences USA* 119(15):e2121808119. doi.org/10.1073/pnas.2121808119.

Kandrashkin, Y., and A. van der Est. 2003. "Light-Induced Electron Spin Polarization in Rigidly Linked, Strongly Coupled Triplet-Doublet Spin Pairs." *Chemical Physics Letters* 379(5,6):574–580. doi.org/10.1016/j.cplett.2003.08.073.

Kandrashkin, Y., and A. van der Est. 2004. "Electron Spin Polarization of the Excited Quartet State of Strongly Coupled Triplet-Doublet Spin Systems." *Journal of Chemical Physics* 120(10):4790–4799. doi.org/10.1063/1.1645773.

Kandrashkin, Y. E., M. S. Asano, and A. van der Est. 2006a. "Light-Induced Electron Spin Polarization in Vanadyl Octaethylporphyrin: I. Characterization of the Excited Quartet State." *Journal of Physical Chemistry A* 110(31):9607–9616. doi.org/10.1021/jp0620365.

Kandrashkin, Y. E., M. S. Asano, and A. van der Est. 2006b. "Light-Induced Electron Spin Polarization in Vanadyl Octaethylporphyrin: II. Dynamics of the Excited States." *Journal of Physical Chemistry A* 110(31):9617–9626. doi.org/10.1021/jp062037x.

Kauranen, M., and A. V. Zayats. 2012. "Nonlinear Plasmonics." *Nature Photonics* 6(11):737–748. doi.org/10.1038/nphoton.2012.244.

Kazmierczak, N. P., and R. G. Hadt. 2022. "Illuminating Ligand Field Contributions to Molecular Qubit Spin Relaxation via T_1 Anisotropy." *Journal of the American Chemical Society* 144(45):20804–20814. doi.org/10.1021/jacs.2c08729.

Kazmierczak, N. P., R. Mirzoyan, and R. G. Hadt. 2021. "The Impact of Ligand Field Symmetry on Molecular Qubit Coherence." *Journal of the American Chemical Society* 143(42):17305–17315. doi.org/10.1021/jacs.1c04605.

Kiktenko, E. O., A. K. Fedorov, A. A. Strakhov, and V. I. Man'ko. 2015. "Single Qudit Realization of the Deutsch Algorithm Using Superconducting Many-Level Quantum Circuits." *Physics Letters A* 379(22–23):1409–1413. doi.org/10.1016/j.physleta.2015.03.023.

Kirk, M. L., D. A. Shultz, P. Hewitt, J. Chen, and A. van der Est. 2022. "Excited State Magneto-Structural Correlations Related to Photoinduced Electron Spin Polarization." *Journal of the American Chemical Society* 144(28):12781–1288. doi.org/10.1021/jacs.2c03490.

Kirk, M. L., D. A. Shultz, A. Reddy Marri, P. Hewitt, and A. van der Est. 2022. "Single-Photon-Induced Electron Spin Polarization of Two Exchange-Coupled Stable Radicals." *Journal of the American Chemical Society* 144(46):21005–21009. doi.org/10.1021/jacs.2c09680.

Kłos, J., and S. Kotochigova. 2020. "Prospects for Laser Cooling of Polyatomic Molecules with Increasing Complexity." *Physical Review Research* 2(1):013384. doi.org/10.1103/PhysRevResearch.2.013384.

Kozyryev, I., L. Baum, K. Matsuda, H. Boerge, and J. M. Doyle. 2016. "Radiation Pressure Force from Optical Cycling on a Polyatomic Molecule." *Journal of Physics B* 49(13):134002. doi.org/10.1088/0953-4075/49/13/134002.

Kozyryev, I., L. Baum, K. Matsuda, and J. M. Doyle. 2016. "Proposal for Laser Cooling of Complex Polyatomic Molecules." *ChemPhysChem* 17(22):3641–3648. doi.org/10.1002/cphc.201601051.

Kozyryev, I., Z. Lasner, and J. M. Doyle. 2021. "Enhanced Sensitivity to Ultralight Bosonic Dark Matter in the Spectra of the Linear Radical SrOH." *Physical Review A* 103(4):043313. doi.org/10.1103/PhysRevA.103.043313.

Kozyryev, I., T. C. Steimle, P. Yu, D.-T. Nguyen, and J. M. Doyle. 2019. "Determination of CaOH and $CaOCH_3$ Vibrational Branching Ratios for Direct Laser Cooling and Trapping." *New Journal of Physics* 21(5):052002. doi.org/10.1088/1367-2630/ab19d7.

Krzyaniak, M. D., L. Kobr, B. K. Rugg, B. T. Phelan, E. A. Margulies, J. N. Nelson, R. M. Young, and M. R. Wasielewski. 2015. "Fast Photo-Driven Electron Spin Coherence Transfer: The Effect of Electron-Nuclear Hyperfine Coupling on Coherence Dephasing." *Journal of Materials Chemistry C* 3(30):7962–7967. doi.org/10.1039/C5TC01446H.

Kues, M., C. Reimer, P. Roztocki, L. Romero Cortés, S. Sciara, B. Wetzel, Y. Zhang, A. Cino, S. T. Chu, B. E. Little, D. J. Moss, L. Caspani, J. Azaña, and R. Morandotti. 2017. "On-Chip Generation of High-Dimensional Entangled Quantum States and Their Coherent Control." *Nature* 546(7660):622–626. doi.org/10.1038/nature22986.

Kumar, K. S., D. Serrano, A. M. Nonat, B. Heinrich, L. Karmazin, L. J. Charbonnière, P. Goldner, and M. Ruben. 2021. "Optical Spin-State Polarization in a Binuclear Europium Complex Towards Molecule-Based Coherent Light-Spin Interfaces." *Nature Communications* 12:2152. doi.org/10.1038/s41467-021-22383-x

Kumarasamy, E., S. N. Sanders, M. J. Y. Tayebjee, A. Asadpoordarvish, T. J. H. Hele, E. G. Fuemmeler, A. B. Pun, L. M. Yablon, J. Z. Low, D. W. Paley, J. C. Dean, B. Choi, G. D. Scholes, M. L. Steigerwald, N. Ananth, D. R. McCamey, M. Y. Sfeir, and L. M. Campos. 2017. "Tuning Singlet Fission in π-Bridge-π Chromophores." *Journal of the American Chemical Society* 139(36):12488–12494. doi.org/10.1021/jacs.7b05204.

Kundu, K., J. Chen, S. Hoffman, J. Marbey, D. Komijani, Y. Duan, A. Gaita-Ariño, J. Stanton, X. Zhang, H.-P. Cheng, and S. Hill. 2023. "Electron-Nuclear Decoupling at a Spin Clock Transition." *Communications Physics* 6(1):38. doi.org/10.1038/s42005-023-01152-w.

Kundu, K., J. R. K. White, S. A. Moehring, J. M. Yu, J. W. Ziller, F. Furche, W. J. Evans, and S. Hill. 2022. "A 9.2-GHz Clock Transition in a Lu(II) Molecular Spin Qubit Arising from a 3,467-MHz Hyperfine Interaction." *Nature Chemistry* 14(4):392–397. doi.org/10.1038/s41557-022-00894-4.

Lao, G., G.-Z. Zhu, C. E. Dickerson, B. L. Augenbraun, A. N. Alexandrova, J. R. Caram, E. R. Hudson, and W. C. Campbell. 2022. "Laser Spectroscopy of Aromatic Molecules with Optical Cycling Centers: Strontium(I) Phenoxides." *Journal of Physical Chemistry Letters* 13(47):11029–11035. doi.org/10.1021/acs.jpclett.2c03040.

Laorenza, D.W., and D.E. Freedman. 2022. "Could the Quantum Internet Be Comprised of Molecular Spins with Tunable Optical Interfaces?" *Journal of the American Chemical Society* 144(48):21810–21825. doi.org/10.1021/jacs.2c07775.

Laorenza, D. W., A. Kairalapova, S. L. Bayliss, T. Goldzak, S. M. Greene, L. R. Weiss, P. Deb, P. J. Mintun, K. A. Collins, D. D. Awschalom, T. C. Berkelbach, and D. E. Freedman. 2021. "Tunable Cr^{4+} Molecular Color Centers." *Journal of the American Chemical Society* 143(50):21350–21363. doi.org/10.1021/jacs.1c10145.

Levi, B. G. 2016. "Making Molecular-Spin Qubits More Robust." *Physics Today* 69(5):17–21. doi.org/10.1063/PT.3.3157.

Liu L. R., J. D. Hood, Y. Yu, J. T. Zhang, N. R. Hutzler, T. Rosenband, and K-K. Ni. 2018. "Building One Molecule from a Reservoir of Two Atoms." *Science* 360 (6391):900–903. doi.org/10.1126/science.aar7797.

Lombardi, F., A. Lodi, J. Ma, J. Liu, M. Slota, A. Narita, W. K. Myers, K. Müllen, X. Feng, and L. Bogani. 2019. "Quantum Units from the Topological Engineering of Molecular Graphenoids." *Science* 366(6469):1107–1110. doi.org/10.1126/science.aay7203.

Lubert-Perquel, D., E. Salvadori, M. Dyson, P. N. Stavrinou, R. Montis, H. Nagashima, Y. Kobori, S. Heutz, and C. W. M. Kay. 2018. "Identifying Triplet Pathways in Dilute Pentacene Films." *Nature Communications* 9(1):1–10. doi.org/10.1038/s41467-018-06330-x.

Luis, F., P. J. Alonso, O. Roubeau, V. Velasco, D. Zueco, D. Aguilà, J. I. Martínez, L. A. Barrios, and G. Aromí. 2020. "A Dissymmetric [Gd_2] Coordination Molecular Dimer Hosting Six Addressable Spin Qubits." *Communications Chemistry* 3(1). doi.org/10.1038/s42004-020-00422-w.

Luis, F., A. Repollés, M. J. Martínez-Pérez, D. Aguilà, O. Roubeau, D. Zueco, P. J. Alonso, M. Evangelisti, A. Camón, J. Sesé, L. A. Barrios, and G. Aromí. 2011. "Molecular Prototypes for Spin-Based CNOT and SWAP Quantum Gates." *Physical Review Letters* 107(11). doi.org/10.1103/physrevlett.107.117203.

Lunghi, A., and S. Sanvito. 2020. "The Limit of Spin Lifetime in Solid-State Electronic Spins." *Journal of Physical Chemistry Letters* 11(15):6273–6278. doi.org/10.1021/acs.jpclett.0c01681.

Luo, J., and P. J. Hore. 2021. "Chiral-Induced Spin Selectivity in the Formation and Recombination of Radical Pairs: Cryptochrome Magnetoreception and EPR Detection." *New Journal of Physics* 23:043032. doi.org/10.1088/1367-2630/abed0b.

Macaluso, E., M. Rubín, D. Aguilà, A. Chiesa, L. A. Barrios, J. I. Martínez, P. J. Alonso, O. Roubeau, F. Luis, G. Aromí, and S. Carretta. 2020. "A Heterometallic [LnLn′Ln] Lanthanide Complex as a Qubit with Embedded Quantum Error Correction." *Chemical Science* 11(38):10337–10343. doi.org/10.1039/d0sc03107k.

Mao, H., G. J. Pažėra, R. M. Young, M. D. Krzyaniak, and M. R. Wasielewski. 2023. "Quantum Gate Operations on a Spectrally Addressable Photogenerated Molecular Electron Spin-Qubit Pair." *Journal of the American Chemical Society* 145(11):6585–6593. doi.org/10.1021/jacs.3c01243.

Matsuda, S., S. Oyama, and Y. Kobori. 2020. "Electron Spin Polarization Generated by Transport of Singlet and Quintet Multiexcitons to Spin-Correlated Triplet Pairs During Singlet Fissions." *Chemical Science* 11(11):2934–2942. doi.org/10.1039/C9SC04949E.

Maylaender, M., S. Chen, E. R. Lorenzo, M. R. Wasielewski, and S. Richert. 2021. "Exploring Photogenerated Molecular Quartet States as Spin Qubits and Qudits." *Journal of the American Chemical Society* 143(18):7050–7058. doi.org/10.1021/jacs.1c01620.

McInnes, E. J. L. 2022. "Molecular Spins Clock In." *Nature Chemistry* 14(4):361–362. doi.org/10.1038/s41557-022-00919-y.
Mi, Q., E. T. Chernick, D. W. McCamant, E. A. Weiss, M. A. Ratner, and M. R. Wasielewski. 2006. "Spin Dynamics of Photogenerated Triradicals in Fixed Distance Electron Donor-Chromophore-Acceptor-TEMPO Molecules." *Journal of Physical Chemistry A* 110(23):7323–7333. doi.org/10.1021/jp061218w.
Mitra, D., Z. D. Lasner, G.-Z. Zhu, C. E. Dickerson, B. L. Augenbraun, A. D. Bailey, A. N. Alexandrova, W. C. Campbell, J. R. Caram, E. R. Hudson, and J. M. Doyle. 2022. "Pathway Toward Optical Cycling and Laser Cooling of Functionalized Arenes." *Journal of Physical Chemistry Letters* 13(30):7029–7035. doi.org/10.1021/acs.jpclett.2c01430.
Mitra, D., N. B. Vilas, C. Hallas, L. Anderegg, B. L. Augenbraun, L. Baum, C. Miller, S. Raval, and J. M. Doyle. 2020. "Direct Laser Cooling of a Symmetric Top Molecule." *Science* 369(6509):1366–1369. doi.org/10.1126/science.abc5357.
Mizuochi, N., Y. Ohba, and S. Yamauchi. 1997. "A Two-Dimensional EPR Nutation Study on Excited Multiplet States of Fullerene Linked to a Nitroxide Radical." *Journal of Physical Chemistry A* 101(34):5966–5968. doi.org/10.1021/jp971569y.
Moreno-Pineda, E., M. Damjanović, O. Fuhr, W. Wernsdorfer, and M. Ruben. 2017. "Nuclear Spin Isomers: Engineering a $Et_4N[DyPc_2]$ Spin Qudit." *Angewandte Chemie International Edition* 56(33):9915–9919. doi.org/10.1002/anie.201706181.
Moreno-Pineda, E., C. Godfrin, F. Balestro, W. Wernsdorfer, and M. Ruben. 2018. "Molecular Spin Qudits for Quantum Algorithms." *Chemical Society Reviews* 47(2):501–513. doi.org/10.1039/C5CS00933B.
Morton, J. J. L., A. M. Tyryshkin, A. Ardavan, K. Porfyrakis, S. A. Lyon, and G. A. D. Briggs. 2007. "Environmental Effects on Electron Spin Relaxation in N@C_{60}." *Physical Review B* 76(8):085418. doi.org/10.1103/PhysRevB.76.085418.
Naaman, R., and D. H. Waldeck. 2012. "Chiral-Induced Spin Selectivity Effect." *Journal of Physical Chemistry Letters* 3(16):2178–2187. doi.org/10.1021/jz300793y.
Nakazawa, S., S. Nishida, T. Ise, T. Yoshino, N. Mori, R. D. Rahimi, K. Sato, Y. Morita, K. Toyota, D. Shiomi, M. Kitagawa, H. Hara, P. Carl, P. Hofer, and T. Takui. 2012. "A Synthetic Two-Spin Quantum Bit: G-Engineered Exchange-Coupled Biradical Designed for Controlled-NOT Gate Operations." *Angewandte Chemie* 51(39):9860–9864. doi.org/10.1002/anie.201204489.
Nelson, J. N., J. Zhang, J. Zhou, B. K. Rugg, M. D. Krzyaniak, and M. R. Wasielewski. 2020. "CNOT Gate Operation on a Photogenerated Molecular Electron Spin-Qubit Pair." *Journal of Chemical Physics* 152(1):014503/1-014503/7. doi.org/10.1063/1.5128132.
Ni, K.-K., T. Rosenband, and D. D. Grimes. 2018. "Dipolar Exchange Quantum Logic Gate with Polar Molecules." *Chemical Science* 9(33):6830–6838. https://doi.org/10.1039/c8sc02355g.
Nielsen, M. A., and I. L. Chuang. 2009. *Quantum Computation and Quantum Information Science: 10th Edition*. New York: Cambridge University Press.
O'Brien, J. L. 2007. "Optical Quantum Computing." *Science* 318(5856):1567–1570. doi.org/10.1126/science.1142892.
Olmschenk, S., D. N. Matsukevich, P. Maunz, D. Hayes, L.-M. Duan, and C. Monroe. 2009. "Quantum Teleportation Between Distant Matter Qubits." *Science* 323(5913):486–489. doi.org/10.1126/science.1167209.
Olshansky, J. H., J. Zhang, M. D. Krzyaniak, E. R. Lorenzo, and M. R. Wasielewski. 2020. "Selectively Addressable Photogenerated Spin Qubit Pairs in DNA Hairpins." *Journal of the American Chemical Society* 142(7):3346–3350. doi.org/10.1021/jacs.9b13398.
Pace, N. A., B. K. Rugg, C. H. Chang, O. G. Reid, K. J. Thorley, S. Parkin, J. E. Anthony, and J. C. Johnson. 2020. "Conversion Between Triplet Pair States is Controlled by Molecular Coupling in Pentadithiophene Thin Films." *Chemical Science* 11(27):7226–7238. doi.org/10.1039/D0SC02497J.
Paquette, M. M., D. Plaul, A. Kurimoto, B. O. Patrick, and N. L. Frank. 2018. "Opto-Spintronics: Photoisomerization-Induced Spin State Switching at 300 K in Photochrome Cobalt–Dioxolene Thin Films." *Journal of the American Chemical Society* 140(44):14990–15000. doi.org/10.1021/jacs.8b09190.
Petiziol, F., A. Chiesa, S. Wimberger, P. Santini, and S. Carretta. 2021. "Counteracting Dephasing in Molecular Nanomagnets by Optimized Qudit Encodings." *npj Quantum Information* 7(1). doi.org/10.1038/s41534-021-00466-3.
Pirandola, S., J. Eisert, C. Weedbrook, A. Furusawa, and S. L. Braunstein. 2015. "Advances in Quantum Teleportation." *Nature Photonics* 9:641–652. doi.org/10.1038/nphoton.2015.154.
Poddutoori, P. K., Y. E. Kandrashkin, P. Karr, and A. van der Est. 2019. "Electron Spin Polarization in an Al(III) Porphyrin Complex with an Axially Bound Nitroxide Radical." *Journal of Chemical Physics* 151(20):204303/1–204303/10. doi.org/10.1063/1.5127760.
Polisseni, C., K. D. Major, S. Boissier, S. Grandi, A. S. Clark, and E. A. Hinds. 2016. "Stable, Single-Photon Emitter in a Thin Organic Crystal for Application to Quantum-Photonic Devices." *Optics Express* 24(5):5615. doi.org/10.1364/oe.24.005615.
Qiu, Y., A. Equbal, C. Lin, Y. Huang, P. J. Brown, R. M. Young, M. D. Krzyaniak, and M. R. Wasielewski. 2022. "Optical Spin Polarization of a Narrow-Linewidth Electron-Spin Qubit in a Chromophore/Stable-Radical System." *Angewandte Chemie International Edition* 62(6). doi.org/10.1002/anie.202214668.

Reta, D., J. G. C. Kragskow, and N. F. Chilton. 2021. "Ab Initio Prediction of High-Temperature Magnetic Relaxation Rates in Single-Molecule Magnets." *Journal of the American Chemical Society* 143(15):5943–5950. doi.org/10.1021/jacs.1c01410.

Rinehart, J. D., and J. R. Long. 2009. "Slow Magnetic Relaxation in a Trigonal Prismatic Uranium(III) Complex." *Journal of the American Chemical Society* 131(35):12558–12559. doi.org/10.1021/ja906012u.

Rugg, B. K., M. D. Krzyaniak, B. T. Phelan, M. A. Ratner, R. M. Young, and M. R. Wasielewski. 2019. "Photodriven Quantum Teleportation of an Electron Spin State in a Covalent Donor-Acceptor-Radical System." *Nature Chemistry* 11(11):981–986. doi.org/10.1038/s41557-019-0332-8.

Rugg, B. K., K. E. Smyser, B. Fluegel, C. H. Chang, K. J. Thorley, S. Parkin, J. E. Anthony, J. D. Eaves, and J. C. Johnson. 2022. "Triplet-Pair Spin Signatures from Macroscopically Aligned Heteroacenes in an Oriented Single Crystal." *Proceedings of the National Academy of Sciences USA* 119(29):e2201879119. doi.org/10.1073/pnas.2201879119.

Sakai, H., R. Inaya, H. Nagashima, S. Nakamura, Y. Kobori, N. V. Tkachenko, and T. Hasobe. 2018. "Multiexciton Dynamics Depending on Intramolecular Orientations in Pentacene Dimers: Recombination and Dissociation of Correlated Triplet Pairs." *Journal of Physical Chemistry Letters* 9(12):3354–3360. doi.org/10.1021/acs.jpclett.8b01184.

Sciara, S., P. Roztocki, B. Fischer, C. Reimer, L. Romero Cortés, W. J. Munro, D. J. Moss, A. C. Cino, L. Caspani, M. Kues, J. Azaña, and R. Morandotti. 2021. "Scalable and Effective Multi-Level Entangled Photon States: A Promising Tool to Boost Quantum Technologies." *Nanophotonics* 10(18):4447–4465. doi.org/10.1515/nanoph-2021-0510.

Serrano, D., S. K. Kuppusamy, B. Heinrich, O. Fuhr, D. Hunger, M. Ruben, and P. Goldner. 2022. "Ultra-Narrow Optical Linewidths in Rare Earth Molecular Crystals." *Nature* 603:241–246.

Sessoli, R., D. Gatteschi, A. Caneschi, and M. A. Novak. 1993. "Magnetic Bistability in a Metal-Ion Cluster." *Nature* 365(6442):141–143. doi.org/10.1038/365141a0.

Sessoli, R., H. L. Tsai, A. R. Schake, S. Wang, J. B. Vincent, K. Folting, D. Gatteschi, G. Christou, and D. N. Hendrickson. 1993. "High-Spin Molecules: $[Mn_{12}O_{12}(O_2CR)_{16}(H_2O)_4]$." *Journal of the American Chemical Society* 115(5):1804–1816. doi.org/10.1021/ja00058a027.

Shannon, C. E. 1948. "A Mathematical Theory of Communication." *The Bell System Technical Journal* 27(3):379–423.

Shapiro, E. A., I. Khavkine, M. Spanner, and M. Y. Ivanov. 2003. "Strong-Field Molecular Alignment for Quantum Logic and Quantum Control." *Physical Review A* 67(1). doi.org/10.1103/physreva.67.013406.

Shiddiq, M., D. Komijani, Y. Duan, A. Gaita-Ariño, E. Coronado, and S. Hill. 2016. "Enhancing Coherence in Molecular Spin Qubits via Atomic Clock Transitions." *Nature* 531(7594):348–351. doi.org/10.1038/nature16984.

Siyushev, P., K. Xia, R. Reuter, M. Jamali, N. Zhao, N. Yang, C. Duan, N. Kukharchyk, A. D. Wieck, R. Kolesov, and J. Wrachtrup. 2014. "Coherent Properties of Single Rare-Earth Spin Qubits." *Nature Communications* 5(1). doi.org/10.1038/ncomms4895.

Smith, M. B., and J. Michl. 2010. "Singlet Fission." *Chemical Reviews* 110(11):6891–6936. doi.org/10.1021/cr1002613.

Smyser, K. E., and J. D. Eaves. 2020. "Singlet Fission for Quantum Information and Quantum Computing: The Parallel JDE Model." *Scientific Reports* 10(1):18480. doi.org/10.1038/s41598-020-75459-x.

Sørensen, M. A., H. Weihe, M. G. Vinum, J. S. Mortensen, L. H. Doerrer, and J. Bendix. 2017. "Imposing High-Symmetry and Tuneable Geometry on Lanthanide Centres with Chelating Pt and Pd Metalloligands." *Chemical Science* 8(5):3566–3575. doi.org/10.1039/C7SC00135E.

Stuhl, B. K., B. C. Sawyer, D. Wang, and J. Ye. 2008. "Magneto-Optical Trap for Polar Molecules." *Physical Review Letters* 101(24):243002. doi.org/10.1103/PhysRevLett.101.243002.

Sun, L., L. Yang, J. H. Dou, J. Li, G. Skorupskii, M. Mardini, K. O. Tan, T. Chen, C. Sun, J. J. Oppenheim, R. G. Griffin, M. Dincă, and T. Rajh. 2022. "Room-Temperature Quantitative Quantum Sensing of Lithium Ions with a Radical-Embedded Metal-Organic Framework." *Journal of the American Chemical Society* 144(41):19008–19016. doi.org/10.1021/jacs.2c07692.

Tame, M. S., K. R. McEnery, S. K. Ozdemir, J. Lee, S. A. Maier, and M. S. Kim. 2013. "Quantum Plasmonics." *Nature Physics* 9(6):329–340. doi.org/10.1038/nphys2615.

Tang, J., M. C. Thurnauer, and J. R. Norris. 1994. "Electron Spin Echo Envelope Modulation Due to Exchange and Dipolar Interactions in a Spin-Correlated Radical Pair." *Chemical Physics Letters* 219:283–290. doi.org/10.1016/0009-2614(94)87059-4.

Tayebjee, M. J. Y., S. N. Sanders, E. Kumarasamy, L. M. Campos, M. Y. Sfeir, and D. R. McCamey. 2017. "Quintet Multiexciton Dynamics in Singlet Fission." *Nature Physics* 13:182–188. doi.org/10.1038/nphys3909.

Teki, Y. 2020. "Excited-State Dynamics of Non-Luminescent and Luminescent π-Radicals." *Chemistry: A European Journal* 26(5):980–996. doi.org/10.1002/chem.201903444.

Tesi, L., E. Lucaccini, I. Cimatti, M. Perfetti, M. Mannini, M. Atzori, E. Morra, M. Chiesa, A. Caneschi, L. Sorace, and R. Sessoli. 2016. "Quantum Coherence in a Processable Vanadyl Complex: New Tools for the Search of Molecular Spin Qubits." *Chemical Science* 7(3):2074–2083. doi.org/10.1039/C5SC04295J.

Thurnauer, M. C., and J. R. Norris. 1980. "An Electron Spin Echo Phase Shift Observed in Photosynthetic Algae: Possible Evidence for Dynamic Radical Pair Interactions." *Chemical Physics Letters* 76:557–561. doi.org/10.1016/0009-2614(80)80667-1.

Toninelli, C., I. Gerhardt, A. S. Clark, A. Reserbat-Plantey, S. Götzinger, Z. Ristanović, M. Colautti, P. Lombardi, K. D. Major, I. Deperasińska, W. H. Pernice, F. H. L. Koppens, B. Kozankiewicz, A. Gourdon, V. Sandoghdar, and M. Orrit. 2021. "Single Organic Molecules for Photonic Quantum Technologies." *Nature Materials* 20(12):1615–1628. doi.org/10.1038/s41563-021-00987-4.

Troiani, F., and M. Affronte. 2011. "Molecular Spins for Quantum Information Technologies." *Chemical Society Reviews* 40(6):3119–3129. doi.org/10.1039/c0cs00158a.

Truppe, S., H. J. Williams, M. Hambach, L. Caldwell, N. J. Fitch, E. A. Hinds, B. E. Sauer, and M. R. Tarbutt. 2017. "Molecules Cooled Below the Doppler Limit." *Nature Physics* 13(12):1173–1176. doi.org/10.1038/nphys4241.

Urtizberea, A., E. Natividad, P. J. Alonso, M. A. Andrés, I. Gascón, M. Goldmann, and O. Roubeau. 2018. "A Porphyrin Spin Qubit and Its 2D Framework Nanosheets." *Advanced Functional Materials* 28(31):1801695. doi.org/10.1002/adfm.201801695.

Van Raden, J. M., D. I. Alexandropoulos, M. Slota, S. Sopp, T. Matsuno, A. L. Thompson, H. Isobe, H. L. Anderson, and L. Bogani. 2022. "Singly and Triply Linked Magnetic Porphyrin Lanthanide Arrays." *Journal of American Chemical Society* 144(19):8693–8706. doi: 10.1021/jacs.2c02084.

Vilas, N. B., C. Hallas, L. Anderegg, P. Robichaud, A. Winnicki, D. Mitra, and J. M. Doyle. 2022. "Magneto-Optical Trapping and Sub-Doppler Cooling of a Polyatomic Molecule." *Nature* 606(7912):70–74. doi.org/10.1038/s41586-022-04620-5.

von Kugelgen, S., M. D. Krzyaniak, M. Gu, D. Puggioni, J. M. Rondinelli, M. R. Wasielewski, and D. E. Freedman. 2021. "Spectral Addressability in a Modular Two Qubit System." *Journal of the American Chemical Society* 143(21):8069–8077. doi.org/10.1021/jacs.1c02417.

Walsh, J. P S., and D. E Freedman. 2016. "Using Supramolecular Chemistry to Build Quantum Logic Gates." *Chem* 1(5):668–669. doi.org/10.1016/j.chempr.2016.10.010.

Wang, D., H. Kelkar, D. Martin-Cano, D. Rattenbacher, A. Shkarin, T. Utikal, S. Götzinger, and V. Sandoghdar. 2019. "Turning a Molecule into a Coherent Two-Level Quantum System." *Nature Physics* 15(5):483–489. doi.org/10.1038/s41567-019-0436-5.

Wang, H., J. Guo, E. D. Bauer, V. A. Sidorov, H. Zhao, J. Zhang, Y. Zhou, Z. Wang, S. Cai, K. Yang, A. Li, P. Sun, Y.-F. Yang, Q. Wu, T. Xiang, J. D. Thompson, and L. Sun. 2019. "Anomalous Connection Between Antiferromagnetic and Superconducting Phases in the Pressurized Noncentrosymmetric Heavy-Fermion Compound $CeRhGe_3$." *Physical Review B* 99(2). doi.org/10.1103/physrevb.99.024504.

Wang, Y., Z. Hu, B. C. Sanders, and S. Kais. 2020. "Qudits and High-Dimensional Quantum Computing." *Frontiers in Physics* 8. doi.org/10.3389/fphy.2020.589504.

Wang, Y.-X., Z. Liu, Y.-H. Fang, S. Zhou, S.-D. Jiang, and S. Gao. 2021. "Coherent Manipulation and Quantum Phase Interference in a Fullerene-Based Electron Triplet Molecular Qutrit." *npj Quantum Information* 7(1):32. doi.org/10.1038/s41534-021-00362-w.

Warner, M., S. Din, I. S. Tupitsyn, G. W. Morley, A. M. Stoneham, J. A. Gardener, Z. Wu, A. J. Fisher, S. Heutz, C. W. M. Kay, and G. Aeppli. 2013. "Potential for Spin-Based Information Processing in a Thin-Film Molecular Semiconductor." *Nature* 503(7477):504–508. doi.org/10.1038/nature12597.

Wasielewski, M.R. 2023. "Light-Driven Spin Chemistry for Quantum Information Science." *Physics Today* 76(3):28–34. doi.org/10.1063/PT.3.5196.

Wasielewski, M. R., M. D. E. Forbes, N. L. Frank, K. Kowalski, G. D. Scholes, J. Yuen-Zhou, M. A. Baldo, D. E. Freedman, R. H. Goldsmith, T. Goodson, M. L. Kirk, J. K. McCusker, J. P. Ogilvie, D. A. Shultz, S. Stoll, and K. B. Whaley. 2020. "Exploiting Chemistry and Molecular Systems for Quantum Information Science." *Nature Reviews Chemistry* 4(9):490–504. doi.org/10.1038/s41570-020-0200-5.

Wedge, C. J., G. A. Timco, E. T. Spielberg, R. E. George, F. Tuna, S. Rigby, E. J. L. McInnes, R. E. P. Winpenny, S. J. Blundell, and A. Ardavan. 2012. "Chemical Engineering of Molecular Qubits." *Physical Review Letters* 108(10):107204. doi.org/10.1103/PhysRevLett.108.107204.

Weiss, L. R., S. L. Bayliss, F. Kraffert, K. J. Thorley, J. E. Anthony, R. Bittl, R. H. Friend, A. Rao, N. C. Greenham, and J. Behrends. 2017. "Strongly Exchange-Coupled Triplet Pairs in an Organic Semiconductor." *Nature Physics* 13:176–181. doi.org/10.1038/nphys3908.

Williams, H. J., L. Caldwell, N. J. Fitch, S. Truppe, J. Rodewald, E. A. Hinds, B. E. Sauer, and M. R. Tarbutt. 2018. "Magnetic Trapping and Coherent Control of Laser-Cooled Molecules." *Physical Review Letters* 120(16):163201. doi.org/10.1103/PhysRevLett.120.163201.

Wright, K., K. M. Beck, S. Debnath, J. M. Amini, Y. Nam, N. Grzesiak, J. S. Chen, N. C. Pisenti, M. Chmielewski, C. Collins, K. M. Hudek, J. Mizrahi, J. D. Wong-Campos, S. Allen, J. Apisdorf, P. Solomon, M. Williams, A. M. Ducore, A. Blinov, S. M. Kreikemeier, V. Chaplin, M. Keesan, C. Monroe, and J. Kim. 2019. "Benchmarking an 11-Qubit Quantum Computer." *Nature Communications* 10(1). doi.org/10.1038/s41467-019-13534-2.

Yan, B., S. Moses, B. Gadway, J. P. Covey, K. R. A. Hazzard, A. M. Rey, D. S. Jin, and J. Ye. 2013. "Observation of Dipolar Spin-Exchange Interactions with Lattice-Confined Polar Molecules." *Nature* 501:521–525. doi.org/10.1038/nature12483.

Yu, C.-J., M. J. Graham, J. M. Zadrozny, J. Niklas, M. D. Krzyaniak, M. R. Wasielewski, O. G. Poluektov, and D. E. Freedman. 2016. "Long Coherence Times in Nuclear Spin-Free Vanadyl Qubits." *Journal of the American Chemical Society* 138(44):14678–14685. doi.org/10.1021/jacs.6b08467.

Yu, C.-J., S. von Kugelgen, D. W. Laorenza, and D. E. Freedman. 2021. "A Molecular Approach to Quantum Sensing." *ACS Central Science* 7(5):712–723. doi.org/10.1021/acscentsci.0c00737.

Yu, P., L. W. Cheuk, I. Kozyryev, and J. M. Doyle. 2019. "A Scalable Quantum Computing Platform Using Symmetric-Top Molecules." *New Journal of Physics* 21(9):093049. doi.org/10.1088/1367-2630/ab428d.

Yu, P., A. Lopez, W. A. Goddard, and N. R. Hutzler. 2022. "Multivalent Optical Cycling Centers: Towards Control of Polyatomics with Multi-Electron Degrees of Freedom." *Physical Chemistry Chemical Physics*. https://doi.org/10.1039/D2CP03545F.

Yunusova, K. M., S. L. Bayliss, T. Chanelière, V. Derkach, J. E. Anthony, A. D. Chepelianskii, and L. R. Weiss. 2020. "Spin Fine Structure Reveals Biexciton Geometry in an Organic Semiconductor." *Physical Review Letters* 125(9):097402. doi.org/10.1103/PhysRevLett.125.097402.

Zadrozny, J. M., A. T. Gallagher, T. D. Harris, and D. E. Freedman. 2017. "A Porous Array of Clock Qubits." *Journal of the American Chemical Society* 139(20):7089–7094. doi.org/10.1021/jacs.7b03123.

Zadrozny, J. M., J. Niklas, O. G. Poluektov, and D. E. Freedman. 2015. "Millisecond Coherence Time in a Tunable Molecular Electronic Spin Qubit." *ACS Central Science* 1(9):488–492. doi.org/10.1021/acscentsci.5b00338.

Zhang, J., P. W. Hess, A. Kyprianidis, P. Becker, A. Lee, J. Smith, G. Pagano, I. D. Potirniche, A. C. Potter, A. Vishwanath, N. Y. Yao, and C. Monroe. 2017. "Observation of a Discrete Time Crystal." *Nature* 543(7644):217–220. doi.org/10.1038/nature21413.

Zhu, G.-Z., D. Mitra, B. L. Augenbraun, C. E. Dickerson, M. J. Frim, G. Lao, Z. D. Lasner, A. N. Alexandrova, W. C. Campbell, J. R. Caram, J. M. Doyle, and E. R. Hudson. 2022. "Functionalizing Aromatic Compounds with Optical Cycling Centres." *Nature Chemistry* 14(9):995–999. doi.org/10.1038/s41557-022-00998-x.

3

Measurement and Control of Molecular Quantum Systems

Key Takeaways

- Time-resolved/transient electron paramagnetic resonance is a powerful tool for investigating QIS properties in spin-based molecular qubits.
- Coherence is important in both organic and biological systems, as demonstrated by ultrafast, multidimensional spectroscopy. Now is the time to expand this area to look for the importance and usefulness of coherence in understanding biological functions.
- The use of quantum light or entangled photons for spectroscopy is a new and important field. Combining research centers to investigate new sources of entangled photons and applications is a new and exciting direction that would benefit from collaboration and shared resources.
- Quantum light could help understand biological systems at a very sensitive level.
- Optically active molecular systems are desirable for quantum science applications, as they can provide high-fidelity quantum-state preparation, detection, and control.
- Recent work with trapped optical cycling centers (OCCs) has demonstrated several DiVincenzo criteria required for quantum computing, including high-fidelity quantum-state readout, detection, and on-demand entanglement of molecular qubits via a two-qubit gate sufficient for universal gate-based quantum computing. With these demonstrations, the time is ripe for investigations of OCCs as new QIS platforms.
- Investigation into the properties of OCCs, including more complex OCCs and their use as functional groups, is a timely and exciting new direction.
- Characterization techniques need to be developed both for measuring and addressing individual qubits within an ensemble.
- Strategies to reduce the costs of advanced spectroscopy techniques, including electron spin resonance, laser sources, and other laboratory-scale instrumentation, are needed so that these techniques can be available to many more chemistry departments.

- Many techniques already exist to measure individual qubits, but they are too costly to be broadly accessible to chemists. Custom instrumentation is being built affordably in individual laboratories, but turnkey industry solutions are lacking.
- A network of regional instrument hubs that augment national facilities is needed. Support should include robust funding for staffing and training at national facilities and regional centers.
- If instrumentation for measuring quantum effects in chemistry was more accessible, more scientists would use these tools to characterize their materials, and increased training opportunities would lead to a more skilled workforce.
- A steady funding stream would support instrumentation at existing large-scale facilities, including dedicated magnets, beamlines, and user end-stations for QIS research by chemists at national laboratories.

As stated in Chapters 1 and 2, the degree of coherence in molecular systems is critical for chemical applications of QIS. Over the past two decades, there has been great progress not only in the discovery of new chemical designs and synthetic approaches but also in the development of new experimental measurements to probe the fundamental processes in molecules important for QIS applications. While there have been many experimental approaches in this regard, the bulk of the reported work in this area consists of magnetic and optical methods. The principle method to probe the spin dynamics and T_1 and T_2 relaxation process in organic and inorganic molecules has involved electron paramagnetic resonance (EPR) or other nuclear magnetic resonance (NMR) effects. Both time-resolved and continuous-wave (CW) EPR methods have had great impacts in the development of QIS chemical systems. A relatively new approach of using optical cycling in small molecules has also gained great attention for addressing and manipulating electron spin states. For optical measurements of coherence and QIS applications, the focus has been directed toward ultrafast, time-resolved, nonlinear optical spectroscopy and spectroscopy with quantum light (e.g., entangled photons). Two-dimensional (2D) time-resolved methods for investigations of organic and biological systems have been developed over the past two decades. In the case of quantum light applications, new methods for higher-yield entangled photon sources, nonlinear spectroscopy and microscopy with entangled photons, and quantum interferometry and the use of microcavities have been developed for chemical applications in QIS.

Chapter 3 provides an overview of the state of the art in the measurement of quantum phenomena using experimental chemical approaches and how quantum tools could be used to study chemical systems. This bidirectional relationship is thread throughout the discussions in the chapter. For example, Sections 3.5–3.7 examine the part of the committee's charge related to "assessing recent and ongoing research in chemistry that draws upon chemistry's unique capabilities in the synthesis, measurement, and modeling of molecular systems to advance QIS." Sections 3.4 and 3.8 focus primarily on "assessing recent and ongoing research in QIS and advances in quantum information processing and technology that have the potential to transform various aspects of chemistry research." This chapter also identifies current emerging tools and future needs in the area of instrumentation. Later, the chapter discusses the state of user facilities at different scales in the United States and recommends opportunities for advancing infrastructure in the near future.

3.1 DEVELOP NEW APPROACHES FOR ADDRESSING AND CONTROLLING MULTIPLE ELECTRON AND NUCLEAR SPINS IN MOLECULAR SYSTEMS

Several main categories of experimental approaches exist for addressing and manipulating electron and nuclear spins in molecules, including NMR, EPR, neutron scattering, and optical control/cycling (for quantum-state initialization and readout). This section discusses the advantages and limitations of each approach and outlines emerging areas of investigation with potential for further development.

3.1.1 NMR and Molecular QIS

NMR has been widely employed by chemists in the molecular magnetism field since the initial demonstration of magnetic bistability in the Mn_{12}-acetate single-molecule magnet (SMM; Sessoli, Gatteschi, et al. 1993), followed by the prized discovery of resonant quantum tunneling of magnetization in the same molecule (Hernandez et al. 1997). Soon after, the technique was applied more broadly in bulk samples, enabling fundamental insights into subtle effects that influence the quantum spin dynamics associated with a variety of different molecular magnetic clusters (Julien et al. 1999; Kubo et al. 2002; Micotti et al. 2006). Of particular importance are the works of Morello and others on the giant spin $S = 10$ Mn_{12} and Fe_8 molecules (Baek et al. 2005; Chakov et al. 2006; Furukawa et al. 2001; Morello and de Jongh 2007; Morello et al. 2004) that led to an initial theoretical framework describing electron-nuclear decoherence processes associated with mesoscopic electron spin moments in molecules (Morello 2008; Stamp and Tupitsyn 2004). A feature common to almost all of the experimental NMR studies was a requirement that measurements be conducted to sub-kelvin temperatures on a wide variety of nuclei (^{1}H, ^{55}Mn, ^{19}F, ^{53}Cr, ^{57}Fe), many of which are quadrupolar. As such, none of these studies are possible using commercial NMR instruments; they require strong collaboration between chemists and physicists as well as unique facilities such as the U.S. National High Magnetic Field Laboratory (NHMFL; Chakov et al. 2006). The need for unique facilities developed by multidisciplinary teams is a common theme in much of the research in this field.

Conventional NMR spectroscopy continues to find utility at the interface between chemistry and QIS as research has shifted toward the study of simpler monometallic systems and their potential use as spin qubits (Gimeno et al. 2021). For example, as illustrated in Figure 3-1, recent studies on a [Yb^{III}(trensal)] molecule demonstrate coherent manipulation of the ^{173}Yb ($I = 5/2$) nucleus at microwave frequencies typically associated with EPR (Hussain et al. 2018). Strong electron–nuclear hyperfine coupling together with a measurable nuclear quadrupolar interaction again necessitated the use of a specialized home-built broadband NMR spectrometer, HyReSpect (Allodi et al. 2005). The novelty of this work lies in the fact that the ^{73}Yb nuclear qudit, with dimension $d = 6$, can encode the effective $S = ½$ electronic moment of the Y^{III} ion with embedded basic error protection. Further development of ultra-broadband instrumentation to enable the application of the most advanced frequency-swept NMR methods (Altenhof et al. 2019) to this area of study is important, as recent EPR investigations (Kundu et al. 2022) highlight cases with nuclear zero-field quadrupole splitting approaching magnitudes typically associated with electrons (i.e., in the gigahertz range). Meanwhile, elegant synthetic work involving substitutional patterning of ^{1}H and $^{79/81}Br$

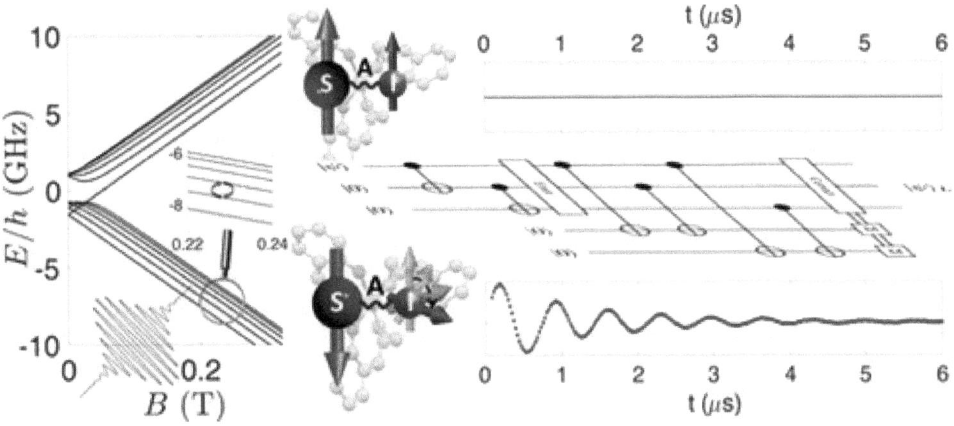

FIGURE 3-1 (left) Multilevel structure of the [Yb^{III}(trensal)] molecule consisting of (center) a coupled electronic doublet (effective $S = ½$) and nuclear sextet ($I = 5/2$). A sizable nuclear quadrupole interaction allows selective excitation of individual nuclear transitions. (right) The multilevel structure of this qudit could be exploited to encode and operate a qubit with embedded basic quantum error correction, with fast gate (Rabi) operation times due to strong electron-nuclear mixing.
SOURCE: Hussain et al. 2018.

nuclear spins on polybrominated catechol ligands and various complexes formed by coordination of the ligands to diamagnetic Ti^{IV} (Johnson, Jackson, and Zadrozny 2020) and magnetic V^{IV} ($S = ½$) (Jackson et al. 2019) ions reveals a pronounced modulation of the proton nuclear spin dynamics in the ligand shell. The V^{IV} example shows how the ^{1}H dynamics and, hence, the nuclear spin patterning can ultimately influence the electronic coherence, suggesting important design principles for molecular spin qubits (see Figure 3-2).

Beyond conventional NMR spectroscopy, several related resonance techniques show tremendous promise for interrogating nuclear spins in areas of interest to chemists working in the QIS field. Several electron and electron-nuclear double resonance (ENDOR) spectroscopies are described in later sections of this report (Goldfarb 2017; Greer et al. 2018; Harmer 2016; Kundu et al. 2023; Sato et al. 2007, 2009; Van Doorslaer 2017; Wang et al. 2018). A particular advantage of these methods is that they often inherit the far greater sensitivity of EPR, in comparison to NMR, and, in some cases, are capable of achieving extreme broadband sensitivity to a multitude of nuclei in a single experiment. Meanwhile, the sensitivity of NMR can also be enhanced greatly by means of dynamic nuclear polarization (DNP; Barnes et al. 2008), whereby electron polarization is dynamically transferred to nuclei via EPR excitation. Although most current applications of DNP-enhanced NMR are in the biophysical research arena (Can, Ni, and Griffin 2015), one can envision a wide range of areas where this blossoming technique can impact research in molecular QIS. First and foremost, DNP enhancement mechanisms (of which there are many) depend intimately on the nature of the coupling between electron and nuclear spins. Thus, DNP-NMR studies can provide valuable insights into decoherence processes associated with molecular spin qubits. Meanwhile, studies have shown that one can use DNP to initialize or control the nuclear environment, leading to a range of intriguing possibilities including massive electron-nuclear entanglement (Simmons et al. 2011), enhanced electronic coherence (Bluhm et al. 2010), and even execution of multiple electron–qubit gate operations (Foletti et al. 2009). While these examples have so far been limited to solid-state semiconductor platforms, application to the molecular QIS field is ripe for investigation. However, this will require cross-fertilization of ideas between chemists working on DNP-NMR and synthesis of molecular qubits, as well as wider access to state-of-the-art magnetic resonance instrumentation and expertise. Finally, DNP offers tremendous prospects for applying NMR to the study of molecules on surfaces (Rossini et al. 2013), where it can provide a means to overcome sensitivity limitations due to low areal concentrations. As illustrated in later sections of this chapter, the study of molecules deposited on solid substrates is an important and growing area of investigation, with potential for the development of scalable device architectures.

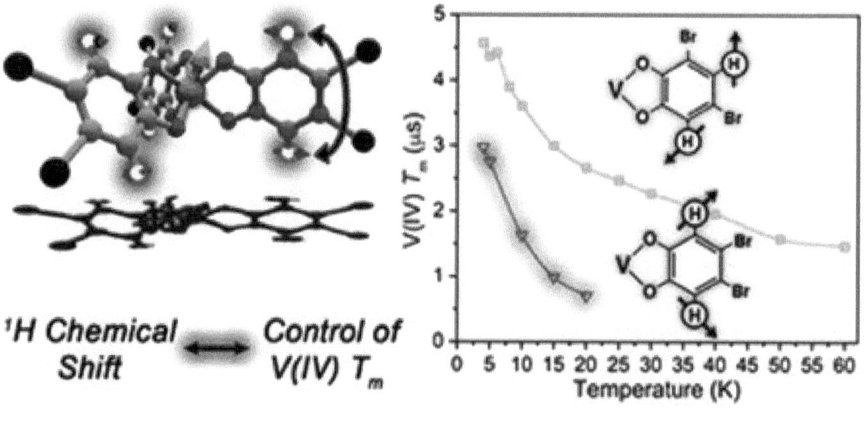

FIGURE 3-2 Nuclear spin-pattern control of the phase memory time in a series of V(IV) complexes. (left) Protons at symmetry equivalent sites have identical chemical and paramagnetic shifts, thus promoting resonant nuclear spin dynamics that cause decoherence of the V(IV) electron spin. (right) Temperature dependence of the electronic phase memory time, T_m, for two complexes with protons patterned at symmetric and asymmetric positions, demonstrating the significantly reduced coherence in the former.
SOURCE: Jackson et al. 2019.

3.1.2 CW EPR and Molecular QIS

EPR has been applied extensively by chemists (and physicists) in the area of molecular quantum spin science, dating back to the very first studies of the Mn_{12} SMM (Gatteschi et al. 2006; Sessoli, Tsai, et al. 1993). Of particular note is the crucial role EPR played in achieving an understanding of the QTM phenomenon (Barra, Gatteschi, and Sessoli 1997; Barra et al. 2007; del Barco et al. 2005; Hill 2013; Hill, Edwards, Jones, et al. 2003; Wilson et al. 2006). As is the case with NMR, the instruments employed in these investigations are often found only in the laboratories of a few expert investigators or at national facilities (Baker et al. 2015). Moreover, even though targeted QIS applications are likely to employ low-field magnets and frequencies matching current communications bands, fundamental spectroscopic investigations often require both high magnetic fields that sometimes exceed those attainable in commercial magnets (Feng et al. 2012; Marriott et al. 2015; Ruamps et al. 2013; Zadrozny et al. 2012) and specialized high-frequency millimeter-wave to terahertz sources and detectors. In particular, the strong spin–orbit coupling and large spin values (giant spin in some polynuclear cases; see Nehrkorn et al. 2021) associated with transition metal and lanthanide complexes can give rise to extremely broad spectral splitting patterns. In comparison to NMR, where chemical shifts are usually measured in parts-per-million, so-called zero-field splitting (ZFS) in EPR can often significantly exceed the unperturbed Larmor frequency (28 GHz/T). This has spurred the development of ultra-wideband EPR instruments that span the millimeter, terahertz, and infrared (IR) spectral ranges (Hassan et al. 2000; Mola et al. 2000; Takahashi and Hill 2005; van Tol, Brunel, and Wylde 2005), including unconventional frequency domain approaches (Blackaby et al. 2022; Hay et al. 2019; Neugebauer et al. 2018; van Slageren et al. 2003) employing synchrotron sources in some cases (Schnegg et al. 2009; Suturina et al. 2017). By comparison, most commercial spectrometers are limited to frequencies below 10 GHz (~3 cm), with a few high-end instruments operating at 34, 94, and 263 GHz.

While much of the QIS-related EPR is currently performed using transient spectrometers (see Section 3.1.3), CW measurements have provided important fundamental chemical insights into this area of investigation in several cases. Motivated by the early quantum computing proposal by Loss and DiVincenzo (1998) involving exchange-coupled quantum dots, several groups investigated engineering coherent interactions within pairs (dimers) of magnetic molecules. The first example involved $[Mn_4]_2$ dimers, where weak hydrogen bonding interactions were crystallographically imposed (Hill, Edwards, Aliaga-Alcalde, and Chrisou 2003). Subsequent synthetic refinements gave rise to covalently linked $[Mn_3]_2$ supramolecular dimers that were shown to retain their coherent quantum properties in dilute frozen solutions (Ghosh et al. 2021; Nguyen et al. 2015), thereby demonstrating that the dimers retain their properties outside of a crystal. Meanwhile, work on $\{Cr_7Ni\}$ qubits demonstrated redox switchable interactions through triangular $\{Ru_2M\}$ linker molecules (M = Zn, Ni, or Co) (Figure 3-3; Ferrando-Soria, Magee, et al. 2016), suggesting possibilities for local switching via a gate electrode or a scanning tunneling microscope (STM) tip. More recent studies have focused on asymmetric lanthanide dimers (Giansiracusa et al. 2018; Luis et al. 2020).

3.1.3 Transient EPR

One of the earliest descriptions of a potentially realizable quantum computer considered localized electronic states in polymers (Lloyd 1993), followed soon after by a proposal to implement Grover's search algorithm using the Hilbert space associated with a giant-spin polynuclear transition metal cluster (Leuenberger and Loss 2001). Common to both of these proposals is the essential role of pulsed EPR, which is the method of choice for coherent quantum manipulation of electron spins in molecules (Sato et al. 2009; Wolfowicz and Morton 2016).

Coherent pulsed EPR experiments have been used to obtain structural details of biologically relevant transition metal complexes as far back as the 1980s (Thomann et al. 1987). However, Ardavan and colleagues (2007) first asked whether coherence times in molecular magnets would be long enough to permit quantum information processing, reporting a phase memory time of several microseconds for a deuterated spin $S = ½$ $\{Cr_7Ni\}$ wheel complex. This marked the dawn of an era that extends to the present day; synthetic inorganic chemists have systematically explored the factors influencing relaxation times in molecular magnets (Gaita-Ariño et al. 2019), with ever-increasing T_2 and T_m being reported from liquid helium temperatures all the way to room temperature

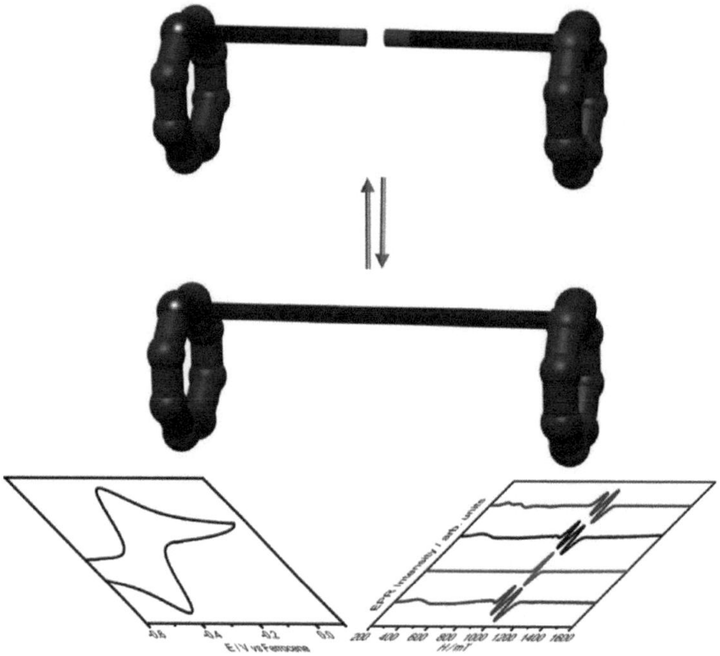

FIGURE 3-3 (top) Dimers of spin $S = ½$ $\{Cr_7Ni\}$ qubits linked by redox switchable $\{Ru_2M\}$ oxo-centered triangles (M = 3d metal, not shown), enabling chemical control of the coupling between the qubits. This control is demonstrated by cyclic voltammetry (lower left) and electron paramagnetic resonance (EPR) (lower right); in the latter case, one observes the coupling via a splitting of the EPR spectrum that can be turned "on" and "off" chemically.
SOURCE: Ferrando-Soria, Magee, et al. 2016.

(Figure 3-4; Atzori et al. 2016; Bader et al. 2014; Bertaina et al. 2008; Warner et al. 2013; Zadrozny et al. 2015). The vast majority of these investigations have been carried out using high-end commercial EPR spectrometers operating in the X-band frequency range (9–10 GHz) down to liquid helium temperatures. One or two studies have been carried out at lower frequencies using both commercial (Zadrozny et al. 2017) and home-built (Collett et al. 2019) instruments. These concerted activities have significantly advanced the understanding of spin-lattice (Kazmierczak, Mirzoyan, and Hadt 2021; Kragskow et al. 2022; Lunghi and Sanvito 2020) and spin-spin (Canarie, Jahn, and Stoll 2020; Chen et al. 2020; Kundu et al. 2023) relaxation in magnetic molecules, leading to the various strategies for enhancing coherence that were described in detail in Chapter 2 of this report: minimization of

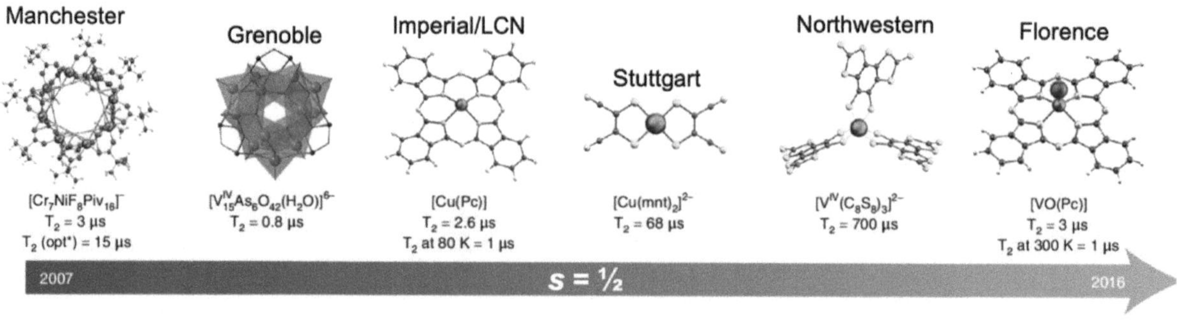

FIGURE 3-4 Some landmarks in the development of molecular complexes for quantum technologies. Timeline for some of the most relevant spin qubits made from transition metal ions in terms of quantum coherence time (T_2).
SOURCE: Gaita-Ariño et al. 2019.

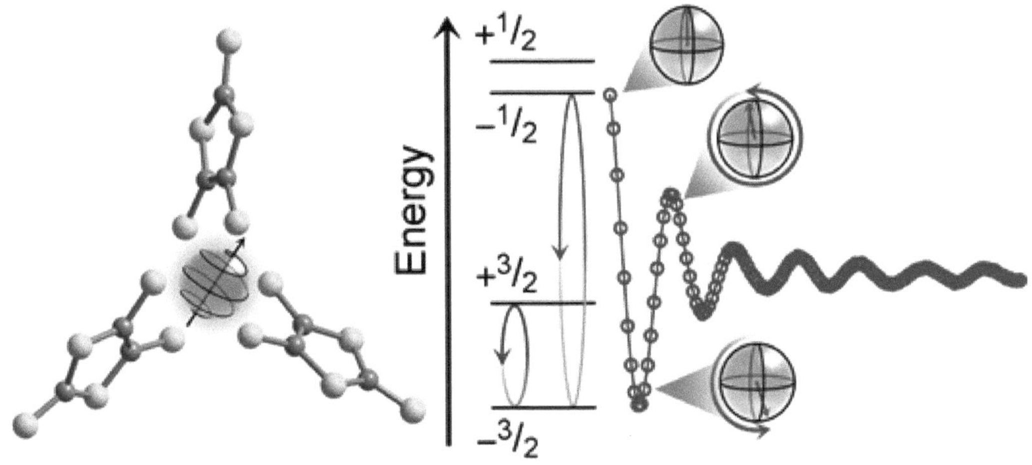

FIGURE 3-5 (left) Molecular structure of the nuclear spin-free $[Cr(C_3S_5)_3]^{3-}$ complex: pink, yellow, and gray spheres represent Cr, S, and C atoms, respectively. (center) Calculated splitting of the M_S energy levels in a 2000 G magnetic field applied along the z axis of the molecule. The circular arrows illustrate formally allowed (red) and forbidden (blue) transitions within the spin $S = 3/2$ Cr^{III} manifold. (right) Rabi oscillations corresponding to one of the formally forbidden transitions.
SOURCE: Fataftah et al. 2016.

spin–orbit coupling (Ariciu et al. 2019; Graham et al. 2014), suppression of spin–vibrational coupling through the design of ligand field symmetry (Kazmierczak, Mirzoyan, and Hadt 2021), use of nuclear spin-free metals (Figure 3-5; Bader et al. 2017; Fataftah et al. 2016) and ligands (Yu et al. 2016), use of nuclear spin patterning (Jackson et al. 2019), and use of clock transitions (Collett et al. 2019; Gaita-Ariño et al. 2019; Harding et al. 2017; Kundu et al. 2022, 2023; McInnes 2022; Shiddiq et al. 2016; Zadrozny et al. 2017).

As with CW EPR, metal-based spin qubits frequently necessitate measurements at frequencies considerably higher than the X-band range. In most cases, this is due to the combined influences of spin–orbit coupling and the ligand field, which gives rise to a large ZFS of the $2S + 1$ spin eigenstates, for $S > ½$ (Fataftah et al. 2020; Takahashi et al. 2009). Even for the $S = ½$ case, a recent lanthanide example demonstrates that the electron-nuclear hyperfine interaction is of a sufficient magnitude that the spectrum cannot be resolved in a commercial X-band instrument (Kundu et al. 2022). Measurements at higher frequencies are challenging due to the lack of available microwave sources with sufficient power to achieve short spin rotation pulses. Although very high-end commercial spectrometers operating at higher frequencies and magnetic fields are available, they have found little utility in this area of research, primarily due to a lack of power and a lack of flexibility for studying metals with broad EPR lines. One notable exception is the case of an Fe_4 molecule with a zero-field excitation that exactly matches the 94 GHz operating frequency (W-band) of a commercial instrument, thus enabling coherent spin manipulation in zero field (Schlegel et al. 2008). However, most if not all other notable high-field/frequency transient EPR studies have been conducted using custom-built hardware that is described in later sections of this chapter. An important early finding from these studies was a significant enhancement in coherence brought about by polarization of the electron spin bath at pumped liquid helium temperatures (Takahashi et al. 2008, 2009, 2011; Wang et al. 2011) (see Figure 3-6). A frequency of 240 GHz is equivalent to a thermal energy of ~11.5 K. Hence observing an EPR signal at low temperatures necessarily involves excitation from the ground spin projection state that is isolated from the first excited state by 11.5 K. Therefore, one can achieve an electron spin polarization exceeding 99.9 percent at a temperature of 1.5 K (pumped ^4He). As a result, electron spin-spin fluctuations are almost completely suppressed, and one can observe coherent electron spin dynamics in a highly concentrated crystalline sample, as has been demonstrated for several polynuclear Fe clusters (Takahashi et al. 2011; Wang et al. 2011).

Besides measuring T_1 and demonstrating the possibility for coherent manipulation of spin states (e.g., Rabi oscillations) in a range of exotic molecular spin systems (see, e.g., Gould et al. 2021; Hu et al. 2018), pulsed

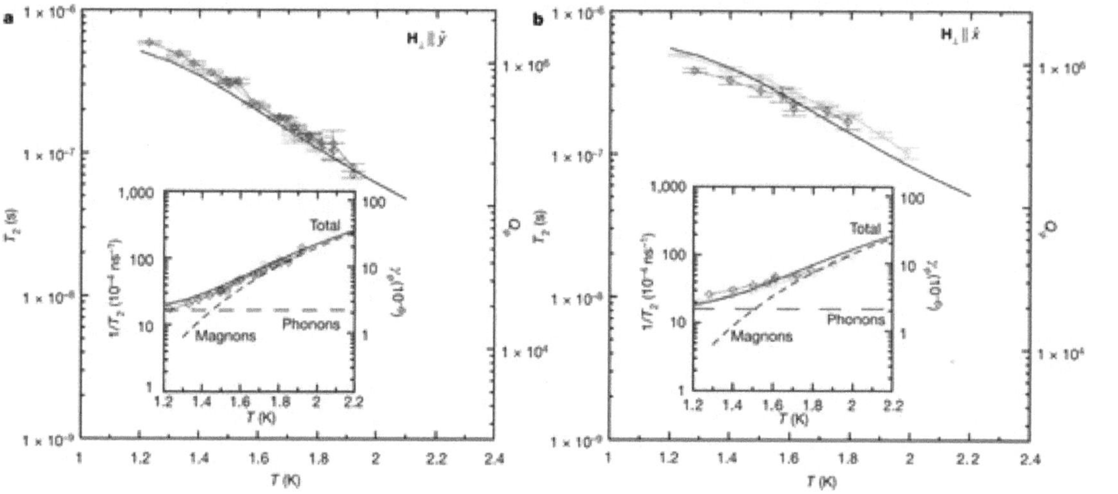

FIGURE 3-6 Temperature dependence of the spin-spin relaxation time, T_2, for two samples (red and green data) of the Fe_8 molecular magnet, with the field applied at two orientations [(a) //x and (b) //y] in the hard plane. The thick blue lines are calculations including phonon and magnon contributions. Insets: partial contributions to $1/T_2$ calculated from magnons (dashed line) and phonons (long-dashed line). The scale on the right side of the main panel indicates the decoherence Q-factor.
SOURCE: Takahashi et al. 2011.

EPR provides access to a wealth of microscopic information via coherent multidimensional pump-probe techniques. In its most basic form, pulsed electron-electron double resonance (ELDOR) involves excitation at one frequency followed by a coherent detection (often by a standard $\pi/2 - \tau - \pi - \tau -$ Hahn echo sequence) at a second frequency within an inhomogeneously broadened EPR spectrum. Such experiments can, for example, provide information on electron spectral diffusion by monitoring changes in polarization at one location/frequency in the spectrum brought about by saturation at another (Wang et al. 2018), thus yielding microscopic insights into electronic relaxation.

A fully coherent pulsed ELDOR sequence, also known as double electron-electron resonance (DEER), involves the execution of an inversion pulse at a frequency ν_2, during either a two-pulse Hahn echo sequence or a three-pulse sequence with a second refocusing π pulse at a frequency ν_1 (Jeschke 2018). This technique, which was developed for structural studies of biomolecules, precisely probes the weak dipolar coupling between two spin labels by monitoring changes in the coherence of one of the spins (at frequency ν_1) by inverting the polarization of the other spin (at frequency ν_2) as a function of delay time, t, within the echo sequence. The result is a modulation of the echo intensity, the Fourier transform of which corresponds to the dipolar coupling in frequency units. The coherent nature of the DEER technique enables the measurement of extremely weak dipolar couplings (~MHz) that would otherwise be buried (i.e., unresolvable) deep within the inhomogeneously broadened CW EPR spectra. Increased spectral resolution at high magnetic fields further permits enhanced orientation selectivity through excitation and detection at spectral locations corresponding to different components of the electron g-tensor. In this way, one obtains information not only on the distances between spins but also on the relative orientations of their g-tensors (Stevens et al. 2016). As depicted in Figure 3-7, DEER has recently been deployed to measure electron–electron coupling within dimers of $\{Cr_7Ni\}$ spin qubits that have been proposed as molecular two-qubit gates (Ardavan et al. 2015), representing a first step toward the demonstration of electron quantum logic in molecules. Indeed, several proposals exist in the literature centered on the use of asymmetric dimers (or oligomers) as two-qubit gates (Aguilà et al. 2014; Collett et al. 2020; Ferrando-Soria, Moreno Pineda, et al. 2016; Uber et al. 2017; Ullah et al. 2022). The asymmetry is necessitated by the requirement that the spins be spectrally distinguishable. However, the demonstration of universal two-qubit logic gates remains a major outstanding goal, for which the primary obstacle is the limited bandwidths (~1 GHz) of current state-of-the-art transient EPR spectrometers (i.e., the limited ability to excite at two well-separated frequencies; Cruickshank et al. 2009). This issue is discussed further in Section 3.11.

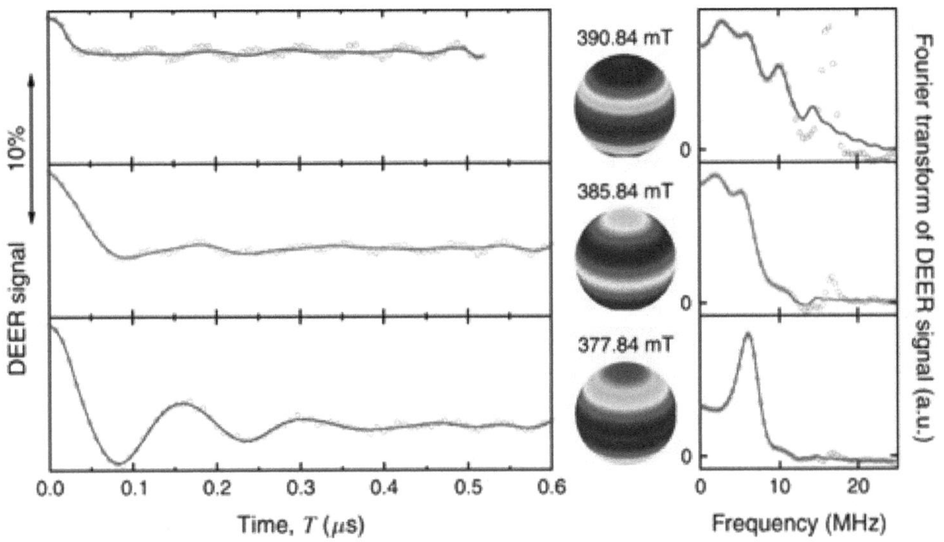

FIGURE 3-7 (left) 2.5 K double electron-electron resonance (DEER) time traces for a dimer of $\{Cr_7Ni\}$ rings connected by rigid organic thread molecules that contain amine binding sites (see Figure 3-4). (center) Different ring orientations relative to the external magnetic field can be excited by the frequencies of the DEER pulses owing to g-factor anisotropy (red: high excitation, blue: low excitation). (right) Fourier transforms of the time-domain data: open circles show raw data; solid lines indicate filtered data. The principal difference between the unfiltered data and the filtered data is the absence in the latter of a component at ~18 MHz arising from the coherent coupling to protons (electron-spin-echo envelope modulation).
SOURCE: Ardavan et al. 2015.

Pulsed EPR can also provide access to electron-nuclear hyperfine and superhyperfine (i.e., coupling of electron density on one atom to nuclear spin on coordinating atoms) interactions. The simplest one-dimensional example is the electron-spin-echo envelope modulation (ESEEM) effect, in which the excitation of formally forbidden electron-nuclear transitions (e.g., the zero-quantum [ZQ], uD → dU; and double-quantum [DQ], dD → uU, resonances, where U and D correspond to up and down and lower- and uppercase denote the nuclear and electron spins) during a two- or three-pulse sequence results in temporal modulations of the echo intensity as a function of delay time on account of the change in nuclear polarization during the coherent sequence (Van Doorslaer 2017). The Fourier transform of the modulation provides information on the coupling of the electron spin to surrounding nuclei; this method has been used recently to demonstrate electron–nuclear decoupling at a clock transition (Kundu et al. 2023). A 2D version of ESEEM—so-called hyperfine sublevel correlation spectroscopy—provides an unprecedented resolution of hyperfine and quadrupolar interactions. This technique is also useful in the study of molecular spin qubits (Ariciu et al. 2019; Atzori et al. 2018), enabling assessment of the degree of electron delocalization from metal to the coordinating ligands.

Multifrequency pulsed ELDOR measurements also provide direct information on NMR frequencies (Goldfarb 2017)—so-called ELDOR-detected NMR—again through excitation of formally forbidden ZQ and DQ electron-nuclear transitions. These transitions become weakly allowed when the electron-nuclear hyperfine interaction breaks rotational symmetry, as is the case, for example, for dipolar coupling. In a 2D experiment, a saturating pump pulse at frequency ν_2 burns multiple holes in the EPR spectrum, with the ZQ and DQ transitions shifted with respect to the detection frequency, ν_1, by an amount $\Delta\nu = \pm\gamma_n B_0 \pm ½A$, where γ_n is the gyromagnetic ratio of the coupled nucleus, B_0 the applied magnetic field strength, and A the hyperfine coupling strength. The major advantage of this technique is that it can provide very wideband (in comparison to conventional NMR) information on the electronic coupling to multiple distinct nuclei in a single experiment, while also inheriting the superior sensitivity of EPR. Again, the method is useful in the study of molecular magnets (Figure 3-8), providing information on electron delocalization in a metal-metal bonded dinuclear Fe-V complex (Greer et al. 2018). In a

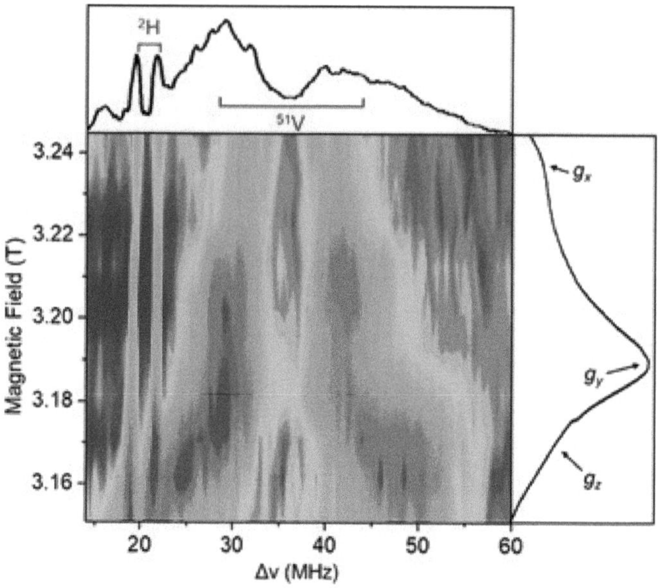

FIGURE 3-8 A two-dimensional color plot of the experimental electron-electron double resonance (ELDOR)-nuclear magnetic resonance (NMR) spectra of an Fe-V metal-metal bonded molecular complex; red and blue correspond to high and low intensities, respectively. The right panel shows the echo-detected electron paramagnetic resonance spectrum recorded at 8 K and 94 GHz. The top panel contains a single ELDOR-NMR trace recorded at ~3.19 T. The isotropic transition centered at Δν ~ 20 MHz accounts for the interaction between the unpaired electron spin density and the deuterated solvent. The axial features centered at Δν ~ 35 MHz arise from the V hyperfine interaction; in the weak-interaction limit, they are centered at the ^{51}V nuclear Larmor frequency and then split by the magnitude of the hyperfine coupling, which is anisotropic (different splitting for excitation at different components of the *g*-tensor).
SOURCE: Greer et al. 2018.

study conducted at the W-band on a custom-built high-power spectrometer, Cruickshank and colleagues (2009) demonstrate the advantages of increased orientation selectivity at high fields, enabling correlation between the electron *g*- and hyperfine coupling tensors. Substitution of one of the ELDOR pulses with a radiofrequency (RF) NMR pulse leads to the ENDOR technique (Harmer 2016), which is useful at the interface between chemistry and QIS (Lutz et al. 2013; Sato et al. 2009; Weiden, Käss, and Dinse 1999). When combined, these multidimensional pump-probe EPR techniques provide access to complementary information about electron-electron and electron-nuclear spin-spin interactions, from which one may infer exquisite microscopic details of physical and electronic structures and spin relaxation behavior. However, more widespread deployment of these methods has likely been limited by the lack of available state-of-the-art transient EPR spectrometers and by the unfamiliarity with advanced (i.e., complex) EPR methodology among the synthetic chemistry community.

3.1.4 Neutron Scattering

The key advance in neutron scattering over the past decade has been the development of a new generation of high-flux cold-neutron time-of-flight spectrometers. They are equipped with arrays of position-sensitive detectors that enable efficient measurement of neutron scattering cross sections as a function of energy and the three components of the momentum transfer vector *Q*, and in vast portions of reciprocal space (Chiesa et al. 2017). The availability of large single crystals of magnetic molecules then permits very detailed four-dimensional (4D) inelastic neutron scattering (INS) spectroscopy, providing unprecedented insight into coherent spin dynamics (Figure 3-9). Indeed, 4D INS enables complete evaluation of the dynamical correlation function, $S(Q,\omega)$, without reliance on any spin Hamiltonian model.

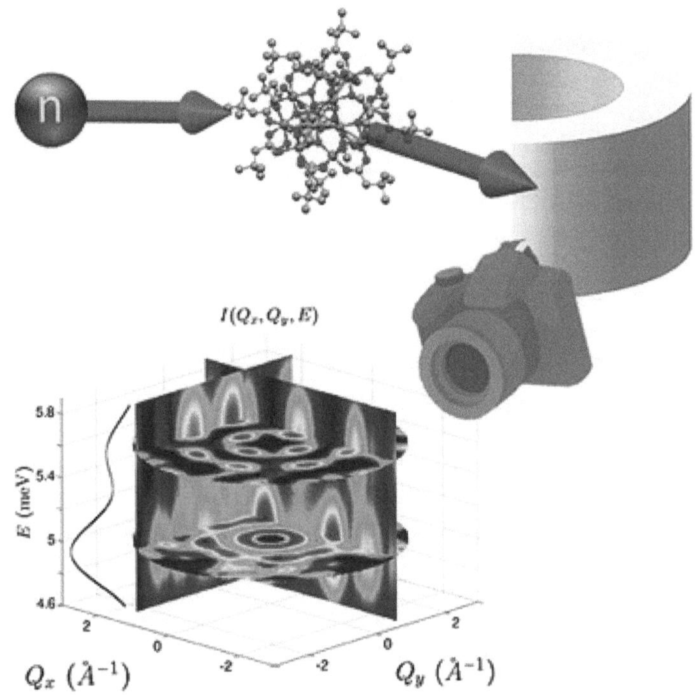

FIGURE 3-9 (top) Illustration of the experimental setup, with incident neutrons scattered from a Mn_{12} molecule into an array of position-sensitive detectors that provide four-dimensional (4D) energy and momentum resolution. (bottom) Snapshot of the 4D inelastic neutron spectrum of Mn_{12}, as a function of the transferred energy E and momentum Q. Here, the scattered neutron intensity is shown as a function of Q_x, Q_y, and E, integrated over the full measured Q_z range.
SOURCE: Garlatti et al. 2019.

These new INS capabilities are starting to impact research in the molecular QIS sphere in dramatic ways. Recent examples include direct evidence for interactions between electronic (i.e., spin) and vibrational degrees of freedom associated with crystals containing vanadyl spin qubits, thus providing crucial information related to spin–phonon coupling strengths (Garlatti et al. 2020), and portrayed entanglement within dimers of $\{Cr_7Ni\}$ molecular spin qubits (Garlatti et al. 2017). One may expect further applications of this powerful new capability over the next 10 years, albeit such measurements are limited to just a handful of large-scale neutron scattering facilities around the world.

3.2 ENHANCE OPTICAL CONTROL IN MOLECULAR SYSTEMS IN QIS

The diversity of molecular species, coupled with the precision available with from molecular synthesis techniques, offer the prospect of building molecular qubits that are uniquely tailored for applications ranging from quantum networks to quantum-enhanced sensing (Wasielewski et al. 2020). As discussed in Chapter 2, the strong interaction of molecules with light (visible to near-infrared [NIR] radiation) in optically active molecules offers a powerful means of control. Molecules with optically addressable spins allow high-fidelity initialization and detection of qubit states (Awschalom et al. 2018). Coherent coupling of molecules with photons allows for long-range entanglement of qubits, opening the possibility of quantum networks over long distances. Separately, photons can be used to transduce quantum information to other physical platforms. Visible/NIR also offers spatial information down to the length scale of the optical wavelength, and potentially below that with the use of near-field techniques.

3.2.1 Optical Quantum-State Initialization

The large number of degrees of freedom (e.g., vibrational, rotational, electronic spin, nuclear spin) present in a molecule or molecular system means that at room temperature, many quantum states are energetically accessible and populated via the Boltzmann distribution. The task of initializing a molecule into a single quantum state involves removing entropy associated with the Boltzmann distribution from a molecular system. Coupling molecular systems to light offers a convenient way to accomplish this: entropy can be transferred to light that propagates away from the system. If this process occurs on a timescale much shorter than that of thermalization with the environment, high-fidelity initialization of molecular qubits can occur *even at room temperature*, which is technologically desirable.

For optical initialization to occur, one first requires optically active molecules—those with high oscillator strength for an electric dipole transition between the ground and some optically excited state at technologically accessible wavelengths. Second, a configuration where certain quantum states are "dark" to incoming light is required. The molecule–light system must be engineered into such a configuration. When this occurs, repeated scattering of incoming photons "pump" molecules into the desired dark quantum state, a technique known as optical pumping. Concretely, the "darkness" required for internal state initialization can be achieved through spectral separation of optical transitions, or through control of the light polarization. Examples can be found in solid-state qubits formed by nitrogen-vacancy (NV) centers in diamond and chemically designed optical Cr^+ spin qubits. The relevant qubit ("spin") states in the ground electronic manifold are split by a large energy difference, allowing qubit states to be spectrally resolved (Awschalom et al. 2018; Jelezko and Wrachtrup 2006; Rodgers et al. 2021). By addressing a specific spin state with light, a qubit can be optically pumped into the unaddressed "dark" states. This form of optical pumping is also known as spectral hole burning. In gas-phase molecules that permit optical cycling (e.g., optical cycling centers [OCCs]), arrays of single diatomic molecules have been initialized through a polarization-sensitive optical pumping scheme (Holland, Lu, and Cheuk 2022).

3.2.2 Optical Quantum-State Readout

By repeatedly exciting a molecular qubit optically and collecting the resulting laser-induced fluorescence, the qubit can be detected. Quantum-state resolution is attained when the optical transitions to different spin states differ in energy and color. Through state-selective optical excitation, isolated OCCs individually trapped in optical tweezer arrays have been detected with quantum-state resolution and with high fidelity (Figure 3-10; Anderegg et al. 2019). In addition to the frequency of the emitted fluorescence, in certain systems, the polarization of an emitted photon can be entangled with the qubit state. This can be harnessed to create quantum networks of distant molecular qubits that are entangled, which can be a resource for quantum communication (Togan et al. 2010).

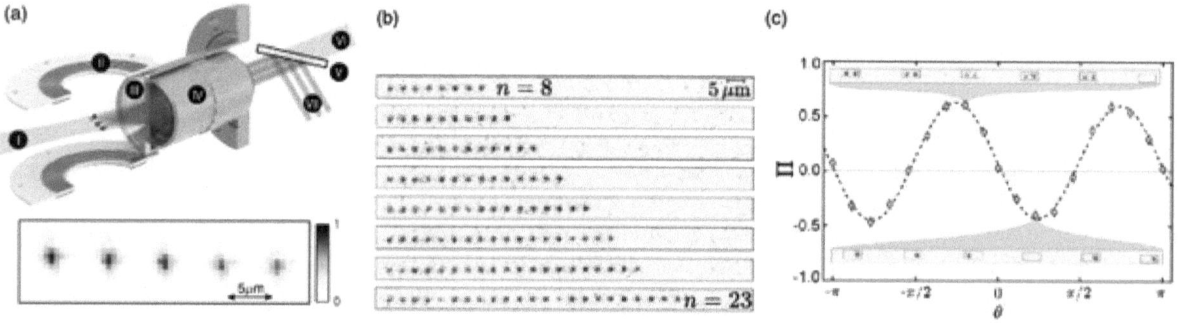

FIGURE 3-10 (a) Schematic of optical tweezer array setup used to trap individual diatomic optical cycling centers (OCCs), and an average fluorescent image of single OCCs in a five-qubit array. (b) Single-shot fluorescent images of defect-free arrays of diatomic OCCs (CaF) held in a reprogrammable array of optical tweezer traps.
SOURCES: (a) Anderegg et al. 2019; (b) Holland, Lu, and Cheuk 2022.

3.2.3 Optical Cycling in Molecular Systems for QIS Applications

Both for optical pumping and quantum-state readout via molecular fluorescence, it is desirable to be able to repeatedly scatter photons off a molecule (i.e., optically cycle photons). For quantum-state initialization through optical pumping, optical cycling is needed since the pumping process can be stochastic, and multiple excitations and decays are needed for a molecule to fall into the desired dark state. For fluorescent quantum-state readout, typical collection efficiencies are much less than unity, implying that multiple photons need to be emitted from a molecule for high-fidelity detection. Achieving optical cycling depends on finding an optically closed system; this can be accomplished by using optically active (high oscillator strength) molecular qubits with atom-like structures or by finding molecules with particularly favorable cycling properties.

Optical cycling, optical initialization of molecular bits, and optical quantum-state (spin) readout have been accomplished in a variety of molecular systems including molecular spin qubits and a variety of gas-phase molecules. A common challenge is decoupling the optical excitations and decays from exciting molecular vibrations.

When a molecule is optically excited, the subsequent decay can be accompanied by a vibrational excitation. That is, the molecule can fall into a different vibrational state that is off-resonant from the initial optical excitation. This problem becomes more severe as the size of a molecule grows and the number of vibrational modes increases. As described in the subsequent section, recent work has shown that this problem can be circumvented in OCCs, in which a very high degree of optical cycling can be achieved.

To provide a point of comparison, the problem of vibrational branching in molecules has an analog in optically active solid-state defects such as NV centers. In these systems, the emission of a photon from an optically excited defect can be accompanied by lattice excitations (phonons), resulting in an emission spectrum containing a narrow line (zero-phonon line [ZPL]) accompanied by a broad background corresponding to various phonon excitations. The latter can be problematic in quantum applications that require indistinguishable photons such as quantum teleportation and quantum networks of entangled spins (Hensen et al. 2015). Owing to conservation of energy, photons emitted in the ZPL can be distinguishable in color from phonon-assisted emission. Even for measurements short enough that the frequency difference is not a concern, phonon-assisted emission can destroy quantum entanglement because the accompanying phonons can lead to rapid decoherence.

3.2.4 Optical Cycling Centers

An area emerging at the intersection between atomic molecular and optical (AMO) physics and chemistry is the identification and control of gas-phase molecules with favorable optical cycling properties, known as OCCs. In the field of AMO physics, full quantum control over molecules at ultracold (< mK) temperatures has been pursued intensely in the past two decades, primarily because of the promise of molecules as a new platform for quantum science. Motivated by the high degree of control achievable in atoms via laser light, initial approaches focused on coherently assembling a molecule out of two ultracold atoms (Ni et al. 2008, 2009; Ospelkaus et al. 2010), whereby one inherits full quantum-state control available with atomic techniques. Although this has been a fruitful path and has led to the observation of reactions modulated by quantum exchange statistics (Ospelkaus et al. 2010) and quantum state–resolved chemistry (J. Liu, Mrozek, et al. 2021; Y. Liu et al. 2021; Liu and Ni 2022), the method of coherent molecular assembly from atoms is difficult to extend beyond diatomic molecules.

An alternative approach generalizable to a large class of molecules has also been pursued. Rather than assembling molecules from their constituent atoms, physicists have sought to directly adapt optical control techniques developed for atoms, such as optical pumping and laser cooling, to molecules. These require optical cycling at a high level of photons (10^4 to 10^5), a task that appeared difficult initially. But with key developments in identifying optically cyclable molecules and devising an optical cycling scheme based on rotational selection rules (Di Rosa 2004; Stuhl et al. 2008), optical cycling of 10^4 to 10^5 photons enabling laser cooling has now been achieved for many gas-phase diatomic molecules (Anderegg et al. 2017; Barry et al. 2014; Collopy et al. 2018; Shuman et al. 2009; Truppe et al. 2017). In these diatomic molecules, the ability to optically cycle has led to many advances that pave the way toward full quantum control. The advances enabled by optical cycling include new techniques to cool and trap large samples of molecules (Anderegg et al. 2018; McCarron 2018), coherent quantum control of rotational states (Williams et al. 2018), and nondestructive fluorescent detection of ultracold molecules (Cheuk et al. 2018).

Notably, with relevance to quantum information processing and quantum simulation, recent work with laser-cooled diatomic molecules has demonstrated high-fidelity detection of arrays of single molecules held in optical tweezer traps (Anderegg et al. 2019; Holland, Lu, and Cheuk 2022). Recent work with molecular tweezer arrays has even demonstrated high-fidelity positioning and quantum-state initialization of single molecules within a large array (Holland, Lu, and Cheuk 2022). Very recently, on-demand entanglement between two molecules spatially separated on the micron scale was demonstrated for the first time, long coherence times on the 100 ms timescale were achieved, and a two-qubit gate sufficient for universal quantum computation was implemented in these arrays (Bao et al. 2022; Holland, Lu, and Cheuk 2022). These demonstrations establish the building blocks needed for quantum information processing and quantum simulation, and make molecular tweezer arrays a promising new platform for quantum science. Looking ahead, the platform of tweezer arrays with optically cyclable molecules could allow high-fidelity individual qubit addressing. By leveraging recent advances in optical tweezer arrays filled with single atoms, these molecular arrays could also be scalable in the near term to hundreds of qubits, as has been demonstrated in their atomic analogs (Ebadi et al. 2021; Scholl et al. 2021).

In the past few years, new work extending beyond diatomic radicals has explored whether optical cycling can be applied to more complex molecules and where new chemical principles can be found that guide the design of these molecules. On the physics front, the search for large molecules with favorable optical cycling properties is motivated by the fact that certain molecules are particularly sensitive to the potential for new fundamental physics beyond the standard model. If these same molecules can be optically controlled at the quantum level, they could offer significant improvements in fundamental physics searches, such as those that look for an electron electric dipole moment (Augenbraun et al. 2020) and ultralight dark matter (Kozyryev, Lasner, and Doyle 2021). The extension of similar techniques to larger molecules could open up possibilities including quantum-enhanced precision measurements and chemistry at ultracold temperatures with entangled matter.

The quest to identify molecules with favorable optical cycling properties along with the discovery of guiding chemical principles is where AMO physics and chemistry intersect. This is a rapidly evolving area. So far, a theme that has guided work on identifying OCCs successfully is the M-O-R motif, where an alkaline earth metal atom (M) is bonded to a halide (Kozyryev, Baum, Matsuda, Boerge, and Doyle 2016; Kozyryev et al. 2019). These systems behave as gas-phase M^+ cation radicals. Optical excitations primarily involve orbitals localized on the metal atom that are minimally coupled to vibrational modes, thereby permitting optical cycling. Based on this principle, theoretical calculations and, in some cases, experiments have shown that more complex M—O—R (where O is oxygen, and R is the organic group) molecules can minimize the issue of vibrational branching and permit optical cycling (Dickerson et al. 2022; Dickerson, Guo, Shin, et al. 2021; Dickerson, Guo, Zhu, et al. 2021; Ivanov et al. 2020; Kłos and Kotochigova 2020; Kozyryev, Baum, Matsuda, and Doyle 2016; Mitra et al. 2022; Zhu et al. 2022). Experimental work has demonstrated optical cycling in the M-O-R polyatomics CaOH and $CaOCH_3$ (Hallas et al. 2023; Mitra et al. 2020; Vilas et al. 2022). Notably, the optical control over CaOH is rapidly approaching that of diatomic OCCs.

For even larger molecules, recent work has discovered new chemical principles that determine optical cycling properties. Theoretical work has shown that Franck–Condon factors in an M-O-R OCC can be tuned through chemical substitution in the R-ligand organic functional group (Figure 3-11; Dickerson, Guo, Shin, et al. 2021). Recent experiments have also verified that OCCs attached to aromatic groups can retain good cycling properties (Mitra et al. 2022; Zhu et al. 2022). Together, these works on M-O-R OCCs have also revealed a possible new chemical principle regarding optical cycling properties—the electron-withdrawing potential of the R ligand correlates well with the molecule's electronic excitation energies and vibrational branching ratios.

The rapidly developing area of large OCCs has also increasingly emphasized the perspective of large OCCs as molecules functionalized with an optical cyclable quantum functional group. Viewing an OCC as a quantum functional group serving as a generic qubit moiety, there are proposals to attach optically controllable qubits to larger molecules and to surfaces (Guo et al. 2021; Zhu et al. 2022). If successful, these endeavors could open up new possibilities such as transducing information among several attached OCCs joined to a large molecule or surface and using attached OCCs as quantum sensors to probe the dynamics of their host molecules (Zhu et al. 2022).

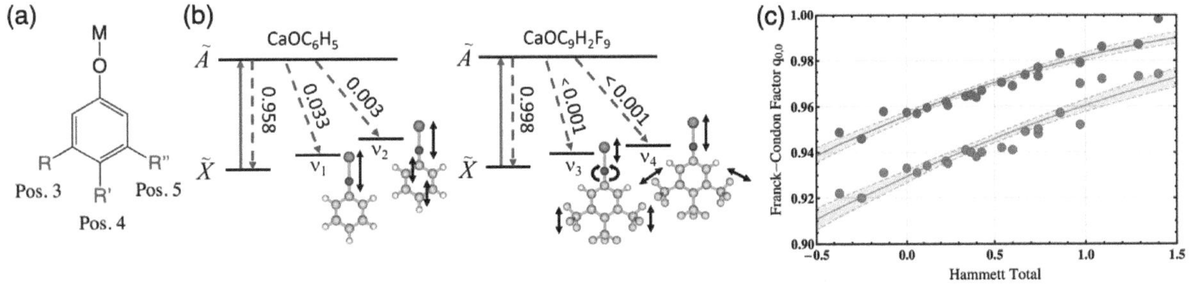

FIGURE 3-11 (a) Diagram illustrating optical cycling centers (OCCs) built upon the M-O-R motif. A metal atom (Ca) is ionically bounded to a molecular fragment (phenyl ring) that can be tuned via chemical substitution (at the R positions). (b) Computed Franck–Condon factors of M-O-R OCCs of Ca-phenoxide and Sr-phenoxide with substitutions. The Franck–Condon factors determine the vibrational branching ratio when the molecule is excited from the ground state (X) to an excited state (A or B). (c) Computed Franck–Condon factors are shown as a function of Hammett's total of the R substituents for CaOR (blue) and SrOR (red) species, illustrating the principle of modulating optical properties of OCCs via chemical substitution. The Hammett total indicates the electron withdrawing strength of the substituent functional group.
SOURCE: Dickerson, Guo, Shin, et al. 2021.

3.3 DEVELOP TECHNIQUES TO PROBE MOLECULAR QUBITS AT COMPLEX INTERFACES TO INFORM THEIR SYSTEMATIC CONTROL

3.3.1 State of the Art in Single-Molecule and Surface NMR

The need to obtain microscopic information concerning nuclear spin dynamics and structural details associated with molecules deposited onto surfaces has driven the development of several novel nuclear resonance methods. For example, ^{57}Fe Mössbauer spectroscopy (sometimes dubbed nuclear gamma resonance), which has long been used in the investigation of iron-based SMMs (see, e.g., Zadrozny et al. 2013), has recently been applied to a monolayer film of an Fe_4 molecular cluster (Cini et al. 2018). Measurements were performed on a 95 percent ^{57}Fe enriched sample using a synchrotron gamma radiation source at the European Synchrotron Radiation Facility. Remarkably, these studies could capture details of structural deformations that escaped detection by conventional synchrotron techniques such as X-ray magnetic circular dichroism. Moreover, the measurements provided information concerning the fluctuation timescale associated with the magnetic moments. Meanwhile, β-detected NMR involves implantation of a magnetically polarized radioactive isotope that undergoes β-decay. By subjecting the sample to a static magnetic field, one can effect a loss of this polarization upon application of RF pulses that are resonant with the Larmor precession of the implanted ions. Analogous to muon spin relaxation (μSR), which has also been applied to the study of molecular spin qubits (Baker et al. 2016), the polarization can be measured from the asymmetry in β-emission from the implanted ions. In a recent example involving Mn_{12} molecules grafted onto a Si substrate, $^{8}Li^{+}$ ions were selectively implanted with different depth profiles based on their implantation energy (Salman et al. 2007). The resulting NMR lineshapes could be used to infer the local distribution of static magnetic fields at different depths relative to the surface, in turn informing on the magnetic properties of the Mn_{12} molecules. These experiments were performed at the Isotope Separator and Accelerator at TRIUMF in Canada.

One of the landmark results in the area of molecular QIS involves the implementation of Grover's quantum search algorithm using a single $I = {}^{3}/_{2}$ nuclear spin carried by a Tb^{III} bis-phthalocyanine complex within a molecular transistor (Godfrin, Ferhat, et al. 2017). As illustrated in Figure 3-12, the method relies on the magnetic bistability of the $J = 6$ spin-orbital moment of the Tb^{III} ion, which can only flip from "up" (↑) to "down" (↓) through quantum tunneling at the crossing point of the $M_J = \pm 6$ Zeeman levels (see Figure 3-12) (Godfrin, Thiele, et al. 2017; Thiele et al. 2014; Vincent et al. 2012). These are split into four sub-pairs due to hyperfine coupling to the $I = {}^{3}/_{2}$ nucleus. Consequently, the electronic moment knows about the quantum state of the nucleus; when the electronic spin flips from "up" to "down," it induces a corresponding jump in the differential conductance of the

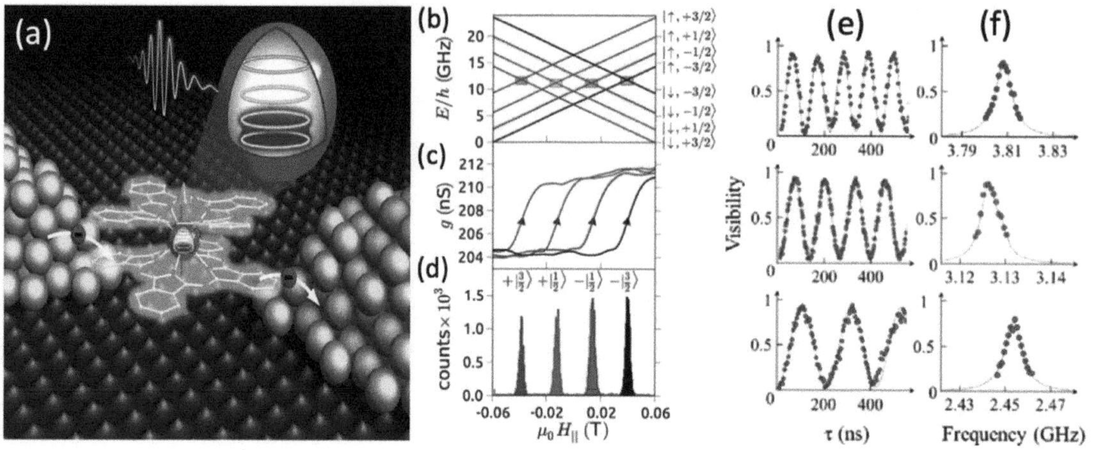

FIGURE 3-12 (a) Artist's view of a nuclear spin qubit transistor based on a single $TbPc_2$ (where Pc is phthalocyanine) molecular magnet, consisting of a Tb^{3+} ion (pink) sandwiched between two Pc ligands (white) and coupled to source and drain electrodes. The four nuclear spin states of the Tb^{3+} (colored circles) can be manipulated with electric field pulses. (b) Zeeman diagram of the $TbPc_2$ molecular magnet, showing the hyperfine split electronic spin ground-state doublet as a function of the external magnetic field. Avoided level crossings (colored rectangles) allow for tunneling of the electron spin. (c) Jumps in conductance read out the state of the Tb^{3+} nuclear spin during field sweeps. (d) Histograms of the conductance jumps. (e) Maximum visibility of the Rabi oscillations. (f) Resonance shape of the three nuclear qubit transitions.
SOURCE: (a and b) Thiele et al. 2014 (c); Godfrin, Ferhat, et al. 2017.

transistor. The location in the magnetic field of this jump, therefore, provides a convenient readout of the nuclear state of the molecule. One can then perform logic operations on the nuclear spin via the application of selective microwave NMR pulses (see Figure 3-13), with the final readout achieved via the transistor. A quality factor can be deduced from the ratio of T_2/T_p (where T_p is the time taken to execute a p-pulse), which gives Q = (300 ms)/(0.12 ms) = 2,500, corresponding to a fidelity of about 99.9 percent. The next step would involve single-qubit gate randomized benchmarking which, to the best of our knowledge, has not been performed for a molecular system to date. These and other related studies (Jo et al. 2006; Perrin, Burzurí, and van der Zant 2015) clearly demonstrate the feasibility of realizing molecular-scale electronic devices for QIS.

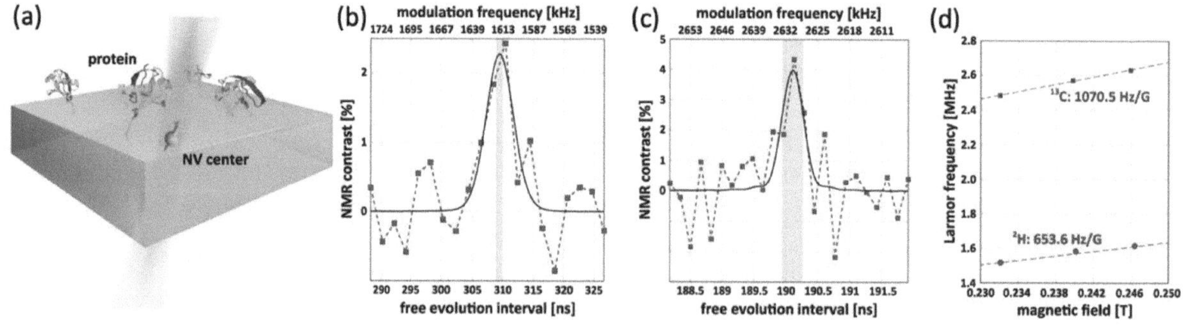

FIGURE 3-13 (a) Schematic of surface nuclear magnetic resonance (NMR) experimental setup. Ubiquitin proteins attached to the diamond surface are probed using a proximal quantum sensor consisting of a nitrogen-vacancy (NV) center electronic spin and its associated ^{15}N nuclear spin. (b) ^{2}H and (c) ^{13}C NMR spectra (red points) recorded at magnetic fields of 2473 and 2457 G, respectively, with Gaussian fits superimposed (black curves). (d) Scalings of resonance frequencies with an applied magnetic field.
SOURCE: Lovchinsky et al. 2016.

Finally, one can turn the single-spin sensing problem on its head by deploying more mature detection platforms such as magnetic resonance force microscopy (MRFM) or the NV defect center in diamond acting as a quantum reporter (discussed in detail in Chapter 2), thereby pushing NMR sensitivity toward the single-spin limit. Nanoscale nuclear magnetic resonance imaging came first (Degen et al. 2009; Hemmer 2013; Mamin et al. 2013; Staudacher et al. 2013), followed shortly thereafter by single-proton detection (Müller et al. 2014; Sushkov et al. 2014). As shown in Figure 3-13, these advances now permit NMR studies of individual proteins (Lovchinsky et al. 2016) and nanoscale nuclear quadrupolar sensing (Lovchinsky et al. 2017), opening up entirely new areas of chemical imaging and spectroscopy (Aslam et al. 2017; Liu et al. 2022; Rugar et al. 2015).

3.3.2 Single-Molecule EPR

Signatures of individual molecular spins within an ensemble were first reported via optically detected EPR in 1993 (Köhler et al. 1993; Wrachtrup et al. 1993). However, the first direct EPR detection of a single localized electron spin was achieved at IBM using MRFM (Rugar et al. 2004), a precursor to the NMR studies described above. A major breakthrough that would eventually lead to the possibility of performing transient EPR on single surface spins (see below) involved the use of an STM both for manipulation and detection of individual Fe adatoms placed on a MgO film (Baumann et al. 2015). This method, which was again developed at IBM, built upon a considerable body of scanning tunneling spectroscopy work on individual surface atoms that mimicked many of the well-known properties of molecular nanomagnets (see, e.g., Heinrich et al. 2004; Loth et al. 2010). Consequently, this work received much attention among chemists working in the area of molecular QIS. The EPR-STM technique has continued to advance at a breathtaking pace, with a number of laboratories around the world developing similar capabilities. These methods hold tremendous promise for exploring next-generation qubit platforms, particularly at the interface between chemistry and QIS. For example, a recent CW EPR investigation published in *Nature Chemistry* explored the surface chemistry of FePc and FePc dimers (where Pc is phthalocyanine) adsorbed on a MgO surface (Zhang et al. 2022). In particular, these studies revealed that the exchange coupling within the FePc dimers depends strongly on molecular geometry, which is dictated by the available surface binding sites. Other groups have looked at organic radicals adsorbed on graphite (Durkan and Welland 2002) and alkali metal dimers on MgO (Kovarik et al. 2022).

As illustrated in Figure 3-14, more recent developments of the EPR-STM method have demonstrated that one can deploy the entire transient EPR toolkit (Chen, Bae, and Heinrich 2022; Willke et al. 2021; Yang et al. 2019). This enables coherent manipulation and readout of individual spins associated with atoms or molecules deposited onto an insulating (typically MgO) film supported on a metallic substrate. RF excitation is achieved either via alternating current (AC) modulation of the tunneling current generated by the STM (the precise mechanism of

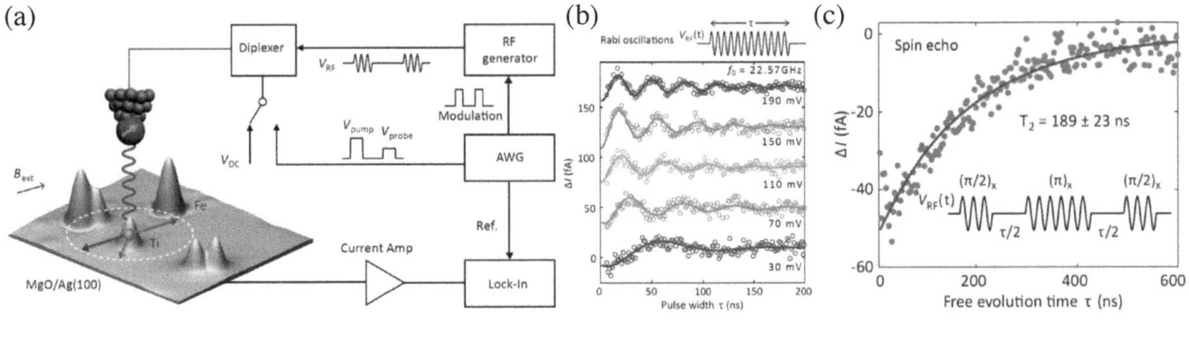

FIGURE 3-14 (a) Experimental setup for a pulsed electron spin resonance–scanning tunneling microscope (STM). The STM is equipped with a radiofrequency (RF) generator and an arbitrary waveform generator. A sequence of RF and direct current pulses is delivered to an oxygen-site Ti atom on MgO/Ag(100) via the STM tip. (b) Rabi oscillations of the Ti spin at different RF powers (V_{RF}). (c) Hahn echo measurements of the Ti spin.
SOURCE: Chen, Bae, and Heinrich 2022.

this driving mode remains somewhat unclear) or via a separate RF antenna close to the tip. Detection is achieved by employing a magnetized STM tip that generates a spin-polarized tunneling current, enabling a projective readout of the state of the spin located below the tip. Averaging then yields the composition of the wave function (i.e., $|\psi\rangle = a|\uparrow\rangle + b|\downarrow\rangle$). By employing different excitation schemes, one can carry out all of the well-known pulsed EPR schemes (e.g., Rabi nutation, Ramsey fringe measurements, Hahn-echo decay measurements of the T_m associated with a single spin, and so on). The technique, which has been pioneered in Andreas Heinrich's group at Ewha Womans University in Seoul, Korea, obviously requires access to high-end STM instruments along with considerable technical expertise. Most demonstrations until now have involved individual surface adatoms as well as pairs of atoms. However, one recent study focused on the coherent control of the spin associated with a single FePc molecule (Willke et al. 2021), emphasizing the tremendous potential of this relatively new method for coherent control of individual molecular qubits. An added powerful advantage of this technique is the possibility of moving atoms and molecules around on surfaces (Loth et al. 2012), enabling a very deliberate design of multispin architectures for studies of the physics associated with relatively simple model spin Hamiltonians. However, a potential drawback is the lack of an obvious pathway to scalable architectures with the possibility to manipulate large numbers of qubits. Nevertheless, EPR-STM offers a remarkably powerful tool for studying the quantum dynamics of individual (or small numbers of) molecules on surfaces. Limited availability of such instruments has resulted in a relatively low number of publications until now. Groups in Europe (Korvarik et al. 2022; Yan et al. 2020) and Asia (Chen, Bae, and Heinrich 2022; Kawaguchi et al. 2023) are increasingly turning to this technique, even though it was first developed in the United States (Baumann et al. 2015). Obviously, less exotic (and less costly) EPR techniques should be used to pre-screen candidate molecules. However, limited access to EPR-STM platforms, particularly in the United States, is preventing research at the intersection of QIS and chemistry to move forward.

3.3.3 Chemical Tailoring of Solid-State Spin Defects

Defects in semiconductors have become powerful hosts for spin-based qubits in the solid state within materials including diamond, silicon carbide, and silicon. By exploiting defects generated via electron or ion radiation as well as naturally occurring defects in these materials, researchers have demonstrated precise quantum control of individual electron and nuclear spins, robust entanglement, electron-nuclear quantum registers, and increasing control of the spin-optical interface at the level of single photons. For example, like NMR, the use of single NV centers close to the surface of diamond nanocrystals or nanopillars to sense single molecular spins deposited on their surfaces shows tremendous promise for chemical EPR investigations. Recent work by Pinto and colleagues (2020) has demonstrated coherent readout and control of an endohedral $^{14}N@C_{60}$ electronic spin via this method, including transient measurements. This approach can also be extended to nanoscale samples at high magnetic fields and frequencies (Fortman et al. 2020). In particular, NV-detected EPR offers the possibility to detect single electron spins and, therefore, to investigate biological processes at the single-molecule level. As such, NV-detected EPR with single-spin sensitivity can potentially eliminate ensemble averaging in heterogeneous systems (Giovannetti 2004).

Meanwhile, driven by predictive theoretical work, several defect-based states have been engineered for practical applications including quantum memories and optical emission in the telecom regime. Impressive quantum sensing of electric, magnetic, and strain fields has been shown along with dramatic improvements in small-volume and single-spin NMR. However, these material systems face considerable challenges for quantum science and engineering, including creating scalable and identical quantum states, mitigating the impact of strain and isotopic variations, as well as controlling decoherence-driven charge fluctuations from interfaces and unintentional dopants.

The selection of appropriate hosts and defects, and their mutual interactions with one another, remains both a challenge and an opportunity for QIS research. There have been considerable advances in understanding the mechanisms of spin decoherence in semiconductors and solid-state nanostructures. State-of-the-art synchrotron spectroscopies, coherent Bragg diffraction imaging, and high-resolution magnetic resonance techniques have proven to be powerful tools to improve the quality of host materials. In addition, improvements in first-principles approaches to understanding and predicting properties (e.g., density functional theory) have led to a deeper physical understanding of quantum decoherence and successful predictions in mitigating many sources of noise, from dynamical decoupling to unique pulse sequences. In addition, there have been successful efforts aimed at harnessing the capabilities of today's electronics technologies for quantum-state control, including electrical control and readout of single quantum states, the creation of decoherence-free subspaces to locally enhance coherence, and the integration of photonics to both enhance the efficiency of quantum emitters and entangle nearby quantum bits. Recently, rare-earth ions in oxide semiconductor hosts—including silicon-compatible materials—have emerged with encouraging properties as single quantum memories, with impressively long coherence times and single-shot photonic readout.

3.4 DEVELOP ENHANCED SPECTROSCOPIC AND MICROSCOPIC TECHNIQUES WITH NONCLASSICAL LIGHT

3.4.1 Quantum Light Spectroscopy and Entangled Photon Sources

As chemists push forward in their quest for deeper knowledge of the important processes and mechanisms of molecular interactions, great pressure is placed upon the nature of the measurements used to probe such processes. Indeed, the measurements themselves are also physical processes, and the accuracy to which these measurements may be performed to help better understand chemical processes is governed by the laws of physics. At the molecular level and for small systems, the processes involved in the measurements are governed by the laws of quantum mechanics that theoretically place limits on the accuracy to which the measurements can be performed. The intrinsic uncertainty of the results of the measurement of complementary observables is derived from the Heisenberg uncertainty relationship. Other quantum constraints combined with these uncertainty relations impose limits on how accurately we can measure quantities with the given physical resources of a measurement tool.

Nonlinear Entanglement Spectroscopy

New high-resolution, noninvasive spectroscopic and imaging methods for chemical processes have been developed over the past two decades. This success has allowed scientists to probe important chemical processes at the sub-femtosecond and sub-micron levels, which was not previously possible. However, better detection and imaging methods are still needed at the nanoscale at very low powers. Several impressive magnetic and optical techniques have been developed for this purpose. In the optical regime, both linear and nonlinear optical methods (see Figure 3-15) have been well illustrated in the literature. For example, nonlinear optical methods, such as two-photon absorption (TPA; and emission), have enjoyed great success in probing (at a very sensitive level) organic

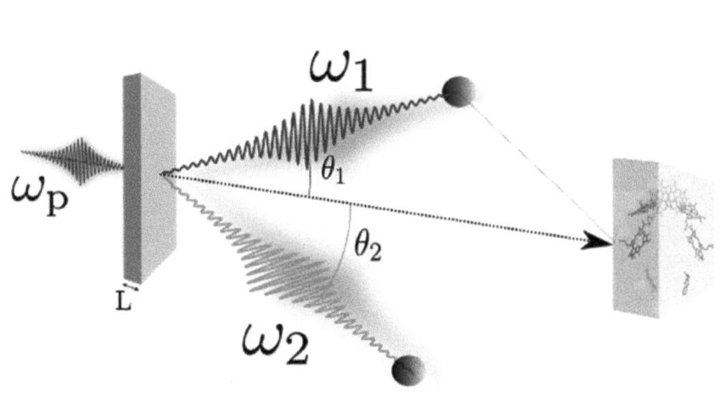

FIGURE 3-15 Schematic diagram of a nonlinear optical experiment with entangled photons. Here, a pump photon of frequency ω_p is down-converted into two photons with frequencies ω_1 and ω_2 and directed onto the sample (bacterial reaction center). SOURCE: Schlawin et al. 2013.

and inorganic molecular systems as well as biological cellular (tissue) processes (Varnavski and Goodson 2020). The quadratic intensity dependence of TPA methods allows for very strong focusing and localization of the optical field, giving a much better effective resolution than their linear optical counterpart. The use of nonclassical sources of light such as squeezed states as well as entangled photons may provide an opportunity to exceed these quantum-imposed limits.

Great enthusiasm exists for exploring quantum information in molecular systems. Constructing a molecular system with the appropriate electronic states as well as inter- and intramolecular interactions for a desired quantum effect is a beneficial goal. This requires substantial advancement in both the synthesis and electronic structure of the proposed molecular systems. However, and perhaps of equal importance, this goal also requires advanced measurements at a very sensitive level for the desired detection of the spectroscopic or microscopic signal. New measurements are now being developed to probe the molecular properties of molecules at an enhanced level, which may one day provide new insights into the molecular design of systems for QIS applications as well as provide a more fundamental understanding of the nature of electronic interactions and mechanisms. In describing quantum phenomena in molecular systems, one usually is seeking behavior that lies outside the norm of classical behavior. For example, one is interested in the processes of coherence and decoherence in optical spectroscopy. These processes are being further defined and developed continually by time-resolved and nonlinear spectroscopic techniques. Chemists are developing the instrumentation and theory for new frontiers in this direction. Molecular systems typically suffer from a loss of coherence (amplitude and phase) at room temperature on ultrafast timescales (femtoseconds). A difficulty in measuring quantum states in molecular systems stems from the fast decoherence or mixing with a homogeneous broadening ensemble environment that is typical in solutions or solids. The development of new experimental techniques to provide new information through nonlinear signals, ultrafast multidimensional systems, and entangled photon interactions could enable improvement in the measurement of quantum-defined processes in molecular systems.

3.4.2 Quantum Light in Molecular Spectroscopy

The use of nonclassical states of light for spectroscopy may provide new approaches toward understanding the physical properties of molecules beyond quantum coherences to include entanglement. There are several ways that one may generate a quantum light source. Photon entanglement occurs between two beams of light when the quantum state of each field cannot be described in the individual parameter space of that field (Mukamel et al. 2020). While there are different parameters of light that can be entangled, the most commonly reported methods involve time and energy, polarization, and position and momentum. Along with the generation of squeezed states of light, these methods provide the bulk of the experimental techniques with nonclassical states of light. The interest in utilizing nonclassical states of light (e.g., entangled photons) stems from the potential advantages they may provide in comparison to classical laser excitation. It may be possible to take advantage of the high degree of temporal and spatial correlations in quantum light resulting from entangled pairs of photons (i.e., the signal and idler). These correlations can provide a sensing tool for chemists to detect potential molecular systems at very low levels of light, which is useful for investigations of chemical and biological systems. The pairs of photons are typically created by the process of spontaneous parametric down-conversion (SPDC) (Figure 3-16). These entangled pairs have shown different and low noise characteristics when compared to classical photon sources. This feature of entangled light can also be utilized in new linear and nonlinear spectroscopic approaches. All light fields have fluctuations in their amplitudes and phases, are subject to a stochastic indeterminacy beyond environmental influences, and involve a quantum indeterminacy. There has been a great deal of progress in producing light fields with fluctuations lower than the shot noise limit (SNL). One of these is amplitude squeezing, which may be produced by an interaction that preferentially removes the large amplitude fluctuations, resulting in a radiation field below the SNL. Both amplitude and photon number squeezing have attracted new applications in sensing and spectroscopy (Ispasoiu and Goodson 2000; Loudon and Knight 1987). Nonlinear four-wave mixing approaches have been carried out to produce bright squeezed states of light. Measurements are achievable with sensitivity beyond the standard classical limit and can be performed at low excitation fluxes for exciting photosensitive materials. With the advent of novel theoretical predictions and models, new measurements have been proposed that may suggest

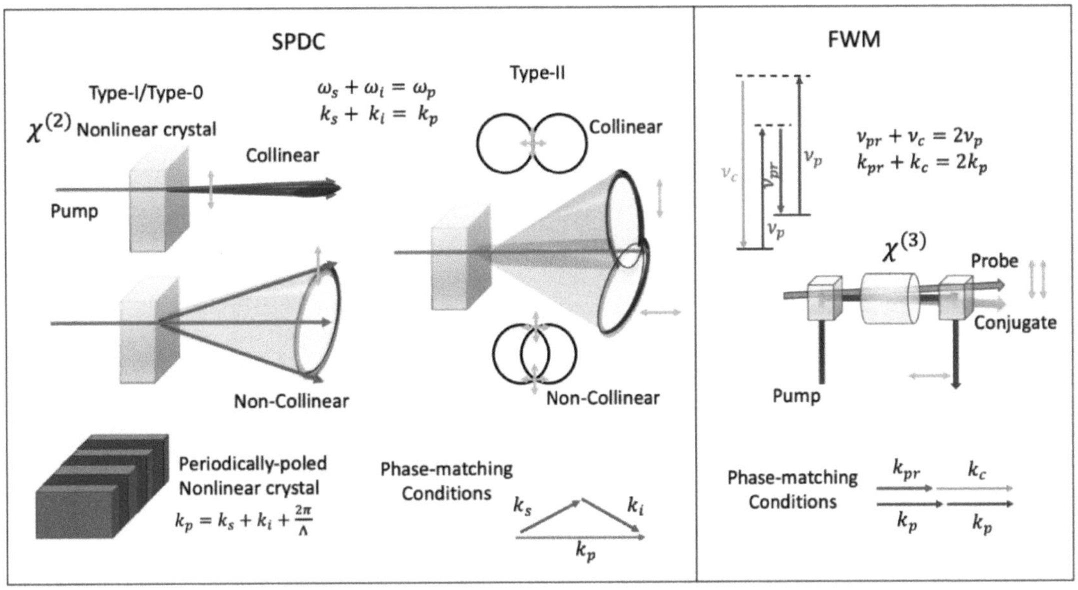

FIGURE 3-16 Diagram of nonclassical light sources. On the left, spontaneous parametric down-conversion (SPDC) with a $\chi^{(2)}$ bulk crystal showing the cone-like spatial arrangement of signal and idler photons for type-I/0 and -II SPDC and a periodically poled crystal (bottom left) with quasi-phase-matching properties. On the right, the phase-matching conditions and diagram of the four-wave mixing (FWM) process for creating squeezed light are shown, as well as an energy diagram of the involved beams. SOURCE: Eshun et al. 2022.

an enhanced spectral resolution. Additionally, by utilizing the unique features of entangled photon pairs, one may also pursue the possibility of targeted entangled or coherent control of photochemical processes (Eshun et al. 2022). In the context of imaging (in biological systems), the highly correlated light provides the opportunity to probe chemical and biological systems at extremely low photon flux, avoiding the typically detrimental light toxicity issue. Entangled photon pairs provide a very powerful way to evaluate fundamental quantum mechanical principles such as Bell's inequalities or in Hong–Ou–Mandel (HOM) photon correlation experiments (Schlawin et al. 2013).

Other intriguing properties of entangled photon spectroscopies include a non-Fourier-related spectral and temporal resolution. The spectral resolution is determined by the linewidth of the down-converted pump, whereas the temporal resolution is determined by the spectral width of the non-degenerate, down-converted photon pairs. This is because only the two entangled photons in a pair will interact with each other in an entangled photon spectroscopy measurement, whereas in a classical laser, any two photons can interact. Along these same lines is the ability to measure one photon using another wavelength or to image an object without spatially detecting the photons hitting the target. These "ghost" measurement techniques rely on the quantum correlations between the two photons in the entangled pair to determine the property of one photon using the other. It is notable that the photons must still be interfered with or measured by coincidence counting circuits to determine these properties. Such techniques are promising because extremes such as measuring an X-ray photon using a visible light photon are possible, all while gaining the quantum advantages of the entangled photons.

As depicted in Figure 3-16, SPDC is the most common method for creating nonclassical, entangled light. In the case of SPDC, a strong pump beam of frequency ω_p interacts with a $\chi^{(2)}$ nonlinear crystal to create a pair of lower-energy photons (the signal ω_s and idler ω_i photon), whose energy adds up to that of the pump, $\omega_s + \omega_i = \omega_p$ (Eshun et al. 2022). In addition to energy, momentum also must be conserved, which leads to "phase-matching" conditions defined by $k_s + k_i = k_p$. The phase-matching conditions determine the spatial correlations, efficiency, bandwidth, and temporal constants of the generated entangled photon pair. The most common nonlinear crystals used for SPDC are bulk β-barium borate (BBO) and potassium dideuterium phosphate (KDP). Because of said phase-matching conditions, bulk crystals usually have efficiencies on the order of 10^{-10} to 10^{-12}. The phase-matching conditions of bulk BBO and KDP also dictate tens of nanometer bandwidths and hundreds of femtosecond correlation times. These parameters put constraints on potential measurements to be performed with entangled photons.

The efficiency, bandwidth, and temporal correlations of an entangled photon source can be improved by using periodic poling of ferroelectric materials to create quasi–phase matching. Materials made popular by telecom entangled photon experiments include potassium titanyl phosphate and lithium niobate. Switching the poling of the crystals on the micrometer scale allows for artificial optimization of the phase-matching conditions. Higher-strength nonlinear tensors can also be used for type 0 SPDC. This results in a dramatic increase of the SPDC efficiency, to as high as 10^{-6}, while also allowing a broader tuning of the central wavelength and bandwidth (and therefore temporal correlations) of the SPDC. Octave-spanning bandwidths with ~10 fs correlation times have been achieved with μW fluxes that are visible by eye. The greater bandwidth is accomplished by the introduction of a chirp in the poling field, which allows more wavelengths to be phase matched while also retaining the benefits associated with longer crystals (Szoke et al. 2021). Quasi–phase matching can be achieved in bulk crystals as well as waveguided sources to provide accessibility in both on-chip and free-space measurements.

A unique aspect of chemistry that is different from quantum information is the ability to create entangled photons over a broad range of central wavelengths. New proposals to use lithium tantalate have been published, offering the ability to pump SPDC down to ~350 nm where lithium niobate begins to suffer from self-absorption issues (Figure 3-17). Of course, even with lithium tantalate, the down-converted entangled photons are still in the 700 nm range, which is too high for most molecular dyes. Recent work has shown the creation of deep ultraviolet entangled photons using $LaBGeO_5$ (LBGO). Entangled photons can also be created in the IR to terahertz regions using lithium niobate (<5 μm) and up to the terahertz range using orientationally patterned GaAs. Future work involves the use of sub-wavelength structures to further enhance the entangled photon generation rate (Price 2022).

As entangled light sources continue to be developed, attention has increased recently toward carrying out spectroscopy with the available entangled photons by current methods. The interest in utilizing entangled photons as spectroscopic tools first emerged in 1990, when scientists theoretically predicted a linear, rather than quadratic, scaling of the TPA rate with the pump photon intensity. One may rationalize this effect by comparing the interaction of the three-level atom with ordinary coherent light and with SPDC entangled light. If the atom interacts with coherent light tuned to the two-photon resonance, the two-photon transition rate may be estimated as the rate of excitation from the (virtual or real) intermediate state multiplied by the probability that the atom is in the intermediate state. At low intensity, both these quantities are proportional to the intensity, and one obtains the usual quadratic dependence of the two-photon rate on intensity. With the SPDC light, on the other hand, the excitation is accomplished in a single step: one photon of the pair promotes the atom to the virtual intermediate state, while its twin immediately (in a time less than the virtual-state lifetime) completes the two-photon transition. One may estimate the two-photon transition rate as the probability of excitation by a photon pair multiplied by the rate of arrival of photon pairs (proportional to intensity). The rate should be linear in intensity. The experimental verification of this effect in atomic and molecular systems established entangled photons as unique light sources for nonlinear spectroscopy with low photon fluxes (Figure 3-18). Measurements in atomic and molecular systems have been carried out for this form of entangled two-photon spectroscopy. Measurements of the entangled two-photon fluorescence with as few as 10^7 photons/sec have been carried out with this method. This gives rise to the possibility of doing spectroscopy with a small number of photons. The exact manner in which the entangled photons interact with particular molecular states and its impact on our understanding of the fundamental photochemical mechanisms could have important ramifications in our ability to excite and possibly control photochemical processes. Theoretical proposals for entangled virtual-state spectroscopy or entanglement-induced two-photon transparency have pointed out the highly unusual bandwidth properties of entangled photons and how this may be used further in understanding and possibly controlling chemical processes (Burdick, Schatz, and Goodson 2021; Eshun et al. 2022; Georgiades et al. 1995; Giri and Schatz 2022; Guzman et al. 2010; Harpham et al. 2009; Kang et al. 2020; Lee and Goodson 2006; Li et al. 2020; Parzuchowski et al. 2021; Raymer et al. 2013; Saleh et al. 1998; Schlawin, Dorfman, and Mukamel 2018; Svidzinsky et al. 2021; Tabakaev et al. 2021, 2022; Upton et al. 2013; Varnavski, Pinsky, and Goodson 2017; Villabona-Monsalve et al. 2018; Ye and Mukamel 2020).

Schlawin, Dorfman, and Mukamel (2018) have been instrumental in making theoretical predictions involving entangled photon spectroscopy. Nonclassical intensity fluctuations can enhance nonlinear optical signals relative to

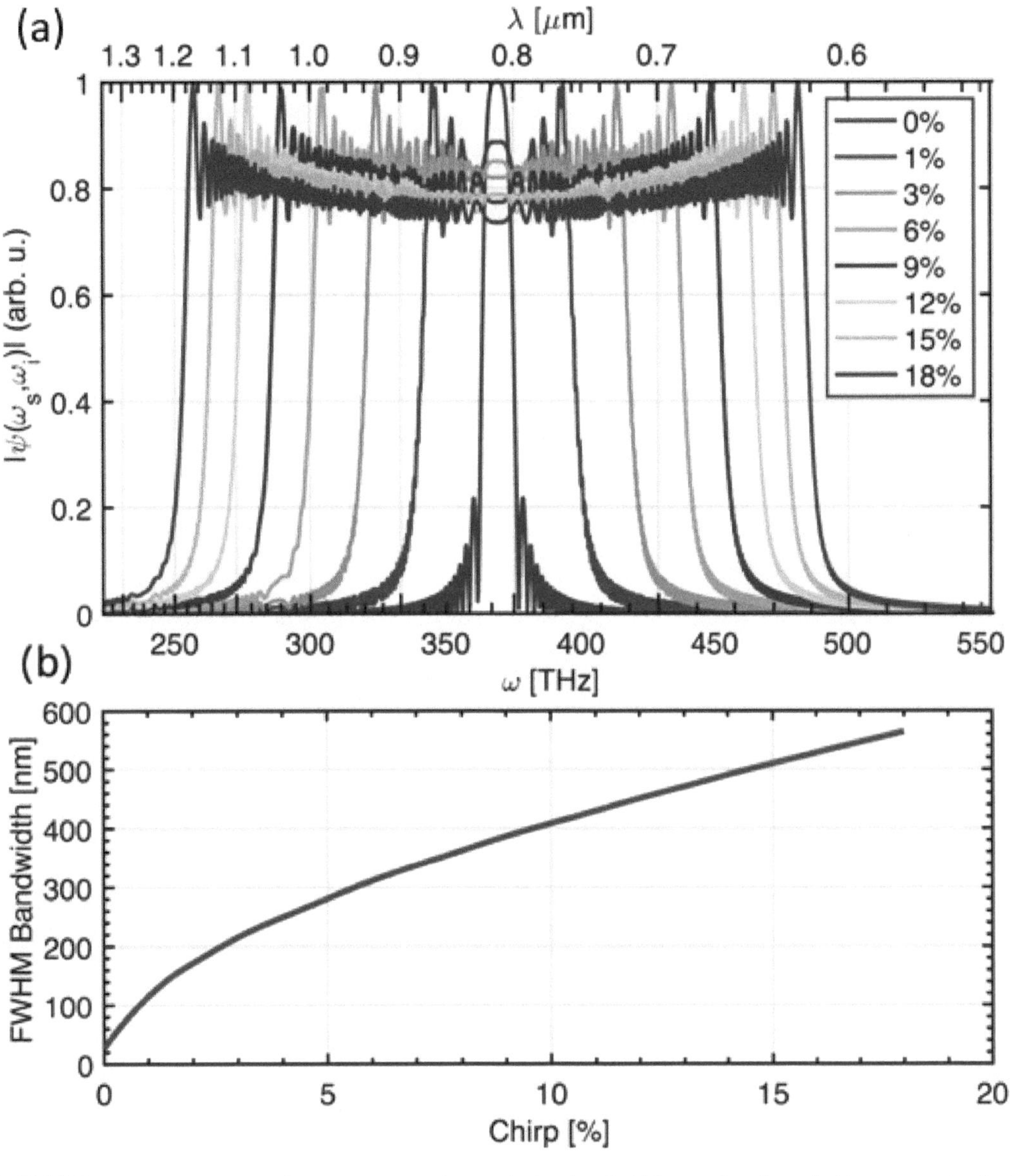

FIGURE 3-17 (a) Spontaneous parametric down-conversion emission bandwidth as a function of chirp parameter for lithium tantalate, defined as a total percentage deviation from the degenerate poling period. (b) Numerically calculated full width at half maximum (FWHM) bandwidth trend as a function of poling period chirp.
SOURCE: Szoke et al. 2021.

linear absorption. This enables nonlinear quantum spectroscopy of, for example, photosensitive biological samples at low light intensities. In addition to the signal's scaling with intensity, the use of entangled photons may also provide a new approach to shaping and controlling excitation pathways in molecular aggregates in ways that cannot be achieved with shaped classical pulses. For example, in some cases the use of entangled photons may suppress background signals and enhance others. In one proposed experiment, time–energy entangled photons produced by SPDC are employed to calculate vibrational hyper-Raman (HR) signals of a conjugated organic chromophore. Compared with classical light, Schlawin, Dorfman, and Mukamel (2018) found that time–energy entanglement can provide selectivity of Liouville-space pathways and will suppress the broad electronic-Raman background arising from one-photon resonances of the intermediate states. They showed that one-photon resonances can significantly

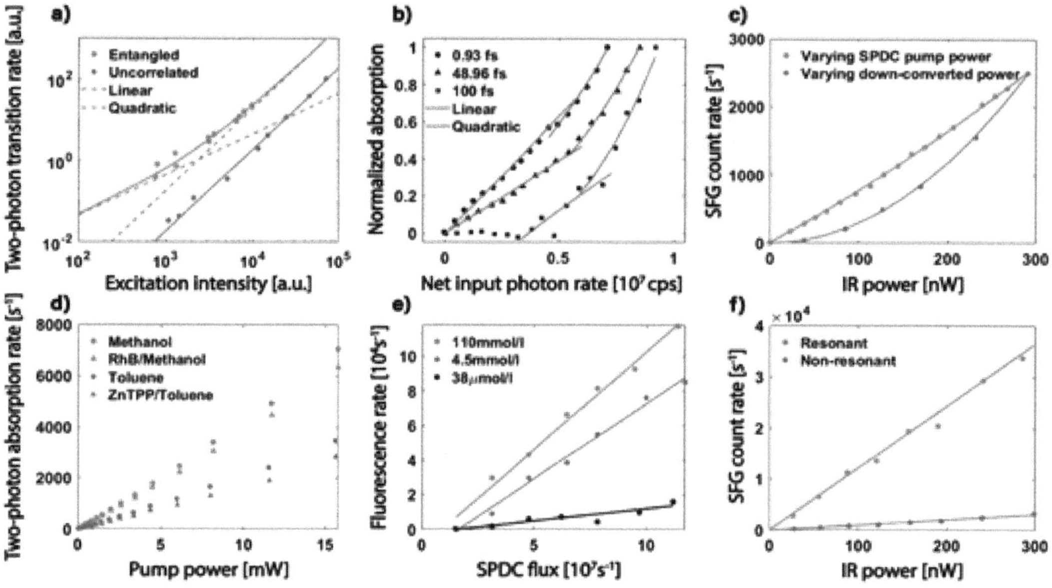

FIGURE 3-18 The linearity of various entangled two-photon absorption (ETPA) processes as a function of power. (a) Two-photon transition rate in trapped cesium with entangled and coherent light. The transition rate for uncorrelated coherent excitation is reduced by a factor of 10 for comparison. (b) Power dependence of ETPA rate in a porphyrin dendrimer under three different entanglement times. (c) Power dependence of sum-frequency generation with entangled photons. (d) ETPA rate for RhB and ZnTPP in solvent. (e) ETPA-induced fluorescence rate for Rh6G in ethanol under different concentrations. (f) Resonantly enhanced sum-frequency generation with entangled photons.
SOURCE: Szoke et al. 2020.

enhance classical HR signals; however, the electronic-Raman background is also introduced. By pathway selectivity, time-energy entangled photons can suppress this background while retaining the intense and narrow HR peaks (Figure 3-19). Another advantage of entangled light for HR spectroscopy is that the signal scales linearly with the pump intensity rather than quadratically. At low field intensities, entangled photons give stronger signals than

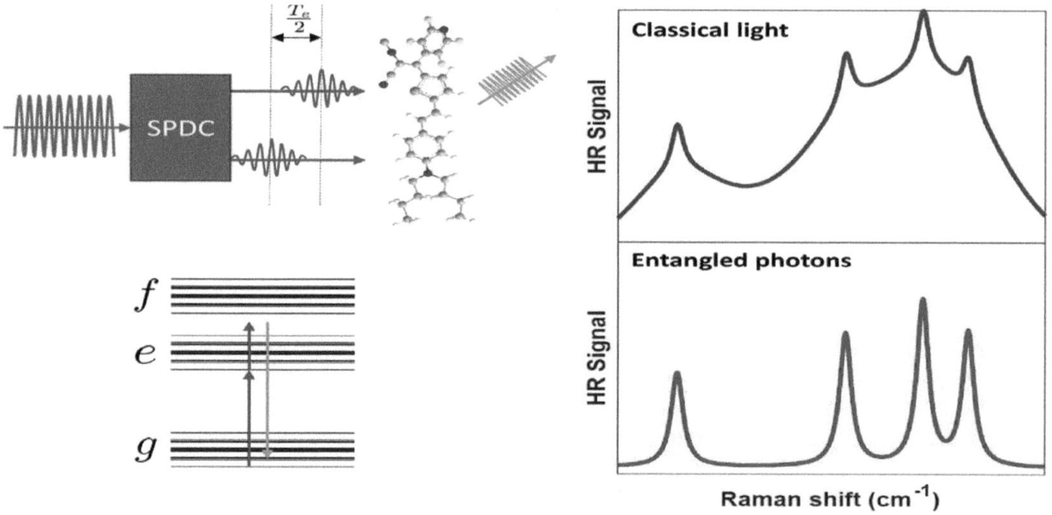

FIGURE 3-19 Schematic of hyper-Raman (HR) spectroscopy with entangled photons generated by spontaneous parametric down-conversion (SPDC).
SOURCE: Chen and Mukamel 2021.

classical light, making this tool valuable for experiments using fragile biological samples (Dorfman, Schlawin, and Mukamel 2014; Roslyak, Marx, and Mukamel 2009; Schlawin, Dorfman, and Mukamel 2018).

Other proposals from Dorfman, Schlawin, and Mukamel (2014) include using interferometric signals for pathway selectivity; applying multidimensional spectroscopy to suppress uncorrelated background signals; and employing interferometric TPA techniques with a triphoton entangled state, which will give stronger spectrally dispersed photon-counting signals than a typical biphoton state. The fundamental idea behind these proposed techniques is two-photon interferometry, which is best described with a HOM interferometer (Hong, Ou, and Mandel 1987). In a HOM interferometer, the entangled photon pairs are separated, and a time delay is introduced between them via differences in their propagation paths. Next, the photons meet and interfere at a 50:50 beam splitter, where single-photon detectors measure whether the photons emerge from the same or opposite sides of the beam splitter. The indistinguishability of the photon pair leads to destructive interference of the indistinguishable paths and a subsequent drop in the coincidence counts known as the "HOM dip" (Figure 3-20). Factors that distinguish the photon pair will affect the shape of the "HOM dip." The dip reflects any matter interactions that the entangled photons encountered, which is an important feature for spectroscopic measurements. The increased sensitivity of quantum interferometers can improve the accuracy of measurements of material linear and nonlinear susceptibilities. This technique was first demonstrated in solid-state crystals and nanostructures with relatively narrow absorption lines. Earlier reports described coherent dynamics and dephasing processes in these systems. More recently, the HOM experiment was utilized to extract a dephasing time in an organic of ~102 fs upon coherent excitation and quantum interference with a path of entangled photons in the HOM interferometer. With theoretical modeling of the coincidence rate of the HOM dip by Eshun and colleagues (2021), the experiment and theory were analyzed and showed good agreement. The proof-of-concept report gives encouragement for further measurements of nonlinear signals with this quantum interferometry tool.

For entangled two-photon absorption (ETPA), and other processes where the photon entanglement is destroyed as a result of the excitation process, uncertainties remain about which proper final-state density to use in Fermi's golden rule for determining the rate of absorption. For classical (unentangled) TPA, this state density is determined by the states of the molecule associated with the doubly excited state; normally, this is dominated by the density

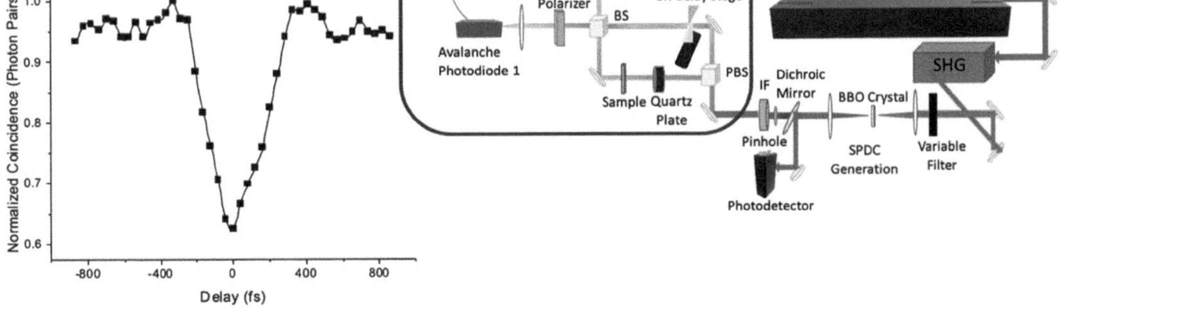

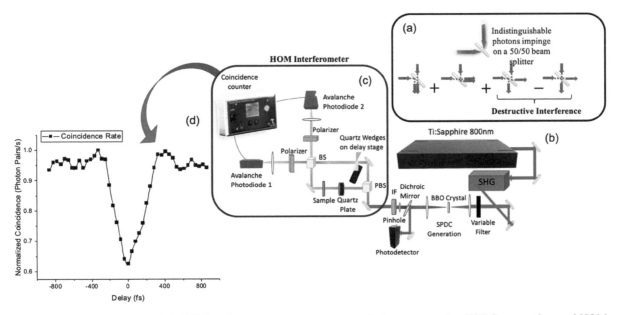

FIGURE 3-20 Hong-Ou-Mandel (HOM) pathways, spontaneous parametric down-conversion (SPDC) generation, and HOM interferometer setup. (a) Schematic of possible HOM interferometer pathways. (b) Femtosecond laser and SPDC generation with type-II SPDC β-barium borate (BBO) crystal. (c) Scheme of HOM interferometer. (d) HOM dip measured without sample. SOURCE: Eshun et al. 2021.

of phonons associated with this state, as this leads to the known rapid dephasing (in ~10 fs) after photoexcitation. However, for entangled photon excitation, Kang and colleagues (2020) proposed that entanglement fidelity can constrain the state density, such that the radiative lifetime of the doubly excited state rather than phonon dephasing determines this density. This result, which previously had been used in atomic physics modeling (Fei et al. 1997; Kojima and Nguyen 2004), leads to a much narrower lineshape for ETPA, and much larger cross sections are obtained from electronic structure calculations that are similar to what has been found in experiments (Kang et al. 2020). However, this assumes that fidelity of the entanglement constrains coupling to phonons, which is an assumption of unknown validity. Further work is needed to relate this result to known electronic and electron-vibrational entanglement properties of molecules (McKemmish et al. 2011; Plasser 2016).

There have been reports of utilizing nonclassical states of light in spectroscopy and microscopy in the linear optical regime. The use of quantum optical approaches in the linear regime could be of great interest to chemists in providing sensitive detection schemes for possible remote detection of chemical and biological systems. For example, there have been reports of quantum imaging with undetected photons. In this quantum interference effect, the photons that pass through the imaged object are never detected. The entangled photons that never interacted with the object provide the image. Here, it is not necessary to detect the photons that illuminate the object at all. This could be very useful for chemical and biological samples, where the material could lead to very opaque and high scattering of the illuminating light (Lemos et al. 2014). There have been other uses of quantum interferometers for linear optical effects in the IR regions. For example, reports have shown that by detecting the signal photons one can acquire the amplitude or phase information of the idler photons. Research has shown that the measured transmission spectrum of a polymer sample is in good agreement with a conventionally measured reference spectrum by detecting the correlated visible light of the SPDC process. Again, this approach could be useful for chemists interested in collecting IR spectral signatures of molecules using a visible laser light source and a silicon-based detector (Lindner et al. 2020; Mukai, Okamoto, Takeuchi 2022). This approach has even been extended to the terahertz region, where terahertz photons interact with a sample in free space and information about the sample thickness is obtained by the detection of visible photons. The ability to do terahertz spectroscopy with detection of visible photons would impact the investigation of thick opaque materials and samples (Kutas et al. 2020).

The use of entangled photons for coherent control has been reported as well by Gu and colleagues (2021). Recent theoretical work revealed that entangled light may provide a coherent control scheme for nuclear wave packets in a one-photon dark excited state of a molecule. This is demonstrated by nonadiabatic conical intersection wave packet dynamics for the *trans–cis* photoisomerization of azobenzene (Figure 3-21) (Casacio et al. 2021). This control leads to a substantial difference in the transient coherences during the passage through the intersection (the transition-state structure of the photochemistry), which is proposed to be detected by a stimulated X-ray Raman signal. Additionally, it was suggested that the photoisomerization yield is noticeably affected by modulating the photon entanglement time. The essential role of energy–time entanglement in the control is clearly demonstrated by contrasting the quantum light–excited wave packets to the wave packets created by classical light. Varying the bandwidth alone for both classical nonchirped and chirped pulses leads to minor differences in the two-photon excitation process. These results may provide a strategy for coherent quantum light control of the photoexcitation of electronic dark states of molecules (Gu et al. 2021).

3.4.3 Use of Quantum Light in Microscopy

One of the most recent applications of nonclassical states of light in chemistry involves microscopy, which is essential in broad areas of science from physics to biology. Using nonclassical states of light, chemists can open possibilities of realizing low-intensity microscopy at intensity levels not achievable with classical resources. For example, using quantum states of light for illumination, precise phase and linear absorption measurements in a microscopic format have been previously reported, and precision beyond the standard quantum limit has been achieved in a microscope (Casacio et al. 2021). These experiments also will be useful in overcoming photodamage, which is important for biological systems. Through the use of quantum correlations in nonclassical light, Casacio and colleagues (2021) demonstrated a highly selective coherent Raman microscope. They observed a 35 percent improved signal-to-noise ratio (SNR) compared to conventional classical light microscopy. This improvement

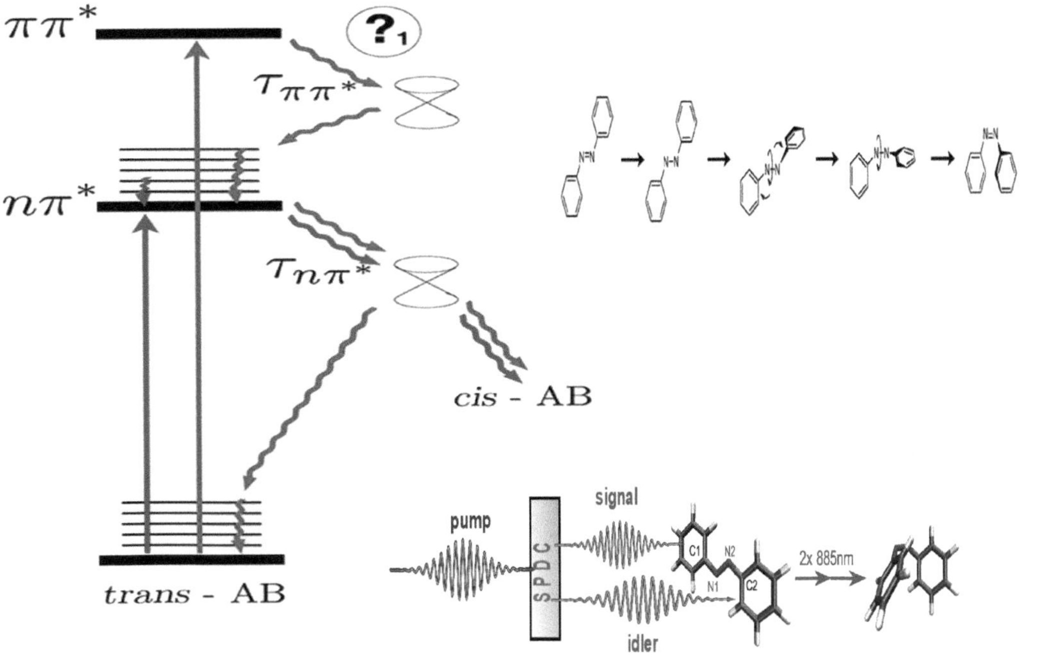

FIGURE 3-21 Azobenzene *trans–cis* isomerization initiated by entangled two-photon absorption. The combined energy of the two photons matches the S_0/S_1 transition at the *trans* geometry. The first reactive nuclear coordinate, q1, is the C1–N1–N2–C2 dihedral angle; the second coordinate, q2, is the symmetric bending of the C1–N1–N2 and N1–N2–C2 angles.
SOURCE: Gu et al. 2021.

translates to a 14 percent enhancement in concentration sensitivity, which reduces phototoxicity levels and allows for increased resolution of otherwise unobservable biological structures. The Raman microscope shows the vibrational spectra of biomolecules. Furthermore, quantum light eases the photon budget because it increases the SNR, as Raman scattering can be observed with even less than a single photon within the measurement time frame. Quantum-enhanced images of samples of both dry polystyrene beads and living *Saccharomyces cerevisiae* yeast were obtained (Figure 3-22) with 23 to 35 percent enhanced SNR, leading to increased image contrast. As such Casacio and colleagues (2021) were able to view features that were uncovered below the SNL.

Another approach toward the use of entangled photons in molecular and biological microscopy involves TPA (Eshun et al. 2022; Mukamel et al. 2020; Varnavski and Goodson 2020; Varnavski et al. 2022). Here, the microscopic image created by the fluorescence selectively excited by the process of the entangled TPA was reported. Entangled two-photon microscopy offers nonlinear imaging capabilities at an unprecedented low excitation intensity, 10^7, which is six orders of magnitude lower than the excitation level for the classical two-photon image. The nonmonotonic dependence of the image on the femtosecond delay between the components of the entangled photon pair is demonstrated. This delay dependence is a result of specific quantum interference effects associated with the entanglement, and this is not observable with classical excitation light. Additionally, fluorescent images of breast cancer cells were captured using entangled TPA in a scanning microscope (Figure 3-23; Varnavski et al. 2022). These images were generated at an excitation intensity orders of magnitude lower than the conditions necessary for classical two-photon microscopy. Quantum-enhanced entangled two-photon microscopy has shown cancer cell imaging capabilities at an unprecedented low excitation intensity of ~3.6×10^7 photons/sec, which is one million times lower than the excitation level for the classical two-photon fluorescence image obtained in the same microscope. The entangled two-photon microscope images resolving specific features of breast cancer cells in different stages of mitosis have been obtained. Varnavski and colleagues (2022) also illustrated the impact of

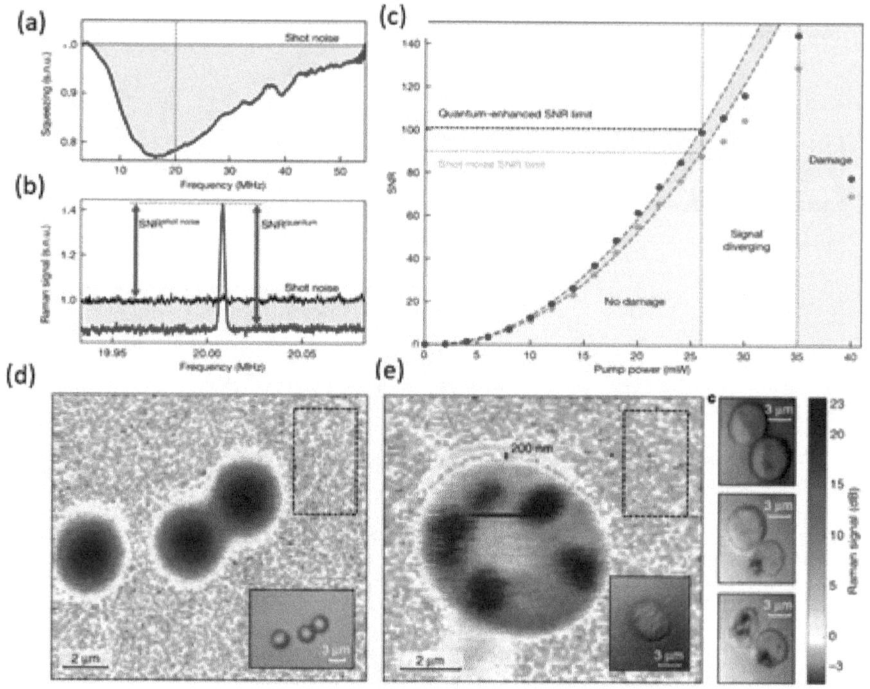

FIGURE 3-22 (a) Squeezed light normalized to shot noises, (b) stimulated Raman signal, and (c) signal-to-noise ratio (SNR) for a polystyrene bead. (d and e) Images of polystyrene beads and a live yeast cell taken with the quantum coherent microscope.
SOURCE: Casacio et al. 2021.

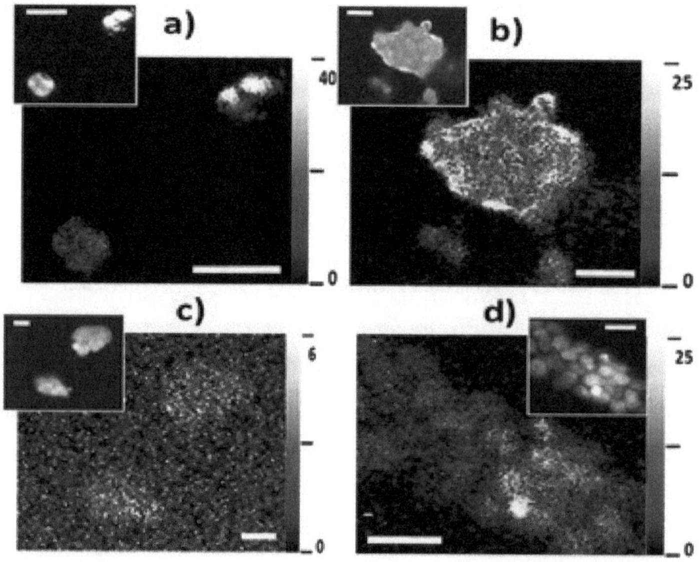

FIGURE 3-23 Entangled two-photon light microscopy images of MCF7 cancer cells stained with the dye Hoechst 34580. (a and c) Cells with different numbers of nuclei. (b) Image of a colony of MCF7 cells. (d) Image of the large cell cluster. Insets: respective cell images created in a fluorescence microscope with the excitation by classical laser light at 405 nm. Spatial scale bar: 20 µm.
SOURCE: Varnavski et al. 2022.

the low light excitation with entangled photons on the photodamage of the cancer cells. This is encouraging in the development of methods with extremely low light probe intensity with entangled two-photon microscopy critical to minimizing photobleaching during repetitive imaging and damage to cells in live-cell applications.

3.4.4 Measurement of Coherence in Molecular Systems

The measurement of coherence in chemical systems has a long history. The development of better methods to resolve more information about the molecular system is a significant part of modern physical chemistry. The instrumentation utilized in measuring the relatively fast dephasing times in molecules was significantly improved with the advent of the femtosecond laser, the use of which is now rather commonplace. As an example, one can utilize short laser pulses to excite ladders of states in phase, thereby making a superposition that can be detected using 2D spectroscopy. In the past decade, the number of research groups with the experience and equipment to carry out electronic and vibrational 2D spectroscopy has significantly increased. In this approach, a cross-peak in the 2D map labels the excited and detected transitions (Figure 3-24), while oscillations in the cross-peaks as a function of pump–probe waiting time reveal coherences involving those transitions occurring together. However, over time the oscillations damp away as a function of waiting time (Figure 3-24c). This spectral characteristic is a consequence of dephasing, giving a lower bound for the decoherence time of the quantum superposition. Figure 3-24a shows an example of electronic coherence for a semiconductor "nanoplatelet" colloid dispersed in

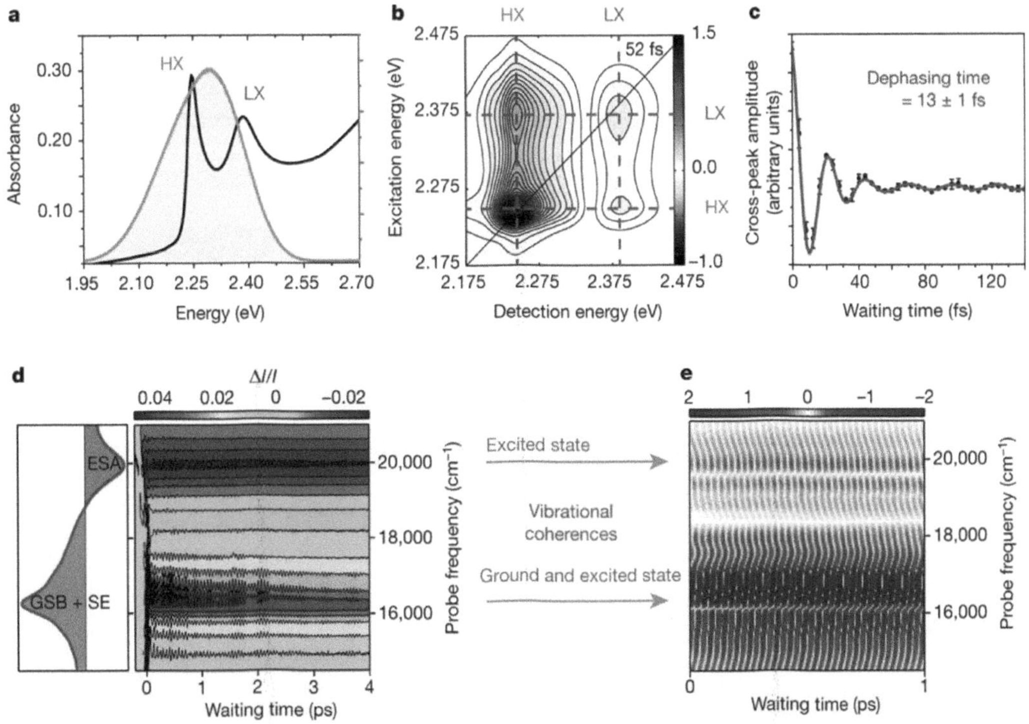

FIGURE 3-24 (a) Absorption spectra of CdSe nanoplatelets (black line) showing the heavy-hole (HX) and light-hole (LX) exciton transitions. The spectrum of the laser pulses used in the two-dimensional (2D) spectroscopy experiments is shaded orange. (b) 2D electronic spectrum recorded at a pump–probe delay time of 52 fs. (c) Amplitude oscillations in the lower cross-peak of the rephasing 2D spectrum for a CdSe/CdZnS nanoplatelet (the real part with population relaxation subtracted) as a function of the waiting time. (d) A contour map of broadband pump–probe data for cresyl violet solution showing the oscillatory modulation on top of the ground- and excited-state population dynamics. (e) Fourier-filtered pump–probe data revealing coherent oscillations of the strong Franck–Condon active modes.
SOURCE: Scholes et al. 2017.

an ambient-temperature solution. Here, the broadband femtosecond pulses (red curve) overlap with the first two exciton states, which are the heavy-hole (HX) and light-hole (LX) exciton. As demonstrated in Figure 3-24b, the amplitudes of the cross-peaks, HX-LX and LX-HX, in the 2D signal map oscillate as a function of excitation-detection time delay, showing that the amplitudes of HX and LX bands indeed are correlated until the superposition dephases. The dephasing of this ensemble happens with a time constant of 13 fs, which is aligned with previous observations for similar systems (Figure 3-24c). Electronic coherences at ambient temperature typically decohere with a time constant of ≤100 fs.

3.4.5 Coherence and Biological Function Is the Future

In the past 10 years, electronic and vibrational coherence experiments have improved significantly with the use of time-resolved nonlinear spectroscopic techniques. Mukamel and colleagues (2020) have been instrumental in the theory and prediction of spectroscopic signals under this context. Chemists have investigated the electronic coherences in organic and biological systems in impressive detail. According to Scholes and colleagues (2017), it might be appropriate for the field to move beyond measuring the coherence lifetimes. The use of multidimensional spectroscopic methods to probe other parameters related to the excited-state properties could provide new insights into the connection between biological function and chemical structure. For example, time-resolved ultrafast optical probes of chiral dynamics may provide a new window to explore how interactions with structured environments drive electronic dynamics. Incorporating optical activity into time-resolved spectroscopies has proven challenging owing to the small signal and large achiral background. Higgins, Allodi, and colleagues (2021) demonstrate that 2D electronic spectroscopy (ES) can be adapted to detect chiral signals and that these signals reveal how excitations delocalize and contract following excitation. They dynamically probe the evolution of the chiral electronic structure in the light-harvesting complex 2 of purple bacteria (Figure 3-25) following photoexcitation by creating a

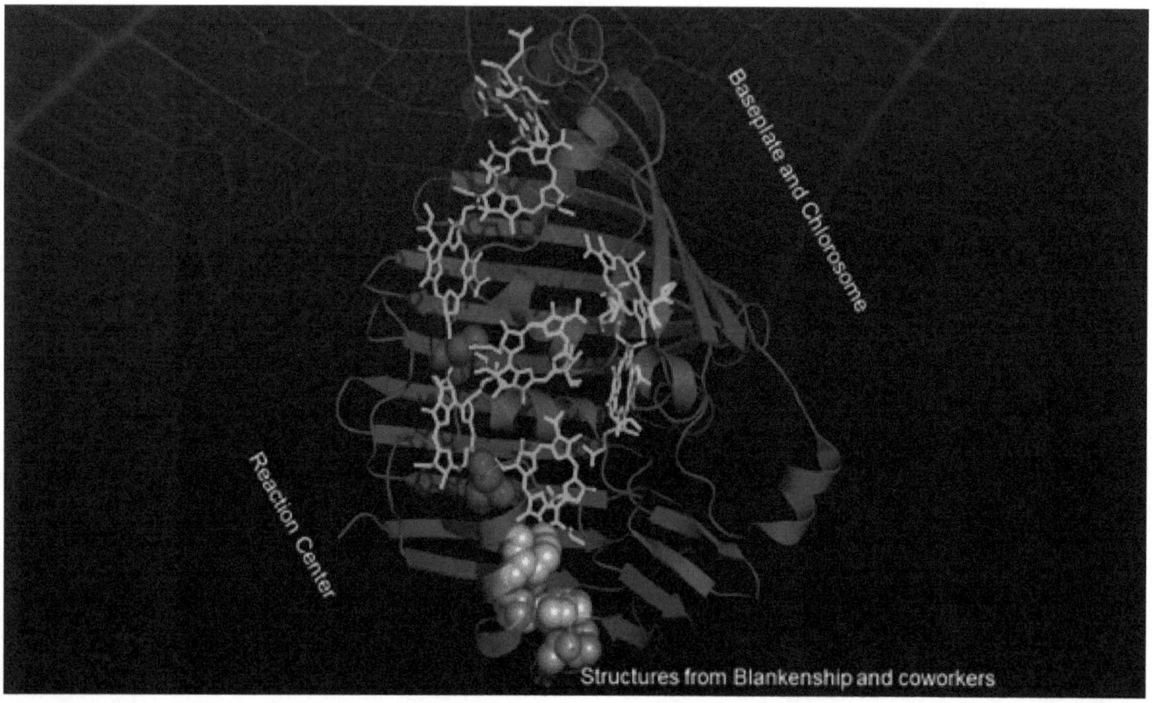

FIGURE 3-25 Structure and model of light-harvesting complex 2 of the purple bacteria reaction center.
SOURCE: Higgins, Allodi, et al. 2021.

chiral 2D mapping. The dynamics of the chiral 2D signal directly report on changes in the degree of delocalization of the excitonic states following photoexcitation. The mechanism of energy transfer in this system may enhance transfer probability owing to the coherent coupling among chromophores while suppressing fluorescence that arises from populating delocalized states. This generally applicable spectroscopy will provide an incisive tool to probe ultrafast transient molecular fluctuations that are obscured in non-chiral experiments. Although coherence has been shown to yield transformative ways for improving function, advances have been confined to pristine matter and coherence was considered fragile. However, recent evidence of coherence in chemical and biological systems suggests that the phenomena are robust and can survive in the face of disorder and noise. Chemists are still assessing the variability of this process and the possibility that coherence experiments and mechanisms can be used effectively in more complex chemical systems. A key missing point is the role of coherence as a design element in realizing function.

Most recently, quantum coherences observed as time-dependent beats in ultrafast spectroscopic experiments arise when light–matter interactions prepare systems in superpositions of states with differing energy (e.g., redox states) and fixed phases across the ensemble. These measurements reveal important variations when compared to previous reports, suggesting that redox conditions tune vibronic coupling in the Fenna–Matthews–Olson (FMO) pigment–protein in green sulfur bacteria; this raises the question of whether redox conditions may also affect the long-lived (>100 fs) quantum coherences observed in this very important complex. Higgins, Lloyd, and colleagues (2021) performed ultrafast 2DES on the FMO complex under both oxidizing and reducing conditions (Figure 3-26). They observed that many excited-state coherences are exclusively present in reducing conditions and are absent or attenuated in oxidizing conditions. Reducing conditions mimic the natural conditions of the complex more closely. Furthermore, the presence of these coherences correlates with the vibronic coupling that produces faster, more efficient energy transfer through the complex under reducing conditions. The growth of coherences across the waiting time and the number of beat frequencies across hundreds of wavenumbers in the power spectra suggest that the beats are excited-state coherences with a mostly vibrational character whose phase relationship is maintained through the energy transfer process. These results suggest that excitonic energy transfer proceeds through a coherent mechanism in this complex and that the coherences may provide a tool to disentangle coherent relaxation from energy transfer driven by stochastic environmental fluctuations.

As this field shifts its focus from confirming the existence of coherence to exploring the connection between coherence and function, new methods are being extended to provide new insights. As expected, this area of investigation will require extensive feedback between theory and experiment, synthesis and measurement, and the development of systematic methods to quantify the influence of coherence in specific processes or devices. Exploring function requires controlled perturbation, and establishing this essential methodology requires new control mechanisms and clear assessment tools. For instance, new and improved experimental techniques will enhance capabilities such as measuring the delocalization of wave functions or measuring the collapse of delocalized states. In this regard, attosecond laser sources have enhanced the ability to study coherent electronic motion, such as charge migration and quantum interference between electrons, as well as electron-nuclear wave packet motion. Similarly, advances in time-resolved X-ray spectroscopies open up new probes of electronic structure by delineating dynamics into element-specific contributions. Inspired by coherent multiple scattering and molecular transport junctions, the reactivity of metal centers might be directed or redox chemistry tuned by interference effects to modify the electron density at an active site of a catalyst. In coherent backscattering, which provides scattering that is twice as bright as diffuse scattering, substantial gains are possible when robust coherence phenomena are exploited. Such gains warrant future research into coherence as a potential force for enhancing function.

3.4.6 High-Finesse Cavities and Nanophotonics for Molecular Qubit Systems

In addition to understanding and designing molecular systems with favorable optical properties, a different and complementary approach is to engineer optical control by modifying the molecule–light coupling, by changing the vacuum field of the light itself. One way to accomplish this is by surrounding a molecule with an optical cavity (Figure 3-27). Light resonant with the cavity recirculates and repeatedly interacts with matter located in the cavity modes, leading to enhanced light–matter coupling. In addition, the cavity can also modify the decay rate of optically excited molecular states by modifying the density of states of the electromagnetic field.

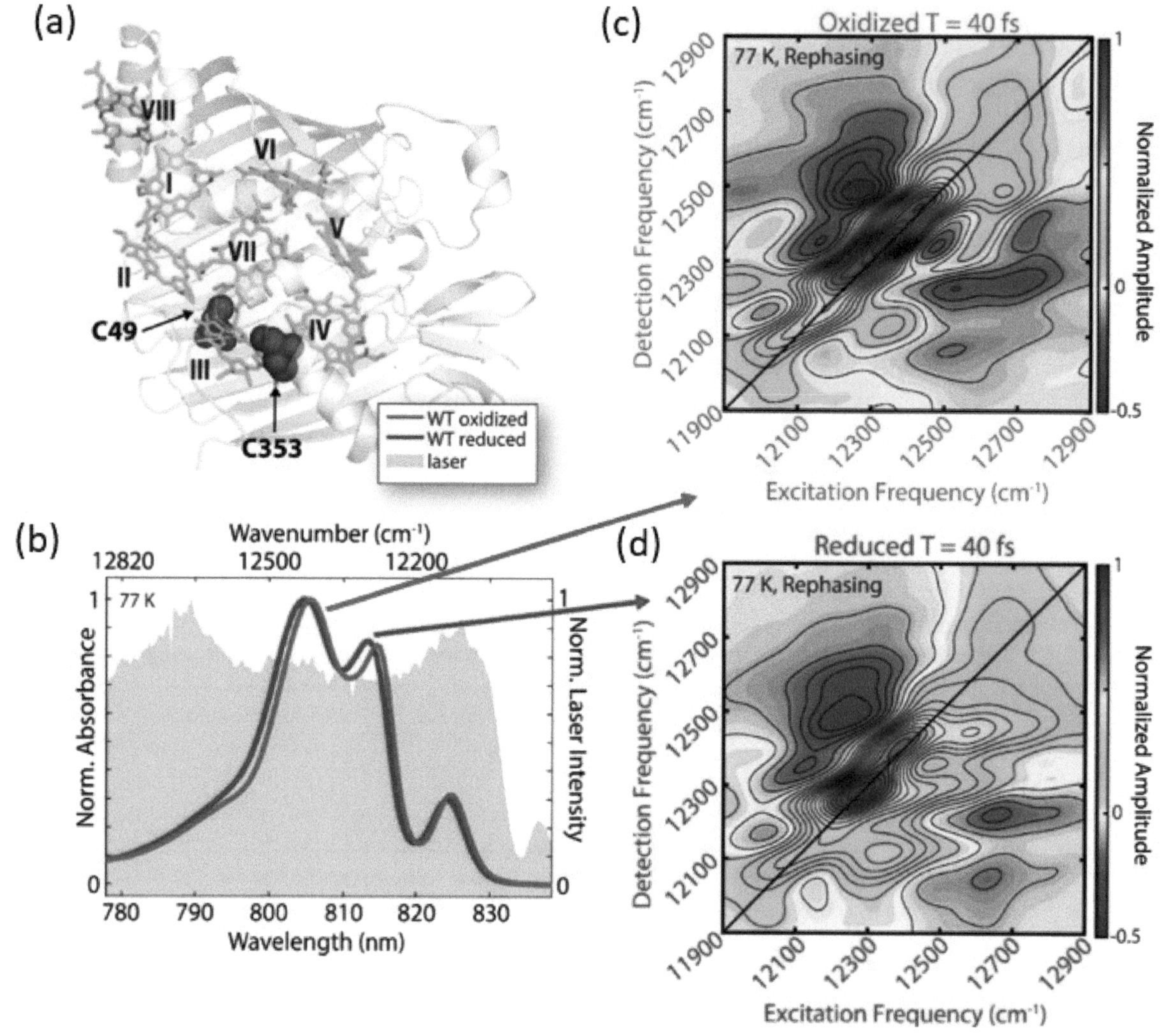

FIGURE 3-26 Redox condition affects excited-state behavior in the wild-type Fenna–Matthews–Olson (FMO) complex. (a) Structure of the FMO complex and the eight bacteriochlorophyll sites held by the protein scaffold (Protein Data Bank ID code 3ENI). Shown in red are the two cysteine residues, C49 and C353, that are known to steer and quench excitations in oxidizing conditions and tune vibronic coupling for enhanced energy transfer in reducing conditions. (b) Linear absorption spectra of the wild-type oxidized (blue) and wild-type reduced (red) FMO complex at 77 K. Shown in gray is the laser spectrum used. (c and d) Rephasing two-dimensional electronic spectra under oxidizing and reducing conditions at waiting time T = 40 fs. Differences in the lower-diagonal cross-peaks between experiments indicate faster, more efficient energy transfer when the complex is reduced.
SOURCE: Higgins, Lloyd, et al. 2021.

The applications in quantum science from surrounding molecules with cavities are many and include building quantum networks (Duan et al. 2001; Kalb et al. 2017), creating photon-mediated interactions between quantum emitters (Evans et al. 2018), and building single-photon sources. Here, we discuss a few examples with relevance to optical control of molecules.

First, with regard to the ZPL problem encountered in solid-state qubits, cavities can enhance emission into the ZPL via the Purcell effect, as has been proposed and experimentally demonstrated (Barclay et al. 2011; Riedel et al. 2017). In the Purcell effect, the decay rate of a transition resonant with an optical cavity is strongly enhanced (Purcell, Torrey, and Pound 1946). This leads to preferential emission into a specific state. State-of-the-art experiments with NV centers in a microcavity have shown enhancement of ZPL emission from 3 to 46 percent, with a

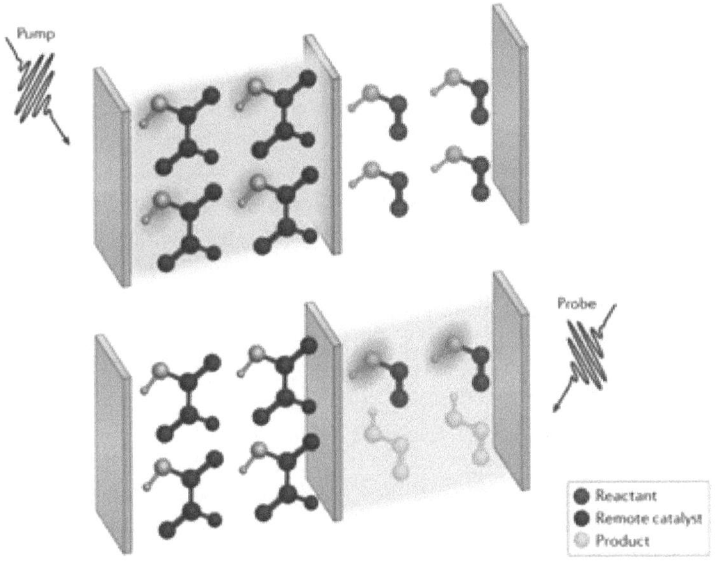

FIGURE 3-27 Illustration of optical microcavities that may one day allow chemists to control chemical reactions.
SOURCE: Du, Ribeiro, and Yuen-Zhou 2019.

viable path toward emission factors exceeding 90 percent (Riedel et al. 2017). Second, optical cavities can provide coherent interactions between matter spaced over large distances. For atomic systems, this has been proposed as a way to produce entanglement and teleportation of quantum information over long distances. The same principles apply to optically active molecular systems.

Although we have focused above on how optical cavities can be used to produce more favorable qubits, combining the QIS technique of optical cavities with molecules also opens a potential new way to control chemical reactions. In the presence of strong light–matter coupling, molecules within a cavity inherit photonic character and form so-called polaritons, which are quantum superpositions of light and matter. This has led to much interest recently in polariton chemistry (see below), where new possibilities abound. With dense highly absorbing molecular ensembles in microfluidic cavities and dye lasers, experiments already have seen polaritons affect chemical reactivity and optoelectronic properties (Ebbesen 2016; Ribeiro et al. 2018). New types of molecular polaritons with unique properties could be synthesized (Gu and Franco 2018), and novel catalysts might be found. For example, with molecules in cavities, it could be possible to control photochemistry remotely simply by tuning a "remote catalyst" in a distant cavity that is coupled to the system of interest via photons (Du, Ribeiro, and Yuen-Zhou 2019). The many new possibilities with molecular systems in optical cavities come with some technical and material challenges, including fabricating cavities with low loss (i.e., high reflectance and low scatter) and integrating cavities with molecular systems.

Analogous to the work with optical cavities described above, demonstrations of strong spin–photon coupling in resonant microwave devices is another area of significant exploration (Bonizzoni, Ghirri, and Affronte 2018; Eddins et al. 2014; Gimeno et al. 2020; Jenkins et al. 2013). All of these investigations have been in a regime far from the quantum limit, involving macroscopic numbers of spins and photons. Nevertheless, strong coupling effects have been reported for a number of different molecular systems and resonator types. This is an area ripe for further development in order to reach a point where entanglement between microwave photons and a small ensemble of molecular spins (or even a single molecular spin) is achieved.

Cavity Polaritons

When strongly coupled, material states and optical cavity modes transform into a *single quantum entity*—a polariton—with its own distinct and modifiable physics and chemistry, fundamentally altering the behavior of both matter and the electromagnetic field. Cavities may provide a nondestructive and versatile way to control energy, charge, and coherence transfer in molecular systems, which underlie reactivity, photochemistry, and catalysis. To establish the fundamental scientific understanding of cavity-enabled molecular control, it is necessary to establish basic principles that can guide future chemical innovation. With these essential concepts identified, the modular nature of external cavity control will usher in the rapid technological development of cavity-enabled chemistry. By leveraging, for example, microfluidic and lab-on-a-chip platforms, it will be possible to carry out complex, high-value chemical transformations that are key to the production of pharmaceutical and industrial chemicals. This could indeed be another avenue for chemistry in QIS.

The use of cavities is already under way in chemical research, with two parallel but related efforts in the polaritonic chemistry community. One thrust has been to apply ultrafast laser spectroscopy methods to monitor the complex energy transfer and relaxation processes in molecular polaritons, both in the vibrational and electronic domains. While considerable basic science interest certainly exists, there are also many possibilities for applications in QIS using molecular polaritons as quantum resources, particularly in the case of quantum transduction. Vibrational polaritons are especially promising in this context because of the relatively long coherence times achievable using molecular vibrations coupled to resonant optical cavities. The use of ultrafast 2D optical spectroscopy, both in the visible and mid-IR spectral regions, has revealed nonlinear optical effects of potential use for optical switching, polaritonic energy ladders, and the exchange of energy between bright polariton states and nearly inactive dark states (Figure 3-28). There are clearly challenges that require advanced cavity designs, as suggested by a recent study showing considerable contributions to the measured two-dimensional infrared (2DIR) response from molecules that are neither polaritons nor dark states but instead comprise a non-polaritonic background (Duan et al. 2021). These fundamental dynamics studies are important because they directly probe the underlying physics of all the components present in a molecule-cavity construct.

The other major thrust in polaritonic chemistry is arguably the most promising avenue for future investigation and discovery. Following the first observation of a cavity-controlled photochemical reaction by Laboratoire des Nanostructures at the University of Strasbourg, there are now several examples of chemical reactions whose kinetics and equilibria can be altered through coupling to a resonant optical cavity (George et al. 2015). An important recent finding showed that the selectivity and rate of bond cleavages in an alkyl silane compound can be modified within a cavity tuned to specific resonances of the reacting species (Nagarajan, Thomas, and Ebbesen 2021). This observation implies that cavities can exert both thermodynamic control (e.g., altering the equilibrium constant of a reaction) as well

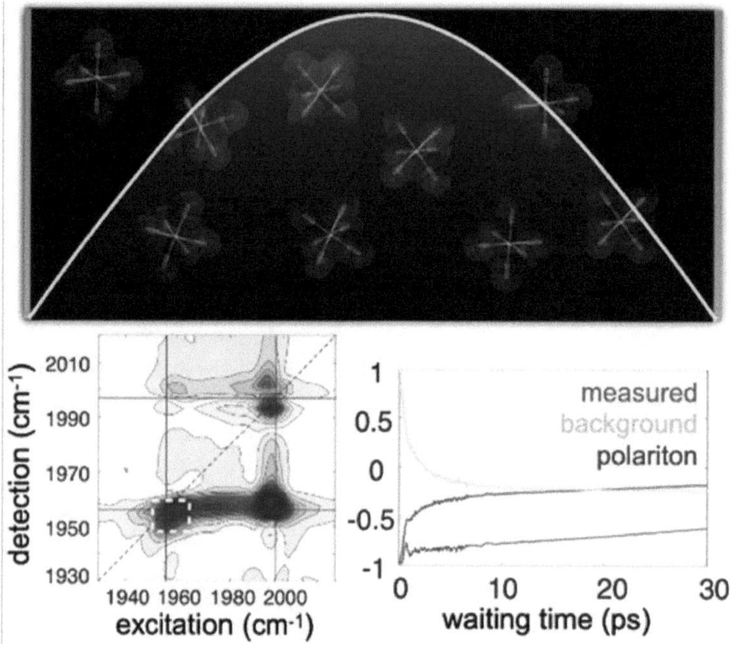

FIGURE 3-28 Two-dimensional infrared spectra of vibrational polaritons.
SOURCE: Duan et al. 2021.

as kinetic control by selectively lowering the barrier for one reaction channel relative to another. Perhaps an exciting recent development is a report that shows that a urethane addition reaction can be controlled by strong coupling to a cavity (Ahn, Herrera, and Simpkins 2022). Though no more fundamental chemically than the alkyl silane example, this latter work is the first by a research group that is completely independent of the University of Strasbourg team. Indeed, challenges in reproducing some cavity results have cast some doubt on the robustness of cavity control, but the recent urethane addition reaction provides ample evidence that cavity control is possible in other laboratories. Theoretically, scientists such as Gu and Mukamel (2021) are working to describe this process in detail.

Perhaps the key outstanding challenge that the field must confront is the lack of a clear set of guiding principles that underlie the mechanism(s) of cavity control. Whereas the physics of strong coupling between molecules and cavity modes are understood, no consensus exists regarding the reasons why creating delocalized states with energies slightly above and below the uncoupled energies would cause reaction rate constants to increase or decrease to the degree reported in the literature. Statistical theories, such as transition-state theory, that routinely apply broadly in chemistry seem to be completely inconsistent with polaritonic effects on reactivity. Dynamical approaches based on quantum electrodynamic density functional theory are feasible but are extremely demanding and are unlikely to be helpful for the broader chemistry community. A set of experimental results on tractable chemically reactive systems that can be studied to provide inspiration and inputs to theoretical analyses will deepen this research area. Ideally, these reactions will be elementary and will not rely on complicated multistep mechanisms. Reactions that are phototriggered will enable direct dynamical investigation rather than more coarse-grained kinetics measurements. In some sense, polaritonic chemistry, especially when coupling vibrational transitions to cavity modes, is the fulfillment of a dream that began with the advent of lasers in the 1960s that we would be able to do chemistry precisely using mode selectivity. A promising aspect of the present situation is that it is highly likely that basic principles will be established quickly—particularly with enhanced funding support—paving the way for accelerated chemical technologies that will offer entirely new ways to conceive of and carry out chemistry.

Another area of research that has received primarily theoretical attention in the chemistry community is the use of quantum light in the context of polaritonic effects in cavities. Gu and Mukamel (2020) have shown that nonlinear quantum light spectroscopy can be employed for probing two-photon excitations in polaritonic systems. Theoretical investigations have identified the combined signatures of a cavity photon mode strongly coupled to molecules and entangled photons on collective bipolariton resonances. Initial studies focused on TPA signals with an entangled photon pair to a polaritonic system consisting of N two-level molecules strongly coupled to a single cavity photon mode. By comparing the TPA spectra with classical and quantum light, Gu and Mukamel (2020) have demonstrated that entangled photons can create two-photon excitations in polaritonic systems that are drastically different from the classical two-photon excitations. Entangled photons can reveal classically dark bipolariton states by modifying the quantum interference among transition pathways leading to TPA (Figure 3-29). These exciting predictions have yet to be examined experimentally and provide a novel opportunity for chemists working in QIS.

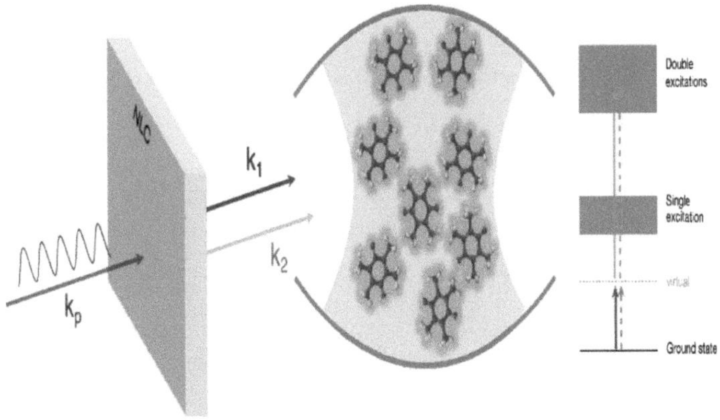

FIGURE 3-29 Illustration of two-photon excitations to bipolariton states created by placing several molecules in an optical cavity excited by quantum light.
SOURCE: Gu and Mukamel 2020.

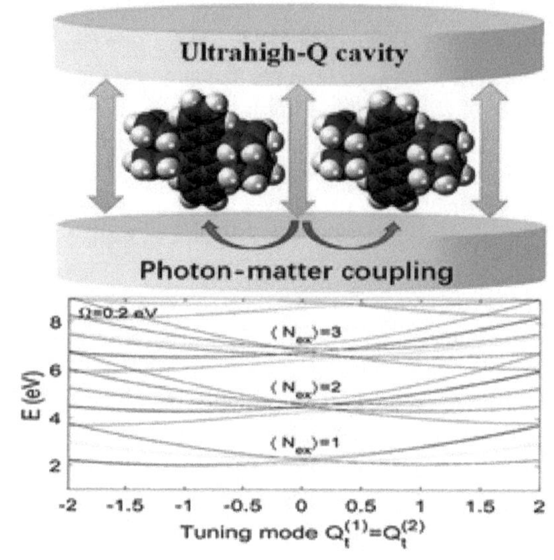

FIGURE 3-30 Illustration of a cavity-manipulated singlet fission system that is mediated by polaritonic conical intersections for both one- and two-molecule systems.
SOURCE: Sun, Gelin, and Zhao 2022.

Other theoretical studies have shown that cavity-manipulated singlet fission (SF) is possible when it is mediated by polaritonic conical intersections for both one- and two-molecule systems (Figure 3-30). The population evolution of the critical state for SF (Triplet-triplet state) and the cavity photons were carefully examined in search of a high fission efficiency via cavity engineering. Several interesting mechanisms were uncovered, such as photon-assisted SF, system localization via a displaced photon state, and collective enhancement of the fission efficiency for the two-molecule system. It was also found that the system localization process in the two-molecule system differs substantially from that in the one-molecule system owing to the appearance of a novel central polaritonic conical intersection in the two-molecule system (Sun, Gelin, and Zhao 2022). The possibility that a cavity-controlled SF process can be switched on and off by controlling the average pumping photon number is another novel opportunity for chemists working in QIS.

3.5 DEVELOP AND EXPLOIT ALTERNATIVE APPROACHES TO SPIN POLARIZATION AND COHERENCE CONTROL

3.5.1 Chirality-Induced Spin Selectivity Effect

Recent experiments have shown that it is possible to generate high spin polarization via electron transport through chiral molecules, even at room temperature. This discovery of the so-called chirality-induced spin selectivity (CISS) effect (Naaman and Waldeck 2012; Ray et al. 1999), whereby spins aligned parallel or antiparallel to the electron transfer displacement vector are preferentially transmitted depending on the chirality of the molecular system, is of huge potential importance for the molecular QIS research endeavor. In particular, the coupling of orbital and spin angular momenta during directional electron transfer processes provides a novel means for manipulating spins over molecular length scales. A detailed discussion of the fundamental chemistry and physics governing the CISS effect can be found in Section 2.4. Nevertheless, it is clear that an area ripe for future investigation involves optimization and exploitation of the CISS effect with potential molecular QIS applications in mind. Such efforts will rely on nanosecond and femtosecond transient absorption spectroscopies in order to probe electron transfer dynamics (Harvey and Wasielewski 2021). In addition, the use (and further development) of time-resolved/transient EPR spectroscopy will be essential to elucidating the CISS-induced spin dynamics.

3.5.2 Electric Field Control

Various efforts have been pursued that employ EPR detection to monitor the response to an external stimulus applied to a sample. For example, slow quantum relaxation of non-equilibrium spin populations has been studied via high-frequency EPR detection (Dressel et al. 2003), as well as over-barrier relaxation due to thermal avalanches triggered by acoustic pulses (Macià et al. 2008). Meanwhile, in situ application of large hydrostatic pressures within a high-field EPR instrument enables fundamental insights into the nature of the microscopic interactions governing the spin physics in molecular materials (Thirunavukkuarasu et al. 2015), while also demonstrating structurally induced switching effects (Prescimone et al. 2012). More recently, several studies have demonstrated magnetoelectric coupling through EPR measurements carried out under static (or quasi-static) electric fields (Boudalis, Robert, and Turek 2018; Fittipaldi et al. 2019; van Slageren 2019). These investigations demonstrated an important milestone on the pathway to molecular QIS by enhancing prospects for local control at the nanometer scale, along with the obvious advantages of the low power consumption needed for the manipulation of quantum systems with electric fields. The first such investigation involved an antiferromagnetic triangular iron-oxo cluster. In this case, competing antiferromagnetic Heisenberg and Dzyaloshinskii-Moriya interactions give rise to spin chirality and the observed magnetoelectric coupling observed as an increase in EPR intensity under the application of a direct current electric field (Figure 3-31; Boudalis, Robert, and Turek 2018). More robust results were obtained subsequently through alternating current electric field modulation of exchange interactions in helical metal-radical chains (Fittipaldi et al. 2019).

Very recently, an entirely new form of coherent pump-probe EPR spectroscopy has been developed by Liu and colleagues (2019) to study spin–electric coupling in molecular qubit candidates. The method is analogous to the pulsed ELDOR and DEER techniques, except that a pump electric field pulse is employed in order to modulate the coherent spin dynamics via the magnetoelectric effect. As highlighted in Section 2.3, perhaps the most exciting

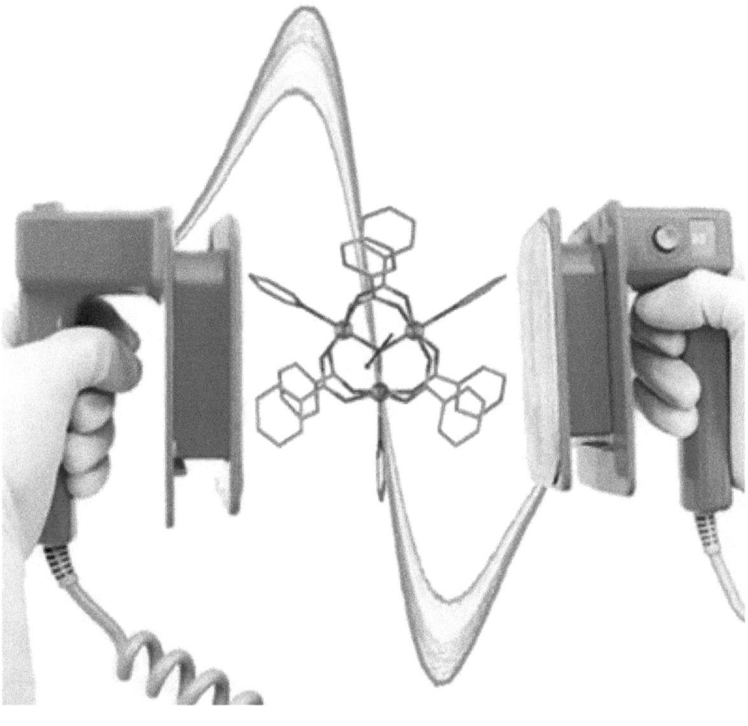

FIGURE 3-31 An electric field applied on the antiferromagnetic spin-chiral complex $[Fe_3O(O_2CPh)_6(py)_3]ClO_4{\cdot}py$ couples to the spin of its ground state and modifies its electron spin resonance spectrum. This magnetoelectric coupling can provide electric control of the spin states of molecular nanomagnets.
SOURCE: Boudalis, Robert, and Turek 2018.

example involves the application to clock transitions associated with a Ho(III) molecular spin qubit (Liu, Mrozek, et al. 2021). Crystals contain two electrically polar molecules related by inversion. Importantly, the structural distortion that gives rise to the electric dipole moment is also responsible for the symmetry breaking that induces the clock transition (the avoided level crossing between the Zeeman levels corresponding to electronic spin-up and -down states)—that is, the clock transition frequency is directly coupled to the molecular dipole moment. In turn, this implies linear coupling of an electric field to the clock transition frequency. However, the frequency shift is opposite for the two inversion-related molecules. Consequently, the application of an electric field pulse during the first half of a Hahn-echo sequence results in an eventual phase-shifted refocusing of the two populations at angles $\pm\phi$ relative to the magnetic pulses within the rotating frame, where the magnitude of ϕ scales linearly with the duration and amplitude of the electric field pulse. Thus, a periodic modulation and decay of the echo is observed in the in-phase echo signal, with no echo detected in the quadrature channel (due to cancelation of the $\pm\phi$ shifted echoes from each subpopulation). The decay is caused by the inhomogeneous response of the ensemble to the electric field pulse. Since these inhomogeneities are static, the spin echo is completely recovered if the electric field pulse spans the second (refocusing) half of the Hahn-echo sequence (see Figure 3-32). It is then possible to perform a 2D experiment, after which the electric field pulse is applied only during the refocusing magnetic π pulse while scanning the frequency of the π pulse. Refocusing is only effective for one of the subpopulations

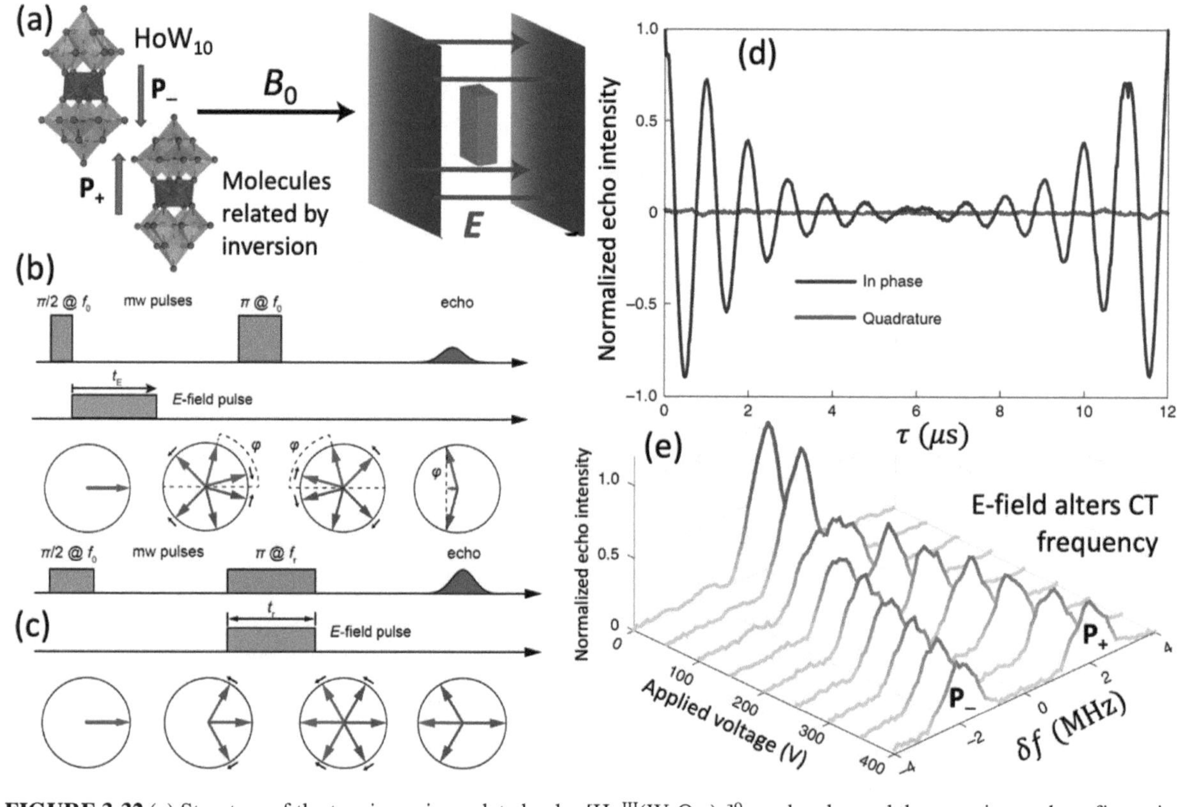

FIGURE 3-32 (a) Structure of the two inversion-related polar $[Ho^{III}(W_5O_{18})_2]^{9-}$ molecules and the experimental configuration in which the sample is subjected to a static applied magnetic field, B_0, and pulsed electric fields, *E* (in addition to microwave electron paramagnetic resonance pulses). (b, c) Sequences of microwave magnetic (yellow) and electric (blue) field pulses and the resulting evolutions of the magnetizations associated with the oppositely polarized molecules, leading eventually to an echo (red). (d) Temporal evolution of the in-phase and quadrature spin-echo signals for the pulse sequence (b); dephasing occurs due to the inhomogeneous response to the electric field pulse. However, this dephasing is refocused if the *E*-field pulse is kept on for the full duration of the Hahn-echo sequence. (e) Applying the *E*-field pulse during the magnetic π pulse (c) breaks the symmetry between the two electrical polarizations.
SOURCE: Liu, Mrozek, et al. 2021.

(i.e., half of the molecules) when the π-pulse frequency matches the shift caused by the electric field. Hence, the echo signal drops in intensity by a factor of two and shifts linearly both up and down in frequency in response to the magnitude of the electric field (see Figure 3-32). In this way, the electric field breaks the inversion symmetry inherent to the crystal, enabling selective excitation of the two inversion-related molecules. More generally, these studies clearly demonstrate coherent spin-electric control in molecular nanomagnets that exhibit long coherence times due to clock transitions (Shiddiq et al. 2016), paving the way toward local electrical control of molecular spin qubits at the nanometer scale with low dissipation (Ullah et al. 2022). These methods are now finding more general utility in the chemical study of magnetoelectric materials (Liu, Mrozek, et al. 2021).

3.6 EXTEND QUANTUM TELEPORTATION AMONG MOLECULAR QUBITS

Quantum communication is a rapidly growing area of QIS and engineering, with quantum network test beds appearing around the world. In contrast to classical communication, quantum communication takes place by sending or receiving information transmitted using the quantum states of a specific quantum system (e.g., a photon). Such "flying qubits" offer built-in security given that the (unwanted) observation of these states collapses the information. A quantum internet is being developed using entangled qubits to create a unique mode of information processing utilizing teleportation. In addition to security, a quantum network may enable quantum computing clusters, connecting multiple quantum computers to create more powerful machines for applications in science and engineering. Ultimately, one imagines a quantum ecosystem where sensors, communications (wiring), and computers exchange quantum information without exposure to the classical world. This new paradigm for communications requires the community to address a growing number of significant challenges, including the creation of quantum repeaters and robust quantum memories. These needs sit at the interface of fundamental physics and chemistry, representing exciting and important scientific research areas.

Today's internet is driven by the existence of repeaters that circumvent our current limits in materials science. Digital optical pulses travel through optical fibers until their amplitude is diminished through scattering processes in the glass. At that point, the signal is read, amplified, and repeated downstream. This process is repeated until the signal reaches its destination. Given the laws of quantum physics, quantum repeaters will require a different operational mode given the inherent information collapse upon observation. To that end, the community currently envisions new concepts such as exploiting and combining different quantum degrees of freedom, as well as mixing photons, phonons, magnons, and spins in an effort to amplify one mode in lieu of another. Devising and constructing atomically engineered materials that enable quantum transduction at the single spin–photon and/or single phonon–photon level may be greatly advanced through molecular chemistry, where advanced theory and synthesis may be used to manipulate local bond strengths and spin–orbit interactions, integrate single ion memories, tune optical levels, and design integration pathways with optical communication pathways (fibers). In addition to a high degree of spatial control and tunable emission energies, there are opportunities to multiplex quantum signals within single structures given molecular length scales and existing organic molecular frameworks. The remarkable precision, placement, and tunability of molecular qubits and potential memories make molecular quantum systems powerful candidates to accelerate research in science and engineering (Baek et al. 2005).

Use Molecular Systems to Teleport Quantum Information Over Distances Greater than 1 μm with High Fidelity

Control of the coherence times of quantum states is currently one of the major challenges in QIS (Wasielewski et al. 2020). Spin-state decoherence can be accelerated greatly by interactions such as spin exchange, hyperfine coupling, spin–orbit coupling, and magnetic dipolar coupling. By controlling the structure and composition of molecular qubits, many of these decoherence sources can be mitigated and/or controlled (Krzyaniak et al. 2015; Yu et al. 2016). For example, purely organic molecular qubits have the advantages of weak spin–orbit coupling and well-defined electron–electron and electron–nuclear spin exchange interactions (Nelson et al. 2017). Moreover, ultrafast photochemical electron transfer within an organic donor-bridge-acceptor (D-B-A) molecule can produce a radical pair that can function as two entangled spin qubits ($D^{\bullet+}$–$A^{\bullet-}$), giving rise to an entangled two-spin singlet or triplet state (Krzyaniak et al. 2015; Yu et al. 2016).

As discussed in Chapter 2, this strategy has been used to achieve electron–spin state teleportation. This is essential to preserve quantum information across long distances (Pirandola et al. 2015) in an ensemble of covalent D-A-R• molecules, in which it is possible to propagate the initially prepared spin state of a stable radical R• to $D^{\bullet+}$ (Pirandola et al. 2015; Rugg et al. 2019). It is desirable that future research focuses on developing molecular systems that can achieve high-fidelity quantum teleportation of electronic states on length scales approaching 1 μm. Such systems could thus serve as quantum interconnects in larger-scale devices. In addition, as discussed in the following section, the development of chemical approaches that take advantage of entangled photons is also of importance for quantum teleportation/communication over ever greater distances.

3.7 DEVELOP MOLECULAR QUANTUM TRANSDUCTION SCHEMES THAT TAKE ADVANTAGE OF ENTANGLED PHOTONS AND ENTANGLED ELECTRON AND NUCLEAR SPINS

Quantum transduction refers to the transfer of quantum states between different forms of quantum subsystems to connect qubits, support quantum networks, or allow for quantum sensing. Current quantum transduction schemes range from electron spins, nuclear spins, bosons, acoustic modes, and microwaves to optical photons. To interface with larger quantum systems or classical input-output schemes, electrical or photon-based transduction is often assumed optimal, but all areas are being explored. Molecules offer a unique approach to quantum transduction because of the tunable interaction between nuclear, rotational, vibrational, and electronic degrees of freedom. Molecular transduction schemes offer more degrees of freedom than ion systems while being more energetically discrete than solid-state approaches. The instrumentation described in this chapter must be developed and combined to probe these potential forms of quantum transduction, as transduction naturally includes measuring the entanglement between multiple degrees of freedom and their modulation simultaneously.

The classic example of quantum transduction is the attempt to couple superconducting transmon qubits by converting electrical or microwave signals into optical photons (flying qubits). While RF and microwave hardware likely will remain central to efforts aimed at coherent control of electron and nuclear spins in molecules, the lack of sensitivity inherent to conventional electromagnetic detection modalities in these frequency bands prompts the exploration of alternative transduction schemes. Optical detection is known to provide single-spin sensitivity in the case of the NV center in diamond (Chen et al. 2017), and efforts focusing on optical sensing of molecular spin qubits were described earlier in Section 3.4 and later in Section 3.10. Another approach allows single-spin detection (Thiele et al. 2014) via electrical readout of a single-molecule transistor (Gehring, Thijssen, and van der Zant 2019; Jo et al. 2006; Pietsch et al. 2016) or chiral-induced spin selectivity and spin chemistry. Entanglement between a nuclear spin and a rotational mode has been measured, creating potential transduction routes through vibrations. Similar coupling to a surface substrate, perhaps through magnons, is also a feasible proposal (Godejohann et al. 2020).

For long-range coupling, there are opportunities to research how spin qubits interact with entangled photons or how photon-based "flying qubits" can be made from molecular qubits. Unique opportunities exist for creating quantum repeaters as optical circuit elements using the inherently spin-controlled optical and vibrational processes of molecules (Dolde et al. 2013). For example, photogenerated electron spins with classical or entangled photons are of great utility because they can be prepared in well-defined quantum states. Quantum transduction between electron and nuclear spins can then be used for storage elements; nuclear spins are highly localized and have ~1,000 times smaller gyromagnetic ratios than electrons, making them less sensitive to their environment and therefore enabling coherence lifetimes orders of magnitude longer than electron spins (Maurer et al. 2012; Saeedi et al. 2013). However, in order to optimize molecular qubit systems that take advantage of the properties of both electron and nuclear spins, three important questions need to be answered through instrumentation development:

1. What are the limits of electron and nuclear spin coherence in designed molecular qubits?
2. How do electron–nuclear spin–spin interactions influence coherence transfer between electron and nuclear spins?
3. What combinations of microwave and optical photons are needed to optimize coherence transfer between multiple such subsystems?

For shorter-range coupling, chemistry enables the design of large molecular structures with atomic precision that can be used as local processing or sensing nodes. Hierarchical structures, such as metal–organic frameworks (Feng et al. 2020) and molecular crystals (Kothe et al. 2021), provide a platform for scaling up single-molecule qubits to functional multiqubit arrays. Molecular qubits can be connected through linkages that are only a few nanometers long, potentially leading to higher-density qubit arrays than those based on isolated atoms or defects. These scaffolds can facilitate the control of the entangling interactions between qubits, such as the spin exchange coupling between two electron spin qubits (Olshansky et al. 2019). Photoactive or switchable bridge molecules can provide a convenient platform for changing the sign and the magnitude of the magnetic exchange couplings between two spin qubits, enabling ultrafast state manipulation, changes in the polarization of individual qubits, and modulation of qubit entanglement. For example, using phenylene bridges, the strength of the coupling can be reduced by a factor of 50 through torsional distortions around the single C–C bonds joining the phenyl groups (Stasiw et al. 2015). Transduction through short-range dipolar-, electronic-, or vibrational-based routes within a dense array, outcoupled by a photon or electrical signal, can then allow for longer-range transduction.

In all of these cases, beyond synthesis needs, measuring quantum transduction requires new tool sets that can evaluate entanglement between multiple quantum subsystems simultaneously—for example, by combining an EPR measurement of spin coherence with a measurement of the emitted entangled optical photon. Control of a single or group of molecular qubits requires transduction, both locally and long range, and linkage to classical external feedback circuits. Methods to measure and realize molecular quantum transduction represent a significant scientific frontier.

3.8 ADVANCE QUANTUM SENSING TECHNIQUES TO FURTHER UNDERSTAND BIOLOGICAL SYSTEMS

As described earlier in this report, the area of quantum sensing is rapidly advancing with early proof-of-concept demonstrations that showcase its breadth. Recent advances in sensor technology suggest that quantum sensors have significant near-term potential. Chemistry is uniquely poised to transform this field. Sensors are inherently analyte specific and environmentally dependent. Developing designer quantum sensors is a challenge perfectly suited for synthetic chemistry. Deep literature exists on modifying chemical compounds for specific environmental compatibility. As an illustrative example, a key question for quantum sensing in biological systems relates to monitoring catalytic turnover in enzymes. Moving from imaging an ensemble of molecules to a single molecule or even hundreds would be transformative for our understanding of catalysis. Since the key steps in a catalytic cycle are very fast, statistically only a small number of molecules in an ensemble are in the critical state at the correct time. As such, rapid measurements that enable sub-ensemble size scales could elucidate the catalytic mechanism for important systems such as the oxygen-evolving complex at the heart of photosystem II. How would a chemical approach be suited to this? Initially, the target would be selected—for instance, readout of the spin state of the dangling Mn ion with light. We would need to use a molecular color center to execute the measurement with an emission frequency that was biologically compatible. Moving toward the lower energy frequency region of the electromagnetic spectrum is generally less biologically damaging and has better depth penetration. The first step would be to design a system suited to sense spin in this regime (Figure 3-33). Subsequently, we would need to introduce a molecule into a biological environment. The molecule would need to be small and biologically compatible, and ideally contain a chemical tether that would bind to the appropriate substrate. Tethering chemicals to biological systems is an old field, with DEER spin labels and fluorescence resonance energy transfer labels as canonical examples. For quantum sensing, there are additional design parameters beyond conventional sensing, coherence time, and temperature dependence (Figure 3-34). The tunability of molecules is very useful for designing these parameters.

3.8.1 Investigating Quantum Effects in Biology with Quantum Tools

Interaction between quantum physicists and biologists can be traced back to the beginning of the quantum era itself. However, as described earlier in Section 3.4, as the sensitivity and detail of experiments have increased, biologists have exhibited a growing interest and goal. Some biologists would like to follow the behavior of

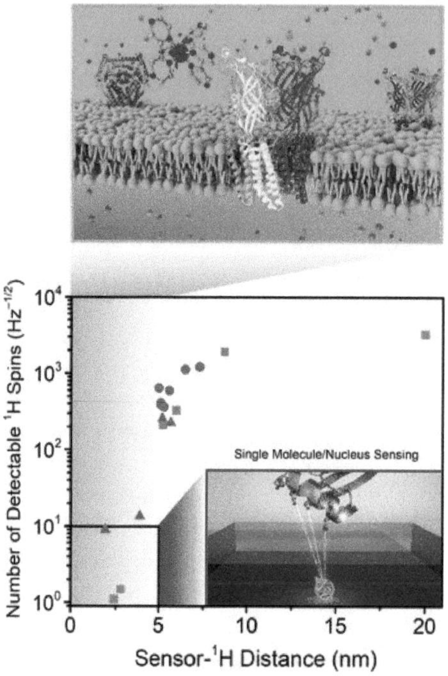

FIGURE 3-33 Calculated sensitivities of nitrogen-vacancy centers to ^{1}H nuclear magnetic moments quantified by the number of detectable ^{1}H spins using a published formula.
SOURCE: Yu et al. 2021.

biomolecules, especially enzymes, in real time. The difficulty arises when one needs to increase the light intensity to levels required to probe the important mechanistic molecules, risking changing their behavior—or even damaging them. In some cases, ordered arrays of quantum sensors that are individually addressable and densely packed in a small volume are used to produce hyper-resolved images or maps that would far exceed the diffraction limit or even room-temperature super-resolution. Creating such an array requires a rigidly organized collection of qubits that are individually optically addressable. These arrays can be organized into a molecular structure, within a single crystal, or in a larger cross-braced scaffold of many solid-state quantum nanoparticles. In other NMR-related experiments, different isotopes such as deuterium, ^{13}C, and ^{15}N are used. Imaging spectroscopies sensitive to molecular vibrations can be used to detect deuterium but have low sensitivity for ^{13}C or ^{15}N. However, ^{13}C and ^{15}N have nuclear magnetic moments that can be detected in conventional NMR spectrometers, raising the intriguing possibility of using high-resolution and high-sensitivity quantum-enhanced magnetometry for detection. Such a tool would enable, for instance, maps of candidate drug distributions through different tissues and, perhaps, even which organelles the drug localizes to, or maps of protein synthesis upregulation in the brain in response to learning or memory formation. To achieve spatially dense measurements with high time resolution using quantum sensors, one will need to integrate the measurements with complementary classical sensing modalities that can provide nanoscale chemical information across an entire cell. These methods will exploit the exquisite sensitivity of enzymes to the local biochemical environment.

3.8.2 Quantum Light and Biological Processes

Other scientists interested in this difficult issue regarding the measurement or tracking of enzymatic activity in biological systems with low light intensities have invoked quantum metrology as a solution. In a proof of concept, conducted by Cimini and colleagues (2019), NOON states were employed to observe and measure enzyme

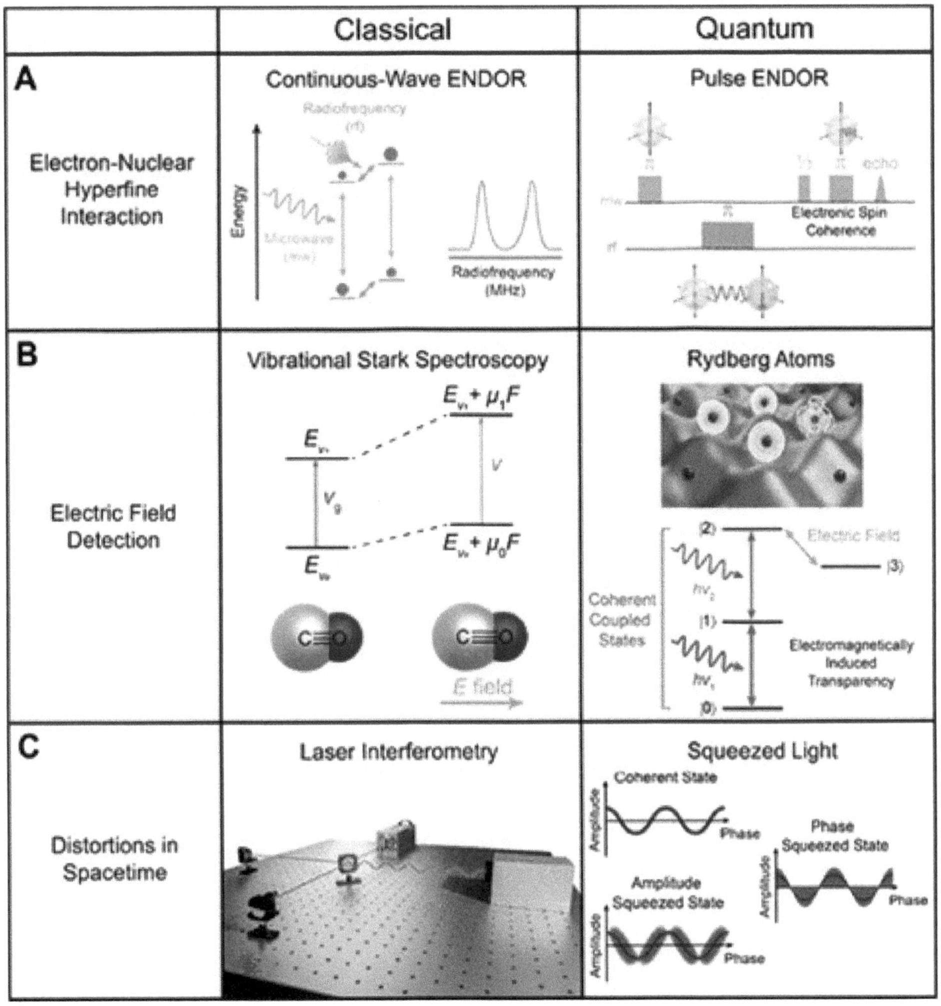

FIGURE 3-34 Examples of existing sensing protocols that leverage quantum mechanical properties such as coherence and entanglement to detect (a) electron–nuclear hyperfine interactions, (b) electric fields, and (c) space–time perturbations. SOURCE: Yu et al. 2021.

activity. First, N00N states that are in circular-polarization basis are created from single photons with orthogonal polarizations (using SPDC) that underwent interference through a polarizing beam splitter (Figure 3-35). This interference is an important step because it generates photons in superposed states, hence exhibiting indistinguishable spatial and temporal degrees of freedom, which allows the creation of the N00N states. After the desired photon state is generated, it is then passed through a solution containing enzymes undergoing catalytic activity. In this example, the enzymes are hydrolyzing sucrose and producing mixtures of chiral products. Due to the chiral states, the molecule's interactions with the photon creates a phase shift between the different N00N states. The extent of the phase shift is a function of the concentration of sucrose. Two avalanche photodiodes were used to measure these shifts. There are still challenges making this technique feasible for use (signal-to-noise detection ratio). However, this proof-of-principle work demonstrated researchers' ability to study biocatalytic reactions to track enzymes' activity in real time by taking advantage of different chiral systems and their distinct optical properties.

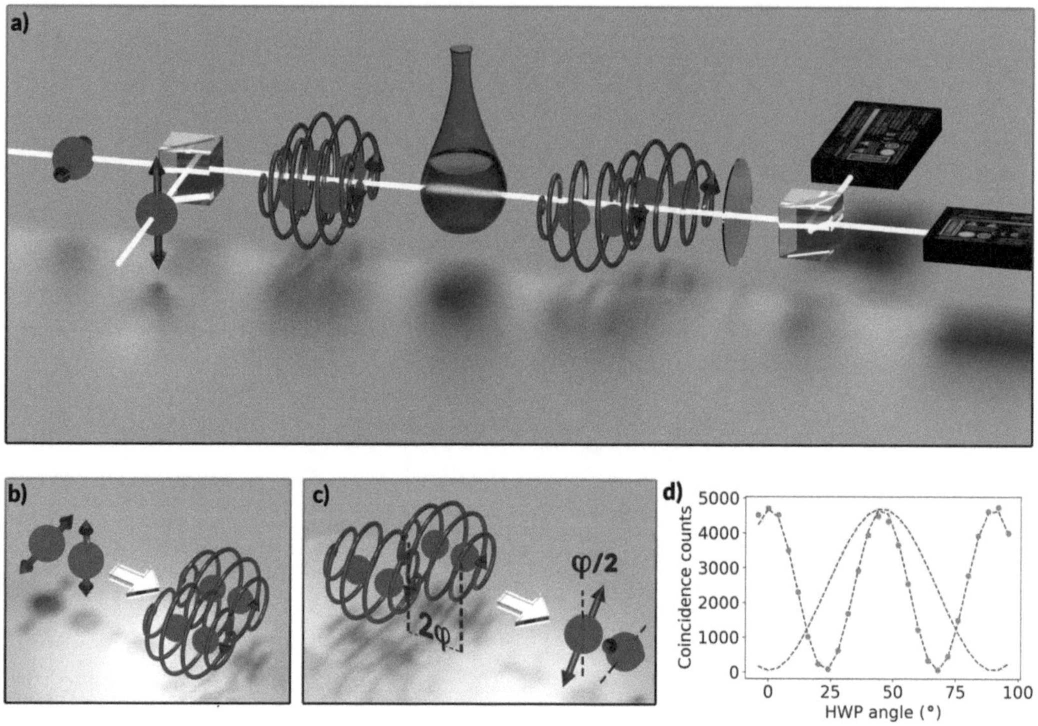

FIGURE 3-35 (a) Schematic of experiment. (b) Two photons produced via spontaneous parametric down-conversion with orthogonal polarizations are sent on a polarizing beam splitter (PBS) to produce a NOON state. (c) 2ϕ between the two components of the NOON state is introduced. When reverting to the linear polarizations, the phase shift corresponds to a rotation by an angle ϕ/2 on both photons. (d) A half-wave plate (HWP) and a second PBS is used to analyze the polarization. The photons are recorded as coincidence counts based on detection using avalanche photodiodes.
SOURCE: Cimini et al. 2019.

3.9 USE BIO-INSPIRED QUANTUM PROCESSES TO DEVELOP NEW QUANTUM TECHNOLOGIES

Strategies to develop platforms for quantum sensing can be found within the realm of biology. Quantum biology is the study of the dynamics of quantum dynamical networks in the presence of an environment. It suggests that biological phenomena such as photosynthesis, enzyme catalysis, avian navigation, and olfaction may utilize fundamental quantum mechanical processes such as coherence, tunneling, and entanglement. Early suggestions for the role of quantum mechanics in biology were made by Schrödinger (2012) and Fröhlich (1968), who proposed that coherence may be the basis of biological oscillators. More recently, superconducting qubit architectures have been used to explore quantum mechanical models of photosynthetic light harvesting inspired by 2DES experiments (Potočnik et al. 2018). The debate over whether quantum physics applies to biological systems centers on the view that biological systems can be seen as noisy and complex from a quantum mechanical perspective (Tegmark 2000).

The suggestion that consciousness depends on biologically relevant coherent quantum processes and that these quantum processes correlate with neuronal synaptic activity (Hameroff and Penrose 2014) has been proposed. Calculation of neural decoherence rates finds that while decoherence timescales are ~10^{-13}–10^{-20} sec, the relevant dynamical timescales (~10^{-3}–10^{-1}sec) for regular neuron firing are much longer, suggesting that the quantum coherence may not be relevant to neural network functions in higher-order organisms. Without a viable qubit specification, the role of quantum phenomena in higher-order neuronal processes remains controversial (Hameroff and Penrose 2014). Strong evidence for the role of coherence, entanglement, and superposition of states, however, has been found in biological photochemical reactions, sensory mechanisms (sight and smell), photosynthesis, and magnetoreception, thus providing fascinating models from biology as to how chemists might design and study quantum sensors for a range of environmental variables.

The role of quantum phenomena in photochemical systems, particularly photosynthetic reaction centers, is perhaps the most well studied. Quantum coherence dictates energy transfer processes in the pigment–protein complex of chlorophyll with decoherence times of 600 fs (Engel et al. 2007; Ishizaki and Fleming 2009; Sarovar et al. 2010; Squire et al. 2013). Coherence in light-harvesting and electronic energy transfer processes in biological systems has been observed on femtosecond timescales, and the phenomena have been modeled to explain the process from classical physics to quantum superposition (Chenu and Scholes 2015).

Important sensory mechanisms such as those of sight and olfactory responses have been proposed to have a quantum origin. Vision depends on the sensing of photons in rods and cones, and sensitivities as high as one to three photons (single-photon counting) have been reported (Rieke and Baylor 1998), which is relevant as inspiration for the development of optical quantum sensors. The sensitivity of olfactory responses to vibrational or phononic processes of molecular systems, beyond simple structural factors, suggests that vibrational modes in molecular structures for olfaction may play a role in sensing through a mechanism analogous to inelastic electronic tunneling spectroscopy (Turin 1996).

Magnetoreception refers to the ability of organisms to utilize external magnetic fields to sense direction (Wiltschko and Wiltschko 2005, 2012). Magnetic field sensing can be with respect to the north or south lines of the geomagnetic field (magnetotaxis) (Rismani Yazdi et al. 2018) or with regard to field strength and inclination involving some combination of the geomagnetic and paleomagnetic field. Magnetoreception has been demonstrated in hundreds of organisms to date, from the orientation of swimming for magnetotactic bacteria, to the orientation of hives for bees and ants, to tunnel direction for mole rats, and to migration over long distances by many species of birds, sea turtles, and invertebrates such as mollusks and crustaceans. The mechanism of magnetoreception can be light dependent, as was found in many species of birds (the "avian compass"), or light independent (sea turtles, bacteria). The light-dependent mechanism is fundamentally a quantum phenomenon that relies on the quantum spin dynamics of transient photoinduced radical pairs that are formed via photoinduced electron transfer between a putative acceptor and donor in a relevant protein (Hiscock et al. 2016; Zoltowski et al. 2019). Coherent spin dynamics between singlet and triplet states are influenced by an external (geomagnetic) field, leading to changes in the quantum yield of the signaling state of the protein. Significant evidence suggests that the critical acceptor–donor pair is comprised of a tryptophan and flavin cofactor found in cryptochromes (Maeda et al. 2012) and the kinetics and quantum yields of photoinduced flavin–tryptophan radical pairs in cryptochrome were found to be magnetically sensitive in field strengths akin to Earth's magnetic field (25–65 μT) (Maeda et al. 2008). The directionality associated with magnetic field detection, however, must include a contribution due to magnetic anisotropy in order to sense "direction" rather than just the magnitude of the Zeeman splitting of states. Neither the nature of the anisotropy (e.g., *g*-anisotropy and hyperfine, on a quantum mechanical level) nor how it is translated within a biological environment to "directionality" are understood. The light-independent mechanism of magnetoreception is a more classical mechanism; biomineralized magnetite Fe_3O_4 or greigite Fe_3S_2 sense changes in the direction and magnitude of external fields to initiate signal transduction pathways. The pathways are not well understood but are believed to involve mechanosensitive ion channels that translate the mechanical motion of the magnetic nanoparticles through the protein matrix to initiate signal transduction. Other magnetic sensing mechanisms based on electromagnetic induction have been evidenced in fish and sharks (Kirschvink, Walker, and Diebel 2001). The mechanisms provided through magnetoreception provide interesting models for the design of chemical systems for magnetic field sensing.

3.10 PROVIDE BROADLY ACCESSIBLE STATE-OF-THE-ART MEASUREMENT TECHNIQUES AND INSTRUMENTATION FOR THE CHEMISTRY COMMUNITY

3.10.1 Emerging Tools and Needs

As the field of chemistry grows into QIS, the tool set already developed for more traditional QIS experiments must be adopted and expanded. In particular, new tools would benefit the following categories: (1) single-molecule qubit measurement, (2) control and detection of ensembles of molecular qubits, and (3) advanced spectroscopy of entangled and classical interactions for molecular qubits. This section discusses these instrumentation classes as emerging tools and needs.

3.10.1a Single-Molecule Qubit Measurement

Currently, molecular qubits and quantum systems are mostly studied as ensembles. While this approach is sufficient for measurements of dephasing or ensemble quantum states, as described in previous sections of this report, a need exists to measure quantum information for individual molecules, between individual parts of a molecule, and between a few molecule systems. The tools used for atomic and ion systems simultaneously allow the isolation of single particles or control over multiple particle systems and their interaction Hamiltonians. The ability to isolate quantum dynamics has led to significant scientific advances in atomic and ion systems, but a suitable technique has yet to be adapted for molecular qubits beyond a few atoms (Chou et al. 2017; Leibfried 2012; Lin et al. 2020; Loh et al. 2013; Shi et al. 2013; Wolf et al. 2016). A similar tool set would benefit molecular systems, where new interactions are possible by chemical synthetic control of qubits and control of multiple molecule systems.

It is important to develop methods for measuring single molecules within an ensemble. This category can be further broken down into the study and readout of the qubit or quantum sensors versus the characterization of molecular properties of the qubit or quantum sensor that lead to its successful operation. Molecular qubits that can be optically read out represent a relatively accessible measurement for most spectroscopy groups. Measuring single-molecule fluorescence is a well-established technique in chemistry and biological imaging (Lelek et al. 2021; Shashkova and Leake 2017). Following the development pathway of the NV center field, molecules can be dispersed on a surface and their fluorescence isolated using a variety of microscopy approaches (Scholten et al. 2021). This has been done, for example, with optically addressable molecular qubits, and a setup is shown in Figure 3-36 (Bayliss et al. 2020). However, further instrumentation development would help a synthetic chemist, for example, rapidly scan candidate molecules. The optics of Figure 3-36 would need to be simplified into a bench-top *entangled* spectrometer, which is an avenue worth exploring if rapid advances in chemistry for QIS are to be made.

Isolating and measuring microwave active molecular qubits is a more challenging task because the microwave wavelength (millimeter and longer) is far from that of visible light (nanometer) and is incommensurate with the size of molecules. A need exists to adapt current atomic resolution methods to be suitable for molecular qubits. Atomic resolution instruments fall into two main categories: scanning probe microscopy and electron imaging techniques. While electron imaging methods like transmission electron microscopy (TEM) have the atomic resolution needed to characterize even small molecules (through cryo–electron microscopy and microcrystal electron diffraction) (Jones et al. 2018), it is more challenging to extract quantum-state information; however, future developments like Lorentz TEM of magnetic fields could prove transformative. On the other hand, scanning probe microscopies, which include instruments like an atomic force microscope and an STM, are already used to measure single-molecule magnetic properties as well as to read out their spin states (see Choi [2019] and Moreno-Pineda and Wernsdorfer [2021] for a review).

Scanning probe techniques can also be adapted to perform nanoscale optical and microwave experiments. Generically speaking, near-field scanning probe microscopes use resonant apertures of various forms, such

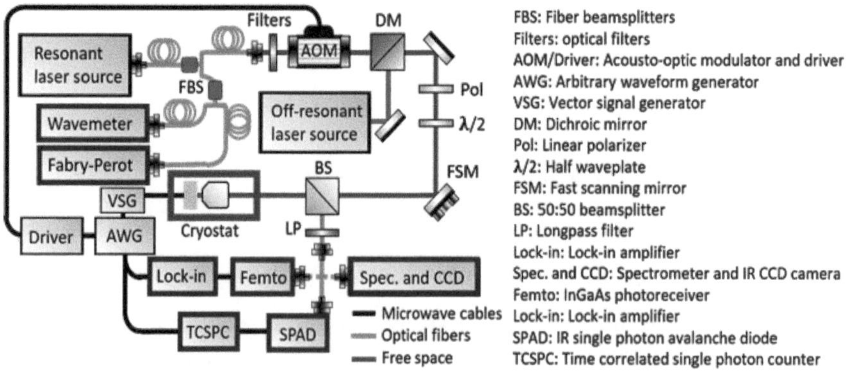

FIGURE 3-36 Experimental setup for characterizing all optical qubits.
SOURCE: Bayliss et al. 2020.

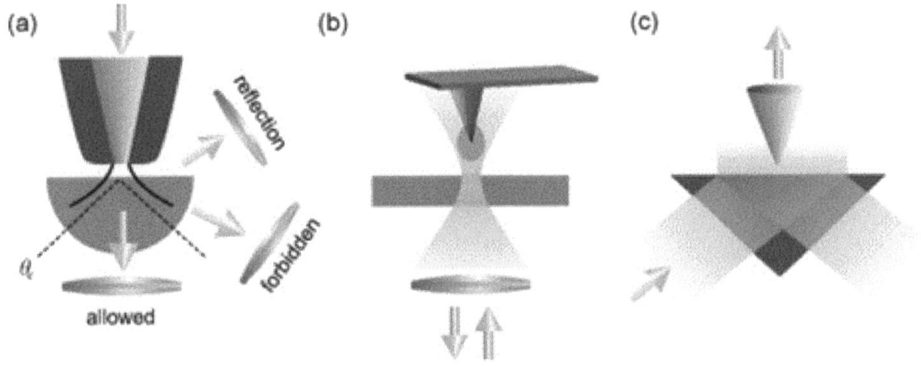

FIGURE 3-37 Different types of scanning near-field optical microscopes: (a) aperture scanning near-field optical microscopy with angular resolved detection, (b) apertureless configuration, and (c) scanning tunneling optical microscope.
SOURCE: Hecht et al. 2000.

as metal tips, to localize an incident electromagnetic field to a sub-wavelength regime (Figure 3-37) (Anlage et al. 2001; Farina and Hwang 2020; Hecht et al. 2000; Mauser and Hartschuh 2014; Rosner and van der Weide 2002; Seo and Kim 2022). Common abbreviations for the techniques are near-field scanning optical microscopy for visible light to terahertz characterization and near-field scanning microwave microscopy for the long-wavelength microwave version. The main difference is the source: visible to terahertz light can be generated by nonlinear frequency generation from a laser, whereas microwaves are created using electronic signal generators and antennas or waveguides when needed. Scanning near-field techniques can reach tens of nanometers in resolution to isolate single molecules or look at interactions between nearest-neighbor molecules. The methods are also easily converted into transient measurements, wherein an optical or microwave pump is used to initiate an excited state of the molecular qubit and then another wavelength region is used as the probe. This allows for measurements of the underlying dephasing mechanisms that are controlling the molecular magnet or qubit operation.

While not as elegant as ion trap measurements, scanning near-field microscopies will be an emergent tool for molecular QIS systems. Multiple companies now sell instruments adaptable for chemistry in QIS work. Depending on the complexity and options selected, the instruments range from the hundreds of thousands to the million-dollar range. The instruments also have multiple applications and therefore can be invested as user facilities. Cutting-edge methods that use STM feedback circuits are now also allowing for sub-nanometer resolution, even with microwave frequencies, and are emergent tools that allow true quantum readout between different spin centers of a molecule (Imtiaz, Wallis, and Kabos 2014). Furthermore, by combining scanning probe microscopy methods with entangled photon spectroscopy, how molecules control or rely on entanglement can be directly measured on the atomic scale.

Combined, the application of single-molecule characterization tools to chemistry in QIS can (1) measure the quantum state of a molecular qubit after being initialized and its decoherence time; (2) use steady-state and ultrafast characterization techniques to measure spin-lattice, spin-spin, and other interactions to understand decoherence mechanisms; and (3) repeat these measurements with nearby molecular qubits to understand coupling mechanisms. These types of measurements would give clues as to how to delineate and achieve multiple qubit interactions in molecular quantum sensors and other QIS systems.

3.10.1b Control and Detection of Ensembles of Molecular Qubits

The application of molecular qubits in quantum scenarios requires the delineation of interactions in an ensemble. Multiple molecular qubits, molecular magnets, or molecular sensors will need to be addressed individually and measured in a complicated environment. While some differentiation can be achieved through synthetic means, as demonstrated by sensing with diamond and other vacancy centers, emerging instrumentation will also play a dominant role.

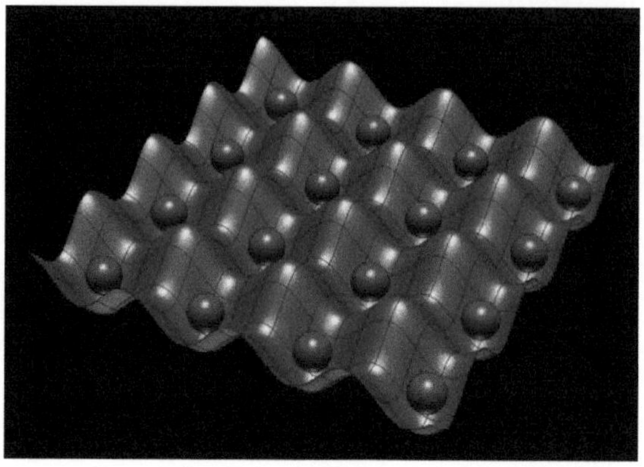

FIGURE 3-38 Schematic of making individual spin qubits controllably interact through the use of an optical lattice approach.
SOURCE: NIST 2009.

First, the topic of constructing defined ensembles of molecular qubits will be discussed. Synthetic methods discussed in Chapter 2 can be used to specifically anchor multiple qubits in ordered fashions; however, more artificial instrumental methods like optical trapping and scanning probe microscopies should also be mentioned. Scanning probe microscopy, the base technique from the previous section, has famously been used to place and order atoms and molecules on a substrate (Gross et al. 2005; Hla and Rieder 2003), even to the point where molecular logic gates have been built (Leisegang et al. 2021). While a tedious technique, copious research has been put into this topic, and artificially creating arrays of molecular qubits could provide a fruitful equivalent to an atomic ion trap. Optical forces can also be used for molecular manipulation. For example, optical tweezers take advantage of the response of a dielectric sphere illuminated by laser light for high levels of precision of molecular control (Bustamante et al. 2021; Moffitt et al. 2008). Applications of optical tweezers were discussed in Section 2.2.2. The downside of this approach is that the molecules must be attached to dielectric spheres in order to provide levitation or multiple ordering. Optical lattices aim to overcome this issue by using complex arrays of interfering laser beams. They are commonly used in atomic, ion, and few-atom molecule work in QIS, although larger molecules are still a technological frontier (Chou et al. 2017; Leibfried 2012; Lin et al. 2020; Loh et al. 2013; Shi et al. 2013; Wolf et al. 2016). Instead of using a dielectric sphere, the system is cooled to ultralow temperatures using a laser or similar methods and then patterned using a complex array of beams with constructive and destructive interference (Figure 3-38). Both optical tweezers and lattices can be considered critical to chemistry in QIS growth because they allow a more free-form and adjustable ordering than is possible with surface manipulation.

Second, instead of trying to create arrays of single molecules, an alternative approach is to isolate a specific qubit(s) within an ensemble. This is especially important for quantum sensing, one of the recommended research priorities of the report, where multiple near-identical qubits will be present in complex physical and biological environments. Inspiration can again be taken from the field of biological imaging, which already addresses this exact challenge. Complex biological microscopes are often already available in commercial forms at campus user centers and are therefore interesting to explore. For example, multiphoton imaging uses nonlinear optical properties to create three-dimensional (3D) images with molecular tags and tagless technologies. Multiphoton microscopy works by overlapping two or more pulsed laser beams within a sample (Hoover and Squier 2013; Lecoq, Orlova, and Grewe 2019). When overlapped in time and space, a multiphoton process, such as two-photon fluorescence, can occur. As the beams are moved throughout a sample, a 3D image is produced, often with up to 1 mm penetration depths and hundreds of nanometers' resolution (Figure 3-39). Light sheets can also be used for rapid, real-time image acquisition. When combined with super-resolution and stimulated depletion techniques, tens of nanometers' spatial resolution is possible (Moneron and Hell 2009). An emergent area of research would be adapting these techniques to isolate and optically address molecular qubits. This would require instrumentation development as well as the creation of appropriate molecular qubit tags that respond to nonlinear or multiphoton activation.

Alternatively, instead of designing qubits to react to multiphoton excitations, tagless methods like stimulated Raman scattering are also now reaching the <100 nm spatial resolution barrier (Li et al. 2021; Qian et al. 2021;

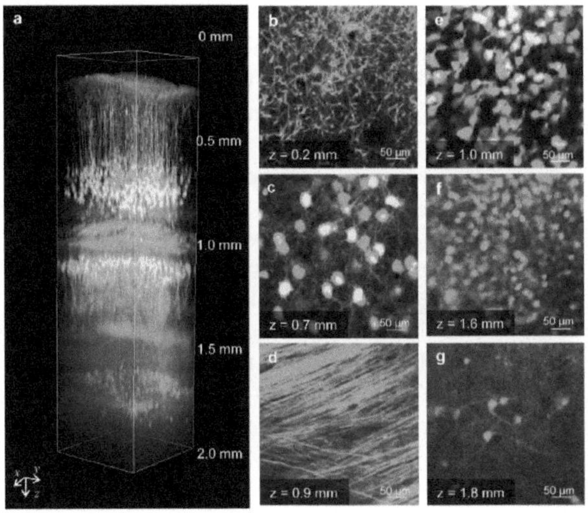

FIGURE 3-39 A rapid optical clearing protocol using 2,2′-thiodiethanol for microscopic observation of a fixed mouse brain. Multiphoton microscopy techniques developed in biology could potentially be used for addressing spin qubits in an array or for measuring multiple molecular qubit sensors in an environment.
SOURCE: Aoyagi et al. 2015.

Shi, Fung, and Zhou 2021). In this case, a molecular qubit would be imaged by its unique Raman spectral profile. Since Raman scattering can measure individual bonds, methods may be developed that can measure decoherence indirectly by vibrational dynamics or, more challengingly, directly through their fine structure. Other nonlinear microscopy techniques, like second and third harmonic generation or sum frequency generation, are also sensitive to local polarizations and vibrational populations and can be used as label-free methods for molecular qubit characterization and interaction (Lim 2019; Rehberg et al. 2011; Wang and Xiong 2021).

3.10.1c Advanced Spectroscopy of Entangled and Classical Interactions for Molecular Qubits

Previous sections have focused on ways to measure or control individual molecular qubits or networks thereof. However, instrumentation development on the ensemble level presents an opportunity to help understand how molecular qubits operate and transduce quantum information. Although microwaves have been used for the control and readout of most molecular qubits to date, how a molecule decoheres and its quantum coupling are often explored using ultrafast lasers. As mentioned throughout this report, techniques like sum frequency generation and stimulated Raman scattering can be used to understand time domain information about bond vibrations; the optical Kerr effect can be used to measure spin polarizations; and multidimensional spectroscopy (2DIR and 2DES) or transient EPR can be used to create and measure molecular spin center coupling within a molecule.

Further development of existing nonlinear optical techniques to operate in conjunction with magnetic fields, microwave access, and optical readout of spin will advance the field. Conventional pulsed EPR cannot measure rapid electron spin relaxation (less than ~5 ns) and does not provide the spatial resolution of optical techniques. Ultrafast spin relaxation processes have been studied in spin-½ systems using all-optical methods such as Faraday rotation to measure optical induction of ground-state magnetization dynamics and subsequent observation of quantum-beat free-induction decay (Furue et al. 2005). While long T_1 and T_2 times are required for quantum information processing applications, measurement of ultrafast spin dynamics still enables fundamental insight into molecular spin–phonon coupling and decoherence mechanisms. A particularly powerful way to explore these mechanisms that need to be developed is probing variable-temperature, variable-field regimes that are typically inaccessible to EPR and AC magnetometry.

Although ensemble based, these techniques remain prominent because high-purity synthesis products can be created. The methods of this section are more mature and accessible, but continued development could further simplify the tools to the level that a synthetic laboratory could use them for rapid characterization without needing a laser spectroscopist collaborator. While these collaborations are highly encouraged and productive, they create a barrier for universities and laboratories that do not have the funds for complex laser tools. **This report concludes that the development of new specialized techniques is needed, but simultaneous efforts to simplify and lower the cost of existing nonlinear or entangled spectroscopy techniques are critical for accelerating molecular qubit design rules.** These developments will also increase ready access to user facilities housing the needed tools.

3.11 IMPROVE INFRASTRUCTURE FOR CHEMICAL MEASUREMENTS IN QIS

Physical resources are required to support the range of established and emerging tools for the experimental characterization of quantum behavior discussed in this chapter. This section reviews available infrastructure in the United States today and makes recommendations for the prioritization of future resources to advance the field of QIS.

This section divides user facilities into three main categories that are based on the scale of the infrastructure. Box 3-1 briefly defines these categories as they are used in the remainder of this section. Although the National Science Foundation (NSF) has established funding ranges (in dollars) that are defined using similar terms, the committee acknowledged the continuously changing costs of various techniques and elected not to assign dollar values to these scale definitions.

Another key element of establishing and maintaining a strong infrastructure for QIS chemistry research in the United States is workforce development. The cultivation of an educated, well-equipped human resources pipeline to support the maintenance and operation of user facilities is discussed in Chapter 5.

3.11.1 Laboratory-Scale, Single–Principal Investigator Centers

A need exists to develop chemistry-oriented QIS laboratories if the field is going to progress. As of now, most single–principal investigator (PI) laboratories do not have access to the basic tools needed to characterize microwave qubit interactions, such as transient EPR, nor the ability to characterize optical qubits. Most chemistry in QIS research has been accomplished by the partnering of synthetic chemistry laboratories with physics or physical chemistry laboratories that specialize in quantum optics or transient EPR techniques. However, these collaborations do not exist at every university and present a barrier to development. Standardized and commercial instruments need to be developed for basic qubit characterization, especially for measuring properties like coherence times or spin-coupling interactions in multiqubit architectures, to allow any synthetic laboratory access to chemistry in QIS research.

Even for quantum optics and optics laboratories, a need for investment remains. Magnetic field and microwave generators are not commonly associated with ultrafast laser optics, and, at the same time, advanced optical capabilities are rarely integrated with transient EPR spectrometers. These capabilities, along with cryostats, represent significant investments. Physical chemists have developed a broad range of nonlinear and linear spectroscopies that can measure almost all of the details of the molecular interaction Hamiltonian. However, these techniques are not yet optimized for measurements of molecular qubits. Ultracold atoms and optical lattice research is currently focused in the physics community, and extension to molecular systems beyond a few atoms will enhance this area of research greatly.

While custom instrument development is needed, multiple tools can already be purchased commercially and adapted to the chemistry in QIS tasks. For example, the multiphoton and tagless Raman imaging microscopes

BOX 3-1
User Facility Scales and Definitions

- **Laboratory-scale or single–**principal investigator (PI) centers: facilities that can operate based on resources provided by a single grant led by one PI.
- **Mid-scale centers:** facilities that serve particular regions of the United States, large enough to require multiple PIs but small enough to be contained within one university.
- **Large-scale centers:** major facilities, such as national laboratories, that may use proposal application processes to prioritize and schedule instrument time for users from visiting institutions.

developed for biology can be used to isolate and measure molecular quantum sensors in complex environments. Optical tweezer systems can be purchased for isolating single-molecular qubits and exploring factors like strain and stiffness on the molecular qubit's dephasing time. Near-field scanning probe microscopies like NSOM and NSMM have reached maturity and can be used for single- and multiple-qubit interaction studies with atomic resolution. Scanning probe methods can also be used to arrange molecules on surfaces. Transient EPR is a dominant tool that is commercially available, albeit at a high cost and on a limited basis. All of these instruments represent significant investments in the hundreds of thousands to several-million-dollar range, but they can be leveraged for multiple experiments across multiple departments to help their adoption.

Driven by communications industries, rapid advances in microwave technologies are such that the hardware employed in transient EPR spectrometers is becoming more advanced, more widely available, and more affordable. Indeed, perhaps with the exception of the magnet, the components needed to assemble quite advanced transient EPR spectrometers can be acquired at a fraction of the current cost of a fully assembled turnkey instrument. Of course, most research groups do not possess the technical expertise to build their own transient EPR spectrometers. Nevertheless, the increased need for these capabilities motivates creative solutions to the significant financial barrier that currently limits the availability of such systems.

There are encouraging signs that several new startups, often supported by Small Business Innovation Research–type federal funding, are entering this scientific space. In turn, this is leading to exciting new routes to reduce costs such as Open Source Hardware,[1] where all relevant information required to produce a given hardware item can be posted and made available to the wider research community at no cost. Indeed, one can already find examples of extremely low-cost components relevant to transient EPR spectroscopy. Another area of rapid progress is the development of advanced on-chip EPR spectrometers that employ tiny permanent magnets (Hassan et al. 2021, 2022; Künstner et al. 2021; Lotfi et al. 2022), with the main spectrometer hardware often having a footprint no larger than a thumb drive. Such systems offer the added advantage of exceptional sample sensitivity. Although still developmental, these instruments are starting to demonstrate some of the capabilities of commercial spectrometers and could soon become commercially available at low cost. Therefore, the prospects for more widespread access to transient EPR capabilities over the next 10 years are promising and should be encouraged by funding agencies, although it is clear that these capabilities will remain somewhat limited in comparison to the most advanced systems (see below). It will be important to ensure that issues related to precision, reproducibility, and safety are addressed if such low-cost instruments become more widely available. However, it should be noted that the international EPR community successfully polices such issues upon the emergence of any new hardware or methodology.

3.11.2 Mid-Scale University User Facilities and Centers

While the prospects for low-cost hardware described in the previous section are exciting, there will always be a need for high-end instrumentation, as well as one-of-a-kind spectrometers offering truly unique capabilities, developed by the leading instrument builders. State-of-the-art commercial spectrometers are essential for benchmarking promising molecular systems and enabling measurements (e.g., of coherence and weak spin–spin interactions) that are simply not possible by other means (Kundu et al. 2023). As noted in a recent article in the International EPR (ESR) Society newsletter (Gallez 2022), "Those universities with modern pulsed EPR systems are currently swamped with demand." However, there are significant barriers to accessing instrumentation in the leading EPR laboratories. Typically, these instruments are operated in support of targeted research programs with relatively narrow scope, and access is limited to collaborators on these projects. The best way to remove these barriers is to locate high-end instrumentation and promote the development of new methodologies (e.g., those discussed in Section 3.11.3) at national user facilities. These organizations are truly open to the wider research community through merit-based proposal processes that are reviewed externally. They support world-leading staff scientists who can optimize the measurement time even for the most inexperienced user (from running the measurements to the data analysis, interpretation, and drafting of manuscripts); they educate their users, thus greatly

[1] The website for Open Source Hardware is www.oshwa.org, accessed April 27, 2023.

expanding intellectual resources within the wider chemistry community; and they enforce policies related to, for example, FAIR data principles.

Although several EPR facilities are scattered across the United States (see Figure 3-40), most of them are supported by the National Institutes of Health, and their mission is focused on biomedical, biochemical, and biophysical research (e.g., Cornell [ACERT]; University of Wisconsin, Milwaukee [National Biomedical EPR Center]; Miami University in Ohio [Ohio Advanced EPR Laboratory]; and the University of Denver [EPR Center]). These facilities generally are not open to chemists working on QIS problems, even though a handful of experiments have been performed using instruments located at these centers (Zadrozny et al. 2017). The Department of Energy also supports facilities focused on environmental and energy-related research (e.g., Argonne National Laboratory, National Renewable Energy Laboratory, Pacific Northwest National Laboratory, and University of California, Davis). The only U.S. EPR user facility that currently supports research at the interface between chemistry and QIS is the NSF Division of Materials Research–funded NHMFL, where the majority of experiments are conducted at magnetic fields and frequencies that are considerably higher than those of commercial instruments (Baker et al. 2015; Morley, Brunel, and van Tol 2008). However, users of the facility also have access to commercial X- (9–10 GHz) and W-band (94 GHz) transient spectrometers as well as a high-power W-band transient spectrometer developed at St. Andrews University (Cruickshank et al. 2009), resources that are highly valued by the user community. The facility is also staffed with internationally renowned EPR support scientists. Therefore, it serves as a model facility for carrying out research on molecular spin qubits of the kinds discussed in Chapter 2, with more than two-thirds of its users drawn from the chemistry community. It is a resource not only for U.S. scientists but also for international researchers, with about 50 percent of its users coming from overseas—particularly from Europe, where a large and active chemistry QIS community already exists. Indeed, the strong demand from overseas users is a testament to the world-leading stature of the NHMFL EPR facility, as similar user centers exist in Europe and Asia. The downside is that limited resources (both in terms of staffing and instrumentation) are making it more difficult for the facility to keep up with user demand, with considerable growth in activity deriving from chemistry-inspired QIS research.

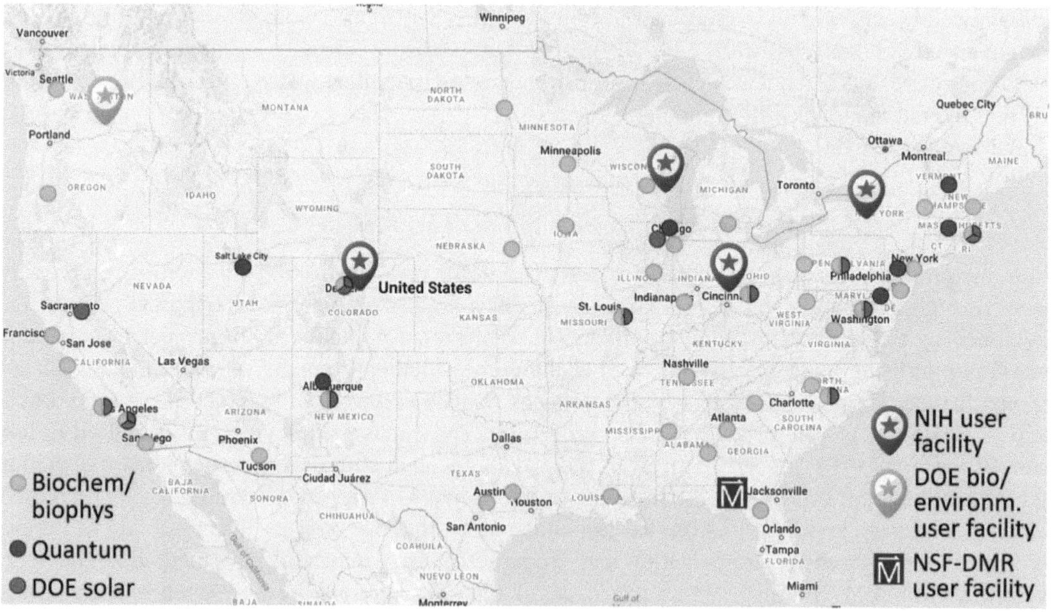

FIGURE 3-40 Locations of electron paramagnetic resonance (EPR) groups, centers, and user facilities in the United States. The pins mark National Institutes of Health (NIH) and Department of Energy (DOE)–funded user facilities that primarily support biochemical and biomedical research. The only major user facility directly supporting QIS research is the National Science Foundation (NSF)–funded National High Magnetic Field Laboratory in Florida. The solid circles denote locations with active EPR research groups.

It is vital for U.S. competitiveness in the growing research area at the interface between chemistry and QIS that the NHMFL and related user facilities in the United States remain at the cutting edge and not lose ground to similar centers overseas. To do so will require continual renewal of instrumentation at these centers. This applies not only to the major infrastructure (e.g., magnets and beamlines; see Section 3.11.3) but perhaps more importantly to the user end-stations (i.e., mid-scale instruments), where rapid technological advances can lead to the situation where a newly developed capability can become uncompetitive (or even obsolete) within a time frame of just a few years. A perfect example is microwave amplifiers and sources where EPR experiments performed today simply were not possible just 10 years ago (Subramanya et al. 2022). Maintaining adequate staffing at these facilities is also vital because of the critical role these people play in much of the research output and the education of users, including junior and early career scientists. Indeed, staff expertise is one of the key factors that fuels the world-leading status for many of the major user facilities in the United States, even though staff are often stretched very thin. Meanwhile, increased efforts should encourage the translation of novel capabilities developed in single-PI laboratories to user facilities, where they can be made available to the wider chemistry community via the knowledgeable staff at these centers. Examples include optical excitation and detection, and integrated scanning probe techniques.

Aside from major user facilities, a strong case can be made for establishing regional user hubs that support mid-scale instrumentation, which the chemistry QIS community relies on for advancing its research—for example, instrumentation that would be out of reach for all but the most research-intensive universities. Indeed, the natural place to locate these hubs would be major R1 academic institutions that have a critical mass of faculty and researchers with relevant expertise. However, it would also be important to staff these hubs with independent research scientists who would primarily serve the users of the instrumentation, in much the same way that a university funds staff to support campus instrumental needs. The hubs would need regular oversight by external entities in order to avoid situations in which the institution absorbs much of the added instrumental capability in support of its own interests. One could imagine national umbrella organizations that handle user applications to the national facilities and regional hubs, a model already employed among the European network of magnet laboratories.[2] National facilities already know how to run user programs, and their insight and oversight could be a model for success of a distributed network of user hubs. This model would alleviate some of the increasing demand at existing national centers, while allowing these major facilities to focus on their core mission (e.g., a focus on high-field measurements at the NHMFL). The addition of hubs would also allow for anticipated growth in research at the interface between chemistry and QIS.

Finally, it is vital to U.S. competitiveness in the area of chemistry QIS research that a steady funding stream exists for the development of truly unique methodologies that overcome the limitations of commercial instrumentation. Given the explosive growth in research at the QIS and chemistry interface, the need for such new developments is becoming increasingly urgent, to the extent that it represents a rate-determining step toward future progress. As an example, most of the current work on spin qubits focuses on single-molecule coherence, whereas the next frontier clearly involves the exploration of entanglement and potential implementation of quantum gates involving multiple qubits (Chiesa et al. 2020; Ferrando-Soria, Pineda et al. 2016; Godfrin, Thiele et al. 2017; Hussain et al. 2018; Ullah et al. 2022). Short-term needs (on a timescale of three to five years) for realizing much of the vision laid out in this chapter are clear. Although recent advances in high-end commercial spectrometers have been breathtaking, such instruments are designed with a broad range of applications in mind (i.e., they are not optimized for specialized chemistry QIS tasks). In particular, bandwidth limitations restrict EPR studies to relatively narrow frequency ranges centered on just one of the various microwave bands (e.g., the L-, S-, X-, and Q-bands at 1, 3, 9, and 34 GHz, respectively). Because of the rapid development of microwave technologies, wideband NMR-type EPR experiments are now becoming routine at frequencies in the 10–50 GHz range, with experiments up to 400 GHz not lagging far behind (Subramanya et al. 2022). As noted in Section 3.1.3, instruments are urgently needed that combine transient EPR and NMR capabilities at multiple widely separated frequencies (from tens of megahertz to hundreds of gigahertz) with optical and other modes of excitation and detection. This will enable qubit initialization and excitation of multiple electron and nuclear spin transitions either within the

[2] The website for the European Magnetic Field Laboratory is https://emfl.eu, accessed April 28, 2023.

same molecule (e.g., a qudit) or among multiple coupled spin centers, thereby facilitating execution of multiqubit gate operations, initially among molecular ensembles. In the longer term (5 to 10 years), efforts would be directed toward arrays of molecules on surfaces. The costs for such developments will be considerable, likely requiring new source/amplifier technologies that extend beyond the current 5G frequency band. Therefore, design criteria must be carefully formulated and based on a viable chemical approach, involving teams of instrument builders, engineers, and synthetic QIS chemists. Significant benchmarking of proposed molecular species represents an important first step in the design process so that future instrument developments do not fail simply because the molecules do not perform as expected. The scale and complexity of these development projects are comparable to those typically associated with national facilities. Moreover, staffing requirements, both during development and upon deployment, suggest that these projects be managed through national facilities. This will ensure availability of necessary expertise, project management, and maximum eventual access.

3.11.3 Large-Scale User Facilities

The committee did not identify any need for new large-scale national facilities. However, barriers to access were clearly identified. Current facilities are oversubscribed, and clearly a need exists to support efforts that increase access—for example, by creating regional instrumentation hubs that relieve national facilities of some of their more routine user demand, increasing availability of magnet/beam time at national facilities to the chemistry QIS community, and developing new beamlines and resonance magnets. However, as noted in the previous section, available user end-station instrumentation also represents a major current barrier to access at national facilities. Therefore, a steady funding stream is needed to develop and sustain support for chemistry QIS instrumentation at national facilities, as has been the case for other large-scale entities such as synchrotrons.

3.11.3a Magnet Laboratories

As noted in earlier sections of this report, strong spin–orbit coupling and ligand field effects can give rise to situations where metal-based spin qubits are EPR-silent at the conventional low frequencies (<50 GHz) and fields (<1.5 T) employed in commercial spectrometers, thus necessitating measurements at high magnetic fields and at frequencies stretching into the terahertz and far-IR regimes. The same is true for NMR, where the most advanced spectrometers capable of targeting quadrupolar nuclei spanning much of the periodic table operate at magnetic fields in excess of 30 T (Gan et al. 2017). For this reason, a large body of work in the molecular QIS field has been performed at large-scale user facilities such as NHMFL.

Much of the high-field research performed today involves powered (resistive) magnets, and the power supplies ultimately limit the amount of time available to the user community at the very highest fields. The current situation at NHMFL is that costs prohibit 24/7 operations. Therefore, some scope exists for increasing user operations without the need for significant investment in infrastructure, although the associated staffing and electricity costs would be quite considerable. Recently, however, the laboratory successfully brought online the first all-superconducting 32 T user magnet, with high-temperature superconducting inner coils (Weijers et al. 2016). Compared to resistive magnet technologies, it provides a lower noise environment for precision measurements, as well as a larger sample space. This magnet is now part of the user NHMFL program, although it is not currently instrumented for EPR or solid-state NMR experiments. Even though this magnet runs independently of the high-power resistive magnets (that require a 56 MW supply), it would not be easy to operate outside of the facility due to a very high helium consumption and complex supporting infrastructure (not to mention that this technology remains developmental). The resonance users of the facility (both EPR and NMR) recognize the tremendous opportunities that such a magnet provides in terms of increased access to the very highest magnetic fields. The current version does not provide the homogeneity needed for high-resolution solid-state NMR, but it is more than adequate for most EPR applications as well as certain specialized NMR applications (and further development of methodologies such as DNP). The development of a second all-superconducting magnet dedicated to resonance applications as well as the future development of high-resolution all-superconducting NMR magnets would serve as a very significant research resource for the chemistry QIS community.

Of course, the availability of new magnets and increased magnet time is not sufficient by itself. Investment in EPR and NMR hardware optimized to these high-field magnets would increase research in chemistry and QIS. Such resources historically have been severely limited within NHMFL's core funding, with most recent new developments being funded through separate grant applications. These activities impose a significant workload on staff, which either subtracts from user support time or imposes barriers that prevent staff from pursuing such development opportunities. Another area of opportunity is increasing involvement of the user community to fund such developments. However, the nature of the user community is often such that users lack the expertise to lead such an effort, instead contributing only to the scientific case. But this shifts the workload back to the scientific staff at the laboratory. This and the obvious limitations in the number of awards made to a given institution currently represent a significant barrier to the development of new experimental hardware for magnetic resonance. Most of this hardware falls under the mid-sale category and is discussed in the previous section. However, we discuss this issue again here, as it clearly represents a major barrier to progress in research at the interface between chemistry and QIS, and also limits access to large-scale facilities.

3.11.3b Light Sources and Synchrotrons

Free electron lasers and synchrotrons are regularly used for the characterization of molecular structure and dynamics (Fukuzawa and Ueda 2020; Heinz et al. 2017; Young et al. 2018). The same approaches apply to molecular qubits, particularly since transient and steady-state X-ray measurements are element-specific and can separate, for example, a metallic spin center from its ligands. Spin sensitivity is also possible through magnetic dichroism and related experiments. However, the development of novel beamlines for molecular qubit research is needed as well.

The first category revolves around the modification of transient absorption and resonant inelastic X-ray scattering beamlines so that they are capable of low-temperature and microwave experiments. Most transient measurements use optical to terahertz pulses, which can excite electronic and vibrational states but not initiate molecular qubits. Using a transient microwave pulse to initialize the qubit and an X-ray pulse to measure the spin-dependent polarization and coupled lattice dynamics would be particularly powerful. Such a beam station could handle molecular jets, liquids, or thin-film samples. Near-field X-ray microscopy using zone plates has also progressed significantly, and its element-specific images could be used for investigating coupled or individual molecular qubits (Shapiro et al. 2020).

The second category of instrumentation revolves around the creation of beamlines to explore how entangled X-rays can be used for novel spectroscopy or characterization of qubits (Defienne et al. 2021; Durbin 2022; Röhlsberger, Evers, and Shwartz 2014; Smith, Wang, and Shih 2020). Similar to the previous section on nonlinear entangled photon spectroscopy (Section 3.4.1), researchers predict that X-rays can linearize multiphoton processes, increase SNRs, and be used for novel "ghost" imaging protocols. This would allow lower-power experiments, solving one of the biggest issues with nonlinear X-ray experiments: sample damage.

For entangled X-ray experiments to be feasible, more efficient sources and optics need to be developed (Röhlsberger, Evers, and Shwartz 2014; Volkovich and Shwartz 2020). Entangled photons are created by SPDC, in which a pump photon couples to a vacuum state to be split into two entangled photons whose energy and momentum sum to that of the pump photon. X-ray processes have low nonlinear coefficients because the refractive index and nonlinear coefficients are near unity. While various nonlinear processes like second harmonic generation, sum frequency generation, and TPA have been achieved using X-ray free-electron lasers, entangled photon generation through SPDC has a million to billions of times lower conversion efficiency than these nonlinear processes. In the hard X-ray regime, phase matching occurs through the nonlocal electron plasma and lattice periodicity even in centrosymmetric materials. Current entangled X-ray photon generation methods rely almost exclusively on diamond for phase matching with conversion efficiencies in the 10^{-12} range (Figure 3-41) (Shwartz et al. 2012).

However, recent reports of using periodically poled materials (similar to visible light entangled photon generation) have shown promising efficiencies in the 10^{-6} range (Shapiro et al. 2020). Metal and dielectric stacks are already employed in extreme ultraviolet optics, so the technology is already in place. For example, using polar materials, extreme ultraviolet (<100 eV) SPDC was recently measured to be one million times more efficient than hard X-rays (Sofer et al. 2019). Free-electron lasers can also natively generate entangled photon and

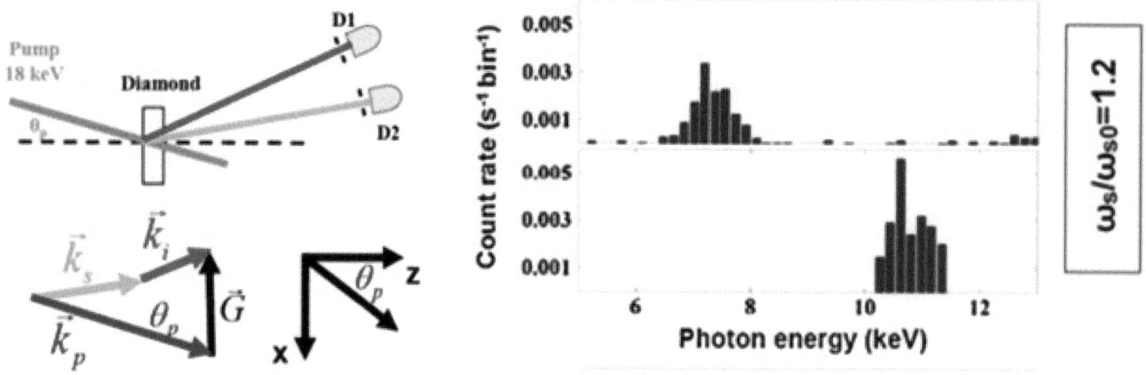

FIGURE 3-41 The generation of X-ray entangled photons by spontaneous parametric down-conversion in diamond has been demonstrated at keV energies at 10^{-12} efficiencies.
SOURCE: Shwartz et al. 2012.

electron–photon pairs (Wong and Kaminer 2021). Given that specialized detectors and long experimental times are needed for most X-ray experiments, the creation of dedicated beamlines could revolutionize nonlinear X-ray source creation as well as exploration of the use of entanglement.

3.11.4 Limited Natural Resources

Applications of rare-earth and actinide elements in the area of optically addressable qubits are discussed in Section 2.2.2 of this report. An important consideration in many of these examples is the availability of naturally occurring isotopes with a variety of nuclear spin quantum numbers, *I*. A perfect example is the $I = ½$ $^{171}Yb^{+}$ ion that is employed in ion trap quantum computers (Wright et al. 2019), where the hyperfine interaction involving the lone unpaired 6*s* electron gives rise to a hydrogenic-like situation and a massive 12.6 GHz clock transition. Similar targets may be anticipated in pursuit of ideal molecular qubit candidates, necessitating isotopic purifications or the production of particular isotopes at nuclear reactor facilities. Access to such isotopes will likely be limited and costly, with sub-milligram syntheses precluding many standard characterization techniques that lack sensitivity. Sample recycling will therefore be essential. Such approaches are already employed—for example, in QIS studies of endohedral metallofullerenes, which are typically extracted and purified via chromatographic techniques. Indeed, several gadolinium endohedral metallofullerenes have recently shown great promise as potential molecular qudits (d-dimensional qubits) (Fu et al. 2022; Hu et al. 2018), where the fullerene cage provides added protection from environmental decoherence sources.

Finally, the ongoing helium crisis represents a very significant challenge for researchers working in the QIS field, where many studies are conducted at temperatures in the sub-10 K range. Moreover, many groups continue to rely on older liquid helium–based cryogenic systems. It goes without saying that continued efforts must be pursued to reduce reliance on such cryogenic approaches, by furnishing researchers with closed-cycle refrigeration systems.

3.12 SUMMARY OF RESEARCH PRIORITIES AND RECOMMENDATION

The following fundamental research priorities have been identified by the committee and extensively discussed in Chapter 3 as those that the Department of Energy and NSF should prioritize within the target research area of "measurement and control of molecular quantum systems."

Research Priorities:

- **Develop new approaches and techniques for addressing and controlling multiple electron and nuclear spins and optical cycling centers in molecular systems.**
- **Develop techniques to probe molecular qubits at complex interfaces to inform their systematic control.**

- **Develop enhanced spectroscopic and microscopic techniques by creating**
    - **i. entangled photon sources with higher yield and better spectral coverage, and**
    - **ii. high-finesse cavities and nanophotonics for molecular qubit systems.**
- **Develop and exploit alternative approaches to spin polarization and coherence control (e.g., chirality-induced spin selectivity and electric field effects).**
- **Use molecular systems to teleport quantum information over distances greater than 1 μm with high fidelity.**
- **Develop molecular quantum transduction schemes that take advantage of entangled photons as well as entangled electrons and nuclear spins.**
- **Advance quantum sensing techniques to further understand biological systems.**
- **Use bio-inspired quantum processes for the development of new quantum technologies.**

In sum, in this chapter the committee discussed infrastructure and instrumentation needs at various levels. The committee provides Recommendation 3-1 to ensure that the infrastructure needs and instrumentation support required to drive chemistry QIS research in the United States are met on a competitive timescale.

Recommendation 3-1. The Department of Energy and the National Science Foundation should support the development of new instrumentation and techniques for the unique needs at the interface of chemistry and quantum information science. Broader access to laboratory-scale and mid-scale instrumentation is needed for the field to progress. For example, investments should be made in time-resolved magnetic resonance and optical spectroscopy. Support is required for professional staff to train users in the operation and utilization of these instruments, as well as to address new technique development and maintenance needs.

REFERENCES

Aguilà, D., L. A. Barrios, V. Velasco, O. Roubeau, A. Repollés, P. J. Alonso, J. Sesé, S. J. Teat, F. Luis, and G. Aromí. 2014. "Heterodimetallic [LnLn′] Lanthanide Complexes: Toward a Chemical Design of Two-Qubit Molecular Spin Quantum Gates." *Journal of the American Chemical Society* 136(40):14215–14222. doi.org/10.1021/ja507809w.

Ahn, W., F. Herrera, and B. Simpkins. 2022. "Modification of Urethane Addition Reaction via Vibrational Strong Coupling." *ChemRxiv*. Cambridge: Cambridge Open Engage. doi.org/10.26434/chemrxiv-2022-wb6vs.

Allodi, G., A. Banderini, R. De Renzi, and C. Vignali. 2005. "HyReSpect: A Broadband Fast-Averaging Spectrometer for Nuclear Magnetic Resonance of Magnetic Materials." *Review of Scientific Instruments* 76(8):083911. doi.org/10.1063/1.2009868.

Altenhof, A. R., A. W. Lindquist, L. D. D. Foster, S. T. Holmes, and R. W. Schurko. 2019. "On the Use of Frequency-Swept Pulses and Pulses Designed with Optimal Control Theory for the Acquisition of Ultra-Wideline NMR Spectra." *Journal of Magnetic Resonance* 309:106612. doi.org/10.1016/j.jmr.2019.106612.

Anderegg, L, B. L. Augenbraun, Y. Bao, S. Burchesky, L. W. Cheuk, W. Ketterle, and J. M. Doyle. 2018. "Laser Cooling of Optically Trapped Molecules." *Nature Physics* 14(9):890–893. https://doi.org/10.1038/s41567-018-0191-z.

Anderegg, L., B. L. Augenbraun, E. Chae, B. Hemmerling, N. R. Hutzler, A. Ravi, A. Collopy, J. Ye, W. Ketterle, and J. M. Doyle. 2017. "Radio Frequency Magneto-Optical Trapping of CaF with High Density." *Physical Review Letters* 119(10):103201. doi.org/10.1103/PhysRevLett.119.103201.

Anderegg, L., L. W. Cheuk, Y. Bao, S. Burchesky, W. Ketterle, K.-K. Ni, and J. M. Doyle. 2019. "An Optical Tweezer Array of Ultracold Molecules." *Science* 365(6458):1156–1158. doi:10.1126/science.aax1265.

Anlage, S. M., D. E. Steinhauer, B. J. Feenstra, C. P. Vlahacos, and F. C. Wellstood. 2001. "Near-Field Microwave Microscopy of Materials Properties." In *Microwave Superconductivity*, edited by H. Weinstock and M. Nisenoff, NATO Science Series, Vol. 375. Dordrecht: Springer. https://doi.org/10.1007/978-94-010-0450-3_10.

Aoyagi, Y., R. Kawakami, H. Osanai, T. Hibi, and T. Nemoto. 2015. "A Rapid Optical Clearing Protocol Using 2,2′-Thiodiethanol for Microscopic Observation of Fixed Mouse Brain." *PLoS ONE* 10(1):e0116280. doi.org/10.1371/journal.pone.0116280.

Ardavan, A., A. M. Bowen, A. Fernandez, A. J. Fielding, D. Kaminski, F. Moro, C. A. Muryn, M. D. Wise, A. Ruggi, E. J. L. McInnes, K. Severin, G. A. Timco, C. R. Timmel, F. Tuna, G. F. S. Whitehead, and R. E. P. Winpenny. 2015. "Engineering Coherent Interactions in Molecular Nanomagnet Dimers." *npj Quantum Information* 1(1):15012. doi.org/10.1038/npjqi.2015.12.

Ardavan, A., O. Rival, J. J. L. Morton, S. J. Blundell, A. M. Tyryshkin, G. A. Timco, and R. E. P. Winpenny. 2007. "Will Spin-Relaxation Times in Molecular Magnets Permit Quantum Information Processing?" *Physical Review Letters* 98(5):057201. doi.org/10.1103/PhysRevLett.98.057201.

Ariciu, A.-M., D. H. Woen, D. N. Huh, L. E. Nodaraki, A. K. Kostopoulos, C. A. P. Goodwin, N. F. Chilton, E. J. L. McInnes, R. E. P. Winpenny, W. J. Evans, and F. Tuna. 2019. "Engineering Electronic Structure to Prolong Relaxation Times in Molecular Qubits by Minimising Orbital Angular Momentum." *Nature Communications* 10(1):3330. doi.org/10.1038/s41467-019-11309-3.

Aslam, N., M. Pfender, P. Neumann, R. Reuter, A. Zappe, F. Fávaro de Oliveira, A. Denisenko, H. Sumiya, S. Onoda, J. Isoya, and J. Wrachtrup. 2017. "Nanoscale Nuclear Magnetic Resonance with Chemical Resolution." *Science* 357(6346):67–71. doi:10.1126/science.aam8697.

Atzori, M., A. Chiesa, E. Morra, M. Chiesa, L. Sorace, S. Carretta, and R. Sessoli. 2018. "A Two-Qubit Molecular Architecture for Electron-Mediated Nuclear Quantum Simulation." *Chemical Science* 9(29):6183–6192. doi.org/10.1039/C8SC01695J.

Atzori, M., L. Tesi, E. Morra, M. Chiesa, L. Sorace, and R. Sessoli. 2016. "Room-Temperature Quantum Coherence and Rabi Oscillations in Vanadyl Phthalocyanine: Toward Multifunctional Molecular Spin Qubits." *Journal of the American Chemical Society* 138(7):2154–2157. doi.org/10.1021/jacs.5b13408.

Augenbraun, B. L., Z. D. Lasner, A. Frenett, H. Sawaoka, C. Miller, T. C. Steimle, and J. M. Doyle. 2020. "Laser-cooled Polyatomic Molecules for Improved Electron Electric Dipole Moment Searches." *New Journal of Physics* 22(2):022003. doi.org/10.1088/1367-2630/ab687b.

Awschalom, D. D., R. Hanson, J. Wrachtrup, and B. B. Zhou. 2018. "Quantum Technologies with Optically Interfaced Solid-State Spins." *Nature Photonics* 12(9):516–527. doi.org/10.1038/s41566-018-0232-2.

Bader, K., D. Dengler, S. Lenz, B. Endeward, S.-D. Jiang, P. Neugebauer, and J. van Slageren. 2014. "Room Temperature Quantum Coherence in a Potential Molecular Qubit." *Nature Communications* 5(1):5304. doi.org/10.1038/ncomms6304.

Bader, K., S. H. Schlindwein, D. Gudat, and J. van Slageren. 2017. "Molecular Qubits Based on Potentially Nuclear-Spin-Free Nickel Ions." *Physical Chemistry Chemical Physics* 19(3):2525–2529. doi.org/10.1039/C6CP08161D.

Baek, S. H., F. Borsa, Y. Furukawa, Y. Hatanaka, S. Kawakami, K. Kumagai, B. J. Suh, and A. Cornia. 2005. "^{57}Fe NMR and Relaxation by Strong Collision in the Tunneling Regime in the Molecular Nanomagnet Fe8." *Physical Review B* 71(21):214436. doi.org/10.1103/PhysRevB.71.214436.

Baker, M. L., S. J. Blundell, N. Domingo, and S. Hill. 2015. "Spectroscopy Methods for Molecular Nanomagnets." In *Molecular Nanomagnets and Related Phenomena*, edited by S. Gao, 231–291. Berlin: Springer Berlin Heidelberg.

Baker, M. L., T. Lancaster, A. Chiesa, G. Amoretti, P. J. Baker, C. Barker, S. J. Blundell, S. Carretta, D. Collison, H. U. Güdel, T. Guidi, E. J. L. McInnes, J. S. Möller, H. Mutka, J. Ollivier, F. L. Pratt, P. Santini, F. Tuna, P. L. W. Tregenna-Piggott, I. J. Vitorica-Yrezabal, G. A. Timco, and R. E. P. Winpenny. 2016. "Studies of a Large Odd-Numbered Odd-Electron Metal Ring: Inelastic Neutron Scattering and Muon Spin Relaxation Spectroscopy of Cr_8Mn." *Chemistry: A European Journal* 22(5):1779–1788. doi.org/10.1002/chem.201503431.

Bao, F., H. Deng, D. Ding, R. Gao, X. Gao, C. Huang, X. Jiang, H-S. Ku, Z. Li, X. Ma, X. Ni, J. Qin, Z. Song, H. Sun, C. Tang, T. Wang, F. Wu, T. Xia, W. Yu, F. Zhang, G. Zhang, X. Zhang, J. Zhou, X. Zhu, Y. Shi, J. Chen, H-H Zhao, and C. Deng. 2022. "Fluxonium: An Alternative Qubit Platform for High-Fidelity Operations." *Physical Review Letters* 129(1):010502. doi.org/10.1103/PhysRevLett.129.010502.

Barclay, P. E., K.-M. C. Fu, C. Santori, A. Faraon, and R. G. Beausoleil. 2011. "Hybrid Nanocavity Resonant Enhancement of Color Center Emission in Diamond." *Physical Review X* 1(1):011007. doi.org/10.1103/PhysRevX.1.011007.

Barnes, A. B., G. De Paëpe, P. C. A. van der Wel, K. N. Hu, C. G. Joo, V. S. Bajaj, M. L. Mak-Jurkauskas, J. R. Sirigiri, J. Herzfeld, R. J. Temkin, and R. G. Griffin. 2008. "High-Field Dynamic Nuclear Polarization for Solid and Solution Biological NMR." *Applied Magnetic Resonance* 34(3):237–263. doi.org/10.1007/s00723-008-0129-1.

Barra, A.-L., A. Caneschi, A. Cornia, D. Gatteschi, L. Gorini, L.-P. Heiniger, R. Sessoli, and L. Sorace. 2007. "The Origin of Transverse Anisotropy in Axially Symmetric Single Molecule Magnets." *Journal of the American Chemical Society* 129(35):10754–10762. doi.org/10.1021/ja0717921.

Barra, A. L., D. Gatteschi, and R. Sessoli. 1997. "High-frequency EPR Spectra of a Molecular Nanomagnet: Understanding Quantum Tunneling of the Magnetization." *Physical Review B* 56(13):8192–8198. doi.org/10.1103/PhysRevB.56.8192.

Barry, J. F., D. J. McCarron, E. B. Norrgard, M. H. Steinecker, and D. DeMille. 2014. "Magneto-Optical Trapping of a Diatomic Molecule." *Nature* 512(7514):286–289. doi.org/10.1038/nature13634.

Baumann, S., W. Paul, T. Choi, C. P. Lutz, A. Ardavan, and A. J. Heinrich. 2015. "Electron Paramagnetic Resonance of Individual Atoms on a Surface." *Science* 350(6259):417–420. doi.org/10.1126/science.aac8703.

Bayliss, S. L., D. W. Laorenza, P. J. Mintun, B. D. Kovos, D. E. Freedman, and D. D. Awschalom. 2020. "Optically Addressable Molecular Spins for Quantum Information Processing." *Science* 370(6522):1309–1312. doi.org/10.1126/science.abb9352.

Bertaina, S., S. Gambarelli, T. Mitra, B. Tsukerblat, A. Müller, and B. Barbara. 2008. "Quantum Oscillations in a Molecular Magnet." *Nature* 453(7192):203–206. doi.org/10.1038/nature06962.

Blackaby, W. J. M., K. L. M. Harriman, S. M. Greer, A. Folli, S. Hill, V. Krewald, M. F. Mahon, D. M. Murphy, M. Murugesu, E. Richards, E. Suturina, and M. K. Whittlesey. 2022. "Extreme g-Tensor Anisotropy and Its Insensitivity to Structural Distortions in a Family of Linear Two-Coordinate Ni(I) Bis-N-heterocyclic Carbene Complexes." *Inorganic Chemistry* 61(3):1308–1315. doi.org/10.1021/acs.inorgchem.1c02413.

Bluhm, H., S. Foletti, D. Mahalu, V. Umansky, and A. Yacoby. 2010. "Enhancing the Coherence of a Spin Qubit by Operating It as a Feedback Loop That Controls Its Nuclear Spin Bath." *Physical Review Letters* 105(21):216803. doi.org/10.1103/PhysRevLett.105.216803.

Bonizzoni, C., A. Ghirri, and M. Affronte. 2018. "Coherent Coupling of Molecular Spins with Microwave Photons in Planar Superconducting Resonators." *Advances in Physics: X* 3(1):1435305. doi.org/10.1080/23746149.2018.1435305.

Boudalis, A. K., J. Robert, and P. Turek. 2018. "First Demonstration of Magnetoelectric Coupling in a Polynuclear Molecular Nanomagnet: Single-Crystal EPR Studies of $[Fe_3O(O_2CPh)_6(py)_3]ClO_4{\cdot}py$ Under Static Electric Fields." *Chemistry: A European Journal* 24(56):14896–14900. doi.org/10.1002/chem.201803038.

Burdick, R. K., G. C. Schatz, and T. Goodson, III. 2021. "Enhancing Entangled Two-Photon Absorption for Picosecond Quantum Spectroscopy." *Journal of the American Chemical Society* 143(41):16930–16934. doi.org/10.1021/jacs.1c09728.

Bustamante, C. J., Y. R. Chemla, S. Liu, and M. D. Wang. 2021. "Optical Tweezers in Single-Molecule Biophysics." *Nature Reviews Methods Primers* 1:25. doi.org/10.1038/s43586-021-00021-6.

Can, T. V., Q. Z. Ni, and R. G. Griffin. 2015. "Mechanisms of Dynamic Nuclear Polarization in Insulating Solids." *Journal of Magnetic Resonance* 253:23–35. doi.org/10.1016/j.jmr.2015.02.005.

Canarie, E. R., S. M. Jahn, and S. Stoll. 2020. "Quantitative Structure-Based Prediction of Electron Spin Decoherence in Organic Radicals." *Physical Review Letters* 11(9):3396–3400. doi.org/10.1021/acs.jpclett.0c00768.

Casacio, C. A., L. S. Madsen, A. Terrasson, M. Waleed, K. Barnscheidt, B. Hage, M. A. Taylor, and W. P. Bowen. 2021. "Quantum-Enhanced Nonlinear Microscopy." *Nature* 594(7862):201–206. doi.org/10.1038/s41586-021-03528-w.

Chakov, N. E., S.-C. Lee, A. G. Harter, P. L. Kuhns, A. P. Reyes, S. O. Hill, N. S. Dalal, W. Wernsdorfer, K. A. Abboud, and G. Christou. 2006. "The Properties of the $[Mn_{12}O_{12}(O_2CR)_{16}(H_2O)_4]$ Single-Molecule Magnets in Truly Axial Symmetry: $[Mn_{12}O_{12}(O_2CCH_2Br)_{16}(H_2O)_4]{\cdot}\ 4CH_2Cl_2$." *Journal of the American Chemical Society* 128(21):6975–6989. doi.org/10.1021/ja060796n.

Chen, F., and Mukamel S. 2021. "Vibrational Hyper-Raman Molecular Spectroscopy with Entangled Photons." ACS Photonics 8(9):2722–27. https://doi.org/10.1021/acsphotonics.1c00777.

Chen, J., C. Hu, J. F. Stanton, S. Hill, H.-P. Cheng, and X.-G. Zhang. 2020. "Decoherence in Molecular Electron Spin Qubits: Insights from Quantum Many-Body Simulations." *Physical Review Letters* 11(6):2074–2078. doi.org/10.1021/Acs.Jpclett.0c00193.

Chen, M., C. Meng, Q. Zhang, C. Duan, F. Shi, and J. Du. 2017. "Quantum Metrology with Single Spins in Diamond Under Ambient Conditions." *National Science Review* 5(3):346–355. doi.org/10.1093/Nsr/Nwx121.

Chen, Y., Y. Bae, and A. J. Heinrich. 2022. "Harnessing the Quantum Behavior of Spins on Surfaces." *Advanced Materials*. doi.org/10.1002/Adma.202107534.

Chenu, A., and G. D. Scholes. 2015. "Coherence in Energy Transfer and Photosynthesis." *Annual Review of Physical Chemistry* 66(1):69–96. doi.org/10.1146/Annurev-Physchem-040214-121713.

Cheuk, L. W., L. Anderegg, B. L. Augenbraun, Y. Bao, S. Burchesky, W. Ketterle, and J. M. Doyle. 2018. "Λ-Enhanced Imaging of Molecules in an Optical Trap." *Physical Review Letters* 121(8):083201. doi.org/10.1103/PhysRevLett.121.083201.

Chiesa, A., T. Guidi, S. Carretta, S. Ansbro, G. A. Timco, I. Vitorica-Yrezabal, E. Garlatti, G. Amoretti, R. E. P. Winpenny, and P. Santini. 2017. "Magnetic Exchange Interactions in the Molecular Nanomagnet Mn_{12}" *Physical Review Letters* 119(21):217202. doi.org/10.1103/Physrevlett.119.217202.

Chiesa, A., E. Macaluso, F. Petiziol, S. Wimberger, P. Santini, and S. Carretta. 2020. "Molecular Nanomagnets as Qubits with Embedded Quantum-Error Correction." *Journal of Physical Chemistry Letters* 11(20):8610–8615. doi.org/10.1021/acs.jpclett.0c02213.

Choi, T. 2019. "Studies of Single Atom Magnets Via Scanning Tunneling Microscopy." *Journal of Magnetism and Magnetic Materials* 481:150–155. doi.org/10.1016/J.Jmmm.2019.03.007.

Chou, C.-W., C. Kurz, D. B. Hume, P. N. Plessow, D. R. Leibrandt, and D. Leibfried. 2017. "Preparation and Coherent Manipulation of Pure Quantum States of a Single Molecular Ion." *Nature* 545(7653):203–207. doi.org/10.1038/Nature22338.

Cimini, V., M. Mellini, G. Rampioni, M. Sbroscia, L. Leoni, M. Barbieri, and I. Gianani. 2019. "Adaptive Tracking of Enzymatic Reactions with Quantum Light." *Optics Express* 27(24):35245. doi.org/10.1364/oe.27.035245.

Cini, A., M. Mannini, F. Totti, M. Fittipaldi, G. Spina, A. Chumakov, R. Rüffer, A. Cornia, and R. Sessoli. 2018. "Mössbauer Spectroscopy of a Monolayer of Single Molecule Magnets." *Nature Communications* 9(1):480. doi.org/10.1038/S41467-018-02840-W.

Collett, C. A., K.-I. Ellers, N. Russo, K. R. Kittilstved, G. A. Timco, R. E. P. Winpenny, and J. R. Friedman. 2019. "A Clock Transition in the Cr_7Mn Molecular Nanomagnet." *Magnetochemistry* 5(1):4. doi.org/10.3390/magnetochemistry5010004.

Collett, C. A., P. Santini, S. Carretta, and J. R. Friedman. 2020. "Constructing Clock-Transition-Based Two-Qubit Gates from Dimers of Molecular Nanomagnets." *Physical Review Research* 2(3):032037. doi.org/10.1103/Physrevresearch.2.032037.

Collopy, A. L., S. Ding, Y. Wu, I. A. Finneran, L. Anderegg, B. L. Augenbraun, J. M. Doyle, and J. Ye. 2018. "3D Magneto-Optical Trap of Yttrium Monoxide." *Physical Review Letters* 121(21):213201. doi.org/10.1103/Physrevlett.121.213201.

Cruickshank, P. A. S., D. R. Bolton, D. A. Robertson, R. I. Hunter, R. J. Wylde, and G. M. Smith. 2009. "A Kilowatt Pulsed 94 Ghz Electron Paramagnetic Resonance Spectrometer with High Concentration Sensitivity, High Instantaneous Bandwidth, and Low Dead Time." *Review of Scientific Instruments* 80(10):103102. doi.org/10.1063/1.3239402.

Defienne, H., B. Ndagano, A. Lyons, and D. Faccio. 2021. "Polarization Entanglement-Enabled Quantum Holography." *Nature Physics* 17(5):591–597. doi.org/10.1038/S41567-020-01156-1.

Degen, C. L., M. Poggio, H. J. Mamin, C. T. Rettner, and D. Rugar. 2009. "Nanoscale Magnetic Resonance Imaging." *Proceedings of the National Academy of Sciences USA* 106(5):1313–1317. doi.org/10.1073/Pnas.0812068106.

del Barco, E., A. D. Kent, S. Hill, J. M. North, N. S. Dalal, E. M. Rumberger, D. N. Hendrickson, N. Chakov, and G. Christou. 2005. "Magnetic Quantum Tunneling in the Single-Molecule Magnet Mn_{12}-Acetate." *Journal of Low Temperature Physics* 140(1):119–174. doi.org/10.1007/s10909-005-6016-3.

Di Rosa, M. D. 2004. "Laser-Cooling Molecules." *European Physical Journal D—Atomic, Molecular, Optical, and Plasma Physics* 31(2):395–402. doi.org/10.1140/Epjd/E2004-00167-2.

Dickerson, C. E., C. Chang, H. Guo, and A. N. Alexandrova. 2022. "Fully Saturated Hydrocarbons as Hosts of Optical Cycling Centers." *Journal of Physical Chemistry A* 126(51):9644–9650. doi.org/10.1021/Acs.Jpca.2c06647.

Dickerson, C. E., H. Guo, A. J. Shin, B. L. Augenbraun, J. R. Caram, W. C. Campbell, and A. N. Alexandrova. 2021. "Franck-Condon Tuning of Optical Cycling Centers by Organic Functionalization." *Physical Review Letters* 126(12):123002. doi.org/10.1103/Physrevlett.126.123002.

Dickerson, C. E., H. Guo, G.-Z. Zhu, E. R. Hudson, J. R. Caram, W. C. Campbell, and A. N. Alexandrova. 2021. "Optical Cycling Functionalization of Arenes." *Physical Review Letters* 12(16):3989–3995. doi.org/10.1021/Acs.Jpclett.1c00733.

Dolde, F., I. Jakobi, B. Naydenov, N. Zhao, S. Pezzagna, C. Trautmann, J. Meijer, P. Neumann, F. Jelezko, and J. Wrachtrup. 2013. "Room-Temperature Entanglement Between Single Defect Spins in Diamond." *Nature Physics* 9(3):139–143. doi.org/10.1038/Nphys2545.

Dorfman, K. E., F. Schlawin, and S. Mukamel. 2014. "Stimulated Raman Spectroscopy with Entangled Light: Enhanced Resolution and Pathway Selection." *Physical Review Letters* 5(16):2843–2849. doi.org/10.1021/Jz501124a.

Dressel, M., B. Gorshunov, K. Rajagopal, S. Vongtragool, and A. A. Mukhin. 2003. "Quantum Tunneling and Relaxation in Mn_{12}-Acetate Studied by Magnetic Spectroscopy." *Physical Review B* 67(6):060405. doi.org/10.1103/Physrevb.67.060405.

Du, M., R. F. Ribeiro, and J. Yuen-Zhou. 2019. "Remote Control of Chemistry in Optical Cavities." *Chem* 5(5):1167–1181. doi.org/10.1016/J.Chempr.2019.02.009.

Duan, L. M., M. D. Lukin, J. I. Cirac, and P. Zoller. 2001. "Long-Distance Quantum Communication with Atomic Ensembles and Linear Optics." *Nature* 414(6862):413–418. doi.org/10.1038/35106500.

Duan, R., J. N. Mastron, Y. Song, and K. J. Kubarych. 2021. "Isolating Polaritonic 2D-IR Transmission Spectra." *Journal of Physical Chemistry Letters* 12(46):11406–11414. doi.org/10.1021/acs.jpclett.1c03198.

Durbin, S. M. 2022. "Proposal for Entangled X-Ray Beams." *Journal of Applied Physics* 131(22):224401. doi.org/10.1063/5.0091947.

Durkan, C., and M. E. Welland. 2002. "Electronic Spin Detection in Molecules Using Scanning-Tunneling-Microscopy-Assisted Electron-Spin Resonance." *Applied Physics Letters* 80(3):458–460. doi.org/10.1063/1.1434301.

Ebadi, S., T. T. Wang, H. Levine, A. Keesling, G. Semeghini, A. Omran, D. Bluvstein, R. Samajdar, H. Pichler, W. W. Ho, S. Choi, S. Sachdev, M. Greiner, V. Vuletić, and M. D. Lukin. 2021. "Quantum Phases of Matter on a 256-Atom Programmable Quantum Simulator." *Nature* 595:227–232. doi.org/10.1038/s41586-021-03582-4.

Ebbesen, T. W. 2016. "Hybrid Light–Matter States in a Molecular and Material Science Perspective." *Accounts of Chemical Research* 49(11):2403–2412. doi.org/10.1021/Acs.Accounts.6b00295.

Eddins, A. W., C. C. Beedle, D. N. Hendrickson, and J. R. Friedman. 2014. "Collective Coupling of a Macroscopic Number of Single-Molecule Magnets with a Microwave Cavity Mode." *Physical Review Letters* 112(12):120501. doi.org/10.1103/Physrevlett.112.120501.

Engel, G. S., T. R. Calhoun, E. L. Read, T.-K. Ahn, T. Mančal, Y.-C. Cheng, R. E. Blankenship, and G. R. Fleming. 2007. "Evidence for Wavelike Energy Transfer Through Quantum Coherence in Photosynthetic Systems." *Nature* 446(7137):782–786. doi.org/10.1038/Nature05678.

Eshun, A., B. Gu, O. Varnavski, S. Asban, K. E. Dorfman, S. Mukamel, and T. Goodson, III. 2021. "Investigations of Molecular Optical Properties Using Quantum Light and Hong–Ou–Mandel Interferometry." *Journal of the American Chemical Society* 143(24):9070–9081. doi.org/10.1021/Jacs.1c02514.

Eshun, A., O. Varnavski, J. P. Villabona-Monsalve, R. K. Burdick, and T. Goodson, III. 2022. "Entangled Photon Spectroscopy." *Accounts of Chemical Research* 55(7):991–1003. doi.org/10.1021/Acs.Accounts.1c00687.

Evans, R. E., M. K. Bhaskar, D. D. Sukachev, C. T. Nguyen, A. Sipahigil, M. J. Burek, B. Machielse, G. H. Zhang, A. S. Zibrov, E. Bielejec, H. Park, M. Lončar, and M. D. Lukin. 2018. "Photon-Mediated Interactions Between Quantum Emitters in a Diamond Nanocavity." *Science* 362(6415):662–665. doi.org/10.1126/Science.Aau4691.

Farina, M., and J. C. M. Hwang. 2020. "Scanning Microwave Microscopy for Biological Applications: Introducing the State of the Art and Inverted SMM." *IEEE Microwave Magazine* 21(10):52–59. doi.org/10.1109/MMM.2020.3008239.

Fataftah, M. S., S. L. Bayliss, D. W. Laorenza, X. Wang, B. T. Phelan, C. B. Wilson, P. J. Mintun, B. D. Kovos, M. R. Wasielewski, S. Han, M. S. Sherwin, D. D. Awschalom, and D. E. Freedman. 2020. "Trigonal Bipyramidal V^{3+} Complex as an Optically Addressable Molecular Qubit Candidate." *Journal of the American Chemical Society* 142(48):20400–20408. doi.org/10.1021/Jacs.0c08986.

Fataftah, M. S., J. M. Zadrozny, S. C. Coste, M. J. Graham, D. M. Rogers, and D. E. Freedman. 2016. "Employing Forbidden Transitions as Qubits in a Nuclear Spin-Free Chromium Complex." *Journal of the American Chemical Society* 138(4):1344–1348. doi.org/10.1021/Jacs.5b11802.

Fei, H-B., B. M. Jost, S. Popescu, B. E. A. Saleh, and M. C. Teich. 1997. "Entanglement-Induced Two-Photon Transparency." *Physical Review Letters* 78(9):1679–1682. doi.org/10.1103/PhysRevLett.78.1679.

Feng, L., K.-Y. Wang, J. Willman, and H.-C. Zhou. 2020. "Hieracrchy in Metal-Organic Frameworks" *Journal of the American Chemical Society* 6(3):359–367. doi.org/10.1021/acscentsci.0c00158

Feng, X., J. Liu, T. D. Harris, S. Hill, and J. R. Long. 2012. "Slow Magnetic Relaxation Induced by a Large Transverse Zero-Field Splitting in a $Mn^{II}re^{IV}(CN)_2$ Single-Chain Magnet." *Journal of the American Chemical Society* 134(17):7521–7529. doi.org/10.1021/Ja301338d.

Ferrando-Soria, J., S. A. Magee, A. Chiesa, S. Carretta, P. Santini, I. J. Vitorica-Yrezabal, F. Tuna, G. F. S. Whitehead, S. Sproules, K. M. Lancaster, A.-L. Barra, G. A. Timco, E. J. L. Mcinnes, and R. E. P. Winpenny. 2016. "Switchable Interaction in Molecular Double Qubits." *Chem* 1(5):727–752. doi.org/10.1016/J.Chempr.2016.10.001.

Ferrando-Soria, J., E. M. Pineda, A. Chiesa, A. Fernandez, S. A. Magee, S. Carretta, P. Santini, I. J. Vitorica-Yrezabal, F. Tuna, G. A. Timco, E. J. L. McInnes, and R. E. P. Winpenny. 2016. "A Modular Design of Molecular Qubits to Implement Universal Quantum Gates." *Nature Communications* 7(1):11377. doi.org/10.1038/ncomms11377.

Fittipaldi, M., A. Cini, G. Annino, A. Vindigni, A. Caneschi, and R. Sessoli. 2019. "Electric Field Modulation of Magnetic Exchange in Molecular Helices." *Nature Materials* 18(4):329–334. doi.org/10.1038/S41563-019-0288-5.

Foletti, S., H. Bluhm, D. Mahalu, V. Umansky, and A. Yacoby. 2009. "Universal Quantum Control of Two-Electron Spin Quantum Bits Using Dynamic Nuclear Polarization." *Nature Physics* 5(12):903–908. doi.org/10.1038/Nphys1424.

Fortman, B., J. Pena, K. Holczer, and S. Takahashi. 2020. "Demonstration of NV-Detected ESR Spectroscopy at 115 Ghz and 4.2 T." *Applied Physics Letters* 116(17):174004. doi.org/10.1063/5.0006014.

Fröhlich, H. 1968. "Long-Range Coherence and Energy Storage in Biological Systems." *International Journal of Quantum Chemistry* 2(5):641–649. doi.org/10.1002/Qua.560020505.

Fu, P.-X., S. Zhou, Z. Liu, C.-H. Wu, Y.-H. Fang, Z.-R. Wu, X.-Q. Tao, J.-Y. Yuan, Y.-X. Wang, S. Gao, and S.-D. Jiang. 2022. "Multiprocessing Quantum Computing Through Hyperfine Couplings in Endohedral Fullerene Derivatives." *Angewandte Chemie International Edition* 61(52):E202212939. doi.org/10.1002/Anie.202212939.

Fukuzawa, H., and K. Ueda. 2020. "X-Ray Induced Ultrafast Dynamics in Atoms, Molecules, and Clusters: Experimental Studies at an X-Ray Free-Electron Laser Facility SACLA and Modelling." *Advances in Physics: X* 5(1):1785327. doi.org/10.1080/23746149.2020.1785327.

Furue, S., T. Kohmoto, M. Kunitomo, and Y. Fukuda. 2005. "Optical Induction of Magnetization and Observation of Fast Spin Dynamics in Aqueous Solutions of Copper Ions." *Physics Letters A* 345(4):415–422. doi.org/10.1016/J.Physleta.2005.07.028.

Furukawa, Y., K. Watanabe, K. Kumagai, F. Borsa, and D. Gatteschi. 2001. "Magnetic Structure and Spin Dynamics of the Ground State of the Molecular Cluster $Mn_{12}O_{12}$ Acetate Studied by ^{55}Mn NMR." *Physical Review B* 64(10):104401. doi.org/10.1103/Physrevb.64.104401.

Gaita-Ariño, A., F. Luis, S. Hill, and E. Coronado. 2019. "Molecular Spins for Quantum Computation." *Nature Chemistry* 11(4):301–309. doi.org/10.1038/S41557-019-0232-Y.

Gallez, B. 2022. "Interview with Professor Graham Smith on the Occasion of His Bruker Prize 2022." *EPR Newsletter*.

Gan, Z., I. Hung, X. Wang, J. Paulino, G. Wu, I. M. Litvak, P. L. Gor'kov, W. W. Brey, P. Lendi, J. L. Schiano, M. D. Bird, I. R. Dixon, J. Toth, G. S. Boebinger, and T. A. Cross. 2017. "NMR Spectroscopy Up to 35.2T Using a Series-Connected Hybrid Magnet." *Journal of Magnetic Resonance* 284:125–136. doi.org/10.1016/J.Jmr.2017.08.007.

Garlatti, E., A. Chiesa, T. Guidi, G. Amoretti, P. Santini, and S. Carretta. 2019. "Unravelling the Spin Dynamics of Molecular Nanomagnets with Four-Dimensional Inelastic Neutron Scattering." *European Journal of Inorganic Chemistry* 2019 (8):1106–1118. doi.org/10.1002/Ejic.201801050.

Garlatti, E., T. Guidi, S. Ansbro, P. Santini, G. Amoretti, J. Ollivier, H. Mutka, G. Timco, I. J. Vitorica-Yrezabal, G. F. S. Whitehead, R. E. P. Winpenny, and S. Carretta. 2017. "Portraying Entanglement Between Molecular Qubits with Four-Dimensional Inelastic Neutron Scattering." *Nature Communications* 8(1):14543. doi.org/10.1038/Ncomms14543.

Garlatti, E., L. Tesi, A. Lunghi, M. Atzori, D. J. Voneshen, P. Santini, S. Sanvito, T. Guidi, R. Sessoli, and S. Carretta. 2020. "Unveiling Phonons in a Molecular Qubit with Four-Dimensional Inelastic Neutron Scattering and Density Functional Theory." *Nature Communications* 11(1):1751. doi.org/10.1038/S41467-020-15475-7.

Gatteschi, D., A. L. Barra, A. Caneschi, A. Cornia, R. Sessoli, and L. Sorace. 2006. "EPR of Molecular Nanomagnets." *Coordination Chemistry Reviews* 250(11):1514–1529. doi.org/10.1016/J.Ccr.2006.02.006.

Gehring, P., J. M. Thijssen, and H. S. J. van der Zant. 2019. "Single-Molecule Quantum-Transport Phenomena in Break Junctions." *Nature Reviews Physics* 1(6):381–396. doi.org/10.1038/S42254-019-0055-1.

George, J., A. Shalabney, J.A. Hutchison, C. Genet, and T.W. Ebbesen. 2015. "Liquid-Phase Vibrational Strong Coupling." *Journal of Physical Chemistry Letters* 6(6):1027–1031. doi.org/10.1021/acs.jpclett.5b00204.

Georgiades, N. Ph., E. S. Polzik, K. Edamatsu, H. J. Kimble, and A. S. Parkins. 1995. "Nonclassical Excitation for Atoms in a Squeezed Vacuum." *Physical Review Letters* 75 (19):3426–3429. doi.org/10.1103/Physrevlett.75.3426.

Ghosh, T., J. Marbey, W. Wernsdorfer, S. Hill, K. A. Abboud, and G. Christou. 2021. "Exchange-Biased Quantum Tunnelling of Magnetization in a $[Mn_3]_2$ Dimer of Single-Molecule Magnets with Rare Ferromagnetic Inter-Mn_3 Coupling." *Physical Chemistry Chemical Physics* 23(14):8854–8867. doi.org/10.1039/D0CP06611G.

Giansiracusa, M. J., E. Moreno-Pineda, R. Hussain, R. Marx, M. Martínez Prada, P. Neugebauer, S. Al-Badran, D. Collison, F. Tuna, J. Van Slageren, S. Carretta, T. Guidi, E. J. L. Mcinnes, R. E. P. Winpenny, and N. F. Chilton. 2018. "Measurement of Magnetic Exchange in Asymmetric Lanthanide Dimetallics: Toward a Transferable Theoretical Framework." *Journal of the American Chemical Society* 140(7):2504–2513. doi.org/10.1021/Jacs.7b10714.

Gimeno, I., W. Kersten, M. C. Pallarés, P. Hermosilla, M. J. Martínez-Pérez, M. D. Jenkins, A. Angerer, C. Sánchez-Azqueta, D. Zueco, J. Majer, A. Lostao, and F. Luis. 2020. "Enhanced Molecular Spin-Photon Coupling at Superconducting Nanoconstrictions." *ACS Nano* 14(7):8707–8715. doi.org/10.1021/Acsnano.0c03167.

Gimeno, I., A. Urtizberea, J. Román-Roche, D. Zueco, A. Camón, P. J. Alonso, O. Roubeau, and F. Luis. 2021. "Broad-Band Spectroscopy of a Vanadyl Porphyrin: A Model Electronuclear Spin Qudit." *Chemical Science* 12(15):5621–5630. doi.org/10.1039/D1SC00564B.

Giovannetti, V. 2004. "Quantum-Enhanced Measurements: Beating the Standard Quantum Limit." *Science* 306(5700):1330–1336. https://doi.org/10.1126/science.1104149.

Giri, S. K., and G. C. Schatz. 2022. "Manipulating Two-Photon Absorption of Molecules Through Efficient Optimization of Entangled Light." *Physical Review Letters* 13(43):10140–10146. doi.org/10.1021/Acs.Jpclett.2c02842.

Godejohann, F., A. V. Scherbakov, S. M. Kukhtaruk, A. N. Poddubny, D. D. Yaremkevich, M. Wang, A. Nadzeyka, D. R. Yakovlev, A. W. Rushforth, A. V. Akimov, and M. Bayer. 2020. "Magnon Polaron Formed by Selectively Coupled Coherent Magnon and Phonon Modes of a Surface Patterned Ferromagnet." *Physical Review B* 102(14):144438. doi.org/10.1103/Physrevb.102.144438.

Godfrin, C., A. Ferhat, R. Ballou, S. Klyatskaya, M. Ruben, W. Wernsdorfer, and F. Balestro. 2017. "Operating Quantum States in Single Magnetic Molecules: Implementation of Grover's Quantum Algorithm." *Physical Review Letters* 119(18):187702. doi.org/10.1103/Physrevlett.119.187702.

Godfrin, C., S. Thiele, A. Ferhat, S. Klyatskaya, M. Ruben, W. Wernsdorfer, and F. Balestro. 2017. "Electrical Read-Out of a Single Spin Using an Exchange-Coupled Quantum Dot." *ACS Nano* 11(4):3984–3989. doi.org/10.1021/Acsnano.7b00451.

Goldfarb, D. 2017. "ELDOR-Detected NMR." *eMagRes* 563:101–114.

Gould, C. A., J. Marbey, V. Vieru, D. A. Marchiori, R. D. Britt, L. F. Chibotaru, S. Hill, and J. R. Long. 2021. "Isolation of a Triplet Benzene Dianion." *Nature Chemistry* 13(10):1001–1005. doi.org/10.1038/S41557-021-00737-8.

Graham, M. J., J. M. Zadrozny, M. Shiddiq, J. S. Anderson, M. S. Fataftah, S. Hill, and D. E. Freedman. 2014. "Influence of Electronic Spin and Spin–Orbit Coupling on Decoherence in Mononuclear Transition Metal Complexes." *Journal of the American Chemical Society* 136(21):7623–7626. doi.org/10.1021/Ja5037397.

Greer, S. M., J. Mckay, K. M. Gramigna, C. M. Thomas, S. A. Stoian, and S. Hill. 2018. "Probing Fe–V Bonding in a C_3-Symmetric Heterobimetallic Complex." *Inorganic Chemistry* 57(10):5870–5878. doi.org/10.1021/Acs.Inorgchem.8b00280.

Gross, L., K.-H. Rieder, F. Moresco, S. M. Stojkovic, A. Gourdon, and C. Joachim. 2005. "Trapping and Moving Metal Atoms with a Six-Leg Molecule." *Nature Materials* 4(12):892–895. doi.org/10.1038/Nmat1529.

Gu, B., and I. Franco. 2018. "Optical Absorption Properties of Laser-Driven Matter." *Physical Review A* 98(6):063412. doi.org/10.1103/Physreva.98.063412.

Gu, B., and S. Mukamel. 2020. "Manipulating Two-Photon-Absorption of Cavity Polaritons by Entangled Light." *Journal of Physical Chemistry Letters* 11(19):8177–8182. doi.org/10.1021/Acs.Jpclett.0c02282.

Gu, B., and S. Mukamel. 2021. "Optical-Cavity Manipulation of Conical Intersections and Singlet Fission in Pentacene Dimers." *Journal of Physical Chemistry Letters* 12(8):2052–2056. doi.org/10.1021/acs.jpclett.0c03829.

Gu, B., D. Keefer, F. Aleotti, A. Nenov, M. Garavelli, and S. Mukamel. 2021. "Photoisomerization Transition State Manipulation by Entangled Two-Photon Absorption." *Proceedings of the National Academy of Sciences USA* 118(47):E2116868118. doi.org/10.1073/Pnas.2116868118.

Guo, H., C. E. Dickerson, A. J. Shin, C. Zhao, T. L. Atallah, J. R. Caram, W. C. Campbell, and A. N. Alexandrova. 2021. "Surface Chemical Trapping of Optical Cycling Centers." *Physical Chemistry Chemical Physics* 23(1):211–218. doi.org/10.1039/D0CP04525J.

Guzman, A. R., M. R. Harpham, Ö. Süzer, M. M. Haley, and T. G. Goodson, III. 2010. "Spatial Control of Entangled Two-Photon Absorption with Organic Chromophores." *Journal of the American Chemical Society* 132(23):7840–7841. doi.org/10.1021/Ja1016816.

Hallas, C., N. B. Vilas, L. Anderegg, P. Robichaud, A. Winnicki, C. Zhang, L. Cheng, and J. M. Doyle. 2023. "Optical Trapping of a Polyatomic Molecule in an l-Type Parity Doublet State." *Physical Review Letters* 130:153202. doi.org/10.1103/PhysRevLett.130.153202.

Hameroff, S., and R. Penrose. 2014. "Consciousness in the Universe: A Review of the 'Orch OR' Theory." *Physics of Life Reviews* 11(1):39–78. doi.org/10.1016/J.Plrev.2013.08.002.

Harding, R. T., S. Zhou, J. Zhou, T. Lindvall, W. K. Myers, A. Ardavan, G. A. D. Briggs, K. Porfyrakis, and E. A. Laird. 2017. "Spin Resonance Clock Transition of the Endohedral Fullerene $^{15}N@C_{60}$." *Physical Review Letters* 119(14):140801. doi.org/10.1103/Physrevlett.119.140801.

Harmer, J. R. 2016. "Hyperfine Spectroscopy–ENDOR." *eMagRes* 5:1493–1514.

Harpham, M. R., Ö. Süzer, C.-Q. Ma, P. Bäuerle, and T. Goodson, III. 2009. "Thiophene Dendrimers as Entangled Photon Sensor Materials." *Journal of the American Chemical Society* 131(3):973–979. doi.org/10.1021/Ja803268s.

Harvey, S. M., and M. R. Wasielewski. 2021. "Photogenerated Spin-Correlated Radical Pairs: From Photosynthetic Energy Transduction to Quantum Information Science." *Journal of the American Chemical Society* 143(38):15508–15529. doi.org/10.1021/Jacs.1c07706.

Hassan, A. K., L. A. Pardi, J. Krzystek, A. Sienkiewicz, P. Goy, M. Rohrer, and L. C. Brunel. 2000. "Ultrawide Band Multifrequency High-Field EMR Technique: A Methodology for Increasing Spectroscopic Information." *Journal of Magnetic Resonance* 142(2):300–312. doi.org/10.1006/Jmre.1999.1952.

Hassan, M. A., T. Elrifai, A. Sakr, M. Kern, K. Lips, and J. Anders. 2021. "A 14-Channel 7 GHz VCO-Based EPR-on-a-Chip Sensor with Rapid Scan Capabilities." 2021 IEEE Sensors, Virtual Conference, Oct. 31–Nov. 3, 2021.

Hassan, M. A., M. Kern, A. Chu, G. Kalra, E. Shabratova, A. Tsarapkin, N. Mackinnon, K. Lips, C. Teutloff, R. Bittl, J. G. Korvink, and J. Anders. 2022. "Towards Single-Cell Pulsed EPR Using VCO-Based EPR-on-a-Chip Detectors." *Frequenz* 76(11–12):699–717. doi.org/10.1515/Freq-2022-0096.

Hay, M. A., A. Sarkar, G. A. Craig, L. Bhaskaran, J. Nehrkorn, M. Ozerov, K. E. R. Marriott, C. Wilson, G. Rajaraman, S. Hill, and M. Murrie. 2019. "In-Depth Investigation of Large Axial Magnetic Anisotropy in Monometallic 3d Complexes Using Frequency Domain Magnetic Resonance and Ab Initio Methods: A Study of Trigonal Bipyramidal Co(II)." *Chemical Science* 10(25):6354–6361. doi.org/10.1039/C9SC00987F.

Hecht, B., B. Sick, U. P. Wild, V. Deckert, R. Zenobi, O. J. F. Martin, and D. W. Pohl. 2000. "Scanning Near-Field Optical Microscopy with Aperture Probes: Fundamentals and Applications." *Journal of Chemical Physics* 112(18):7761–7774. doi.org/10.1063/1.481382.

Heinrich, A. J., J. A. Gupta, C. P. Lutz, and D. M. Eigler. 2004. "Single-Atom Spin-Flip Spectroscopy." *Science* 306(5695):466–469. doi.org/10.1126/Science.1101077.

Heinz, T., O. Shpyrko, D. Basov, N. Berrah, P. Bucksbaum, T. Devereaux, D. Fritz, K. Gaffney, O. Gessner, V. Gopalan, Z. Hasan, A. Lanzara, T. Martinez, A. Millis, S. Mukamel, M. Murnane, K. Nelson, R. Prasankumar, D. Reis, K. Schafer, G. Scholes, Z. X. Shen, A. Stolow, H. Wen, M. Wolf, D. Xiao, L. Young, B. Garrett, L. Horton, H. Kerch, J. Krause, T. Settersten, L. Wilson, K. Runkles, T. Anderson, G. Chui, and E. Rutherford. 2017. "Basic Energy Sciences Roundtable: Opportunities for Basic Research at the Frontiers of XFEL Ultrafast Science." Office of Scientific and Technical Information (OSTI). doi.org/10.2172/1616251.

Hemmer, P. 2013. "Toward Molecular-Scale MRI." *Science* 339(6119):529–530. doi.org/10.1126/Science.1233222.

Hensen, B., H. Bernien, A. E. Dréau, A. Reiserer, N. Kalb, M. S. Blok, J. Ruitenberg, R. F. L. Vermeulen, R. N. Schouten, C. Abellán, W. Amaya, V. Pruneri, M. W. Mitchell, M. Markham, D. J. Twitchen, D. Elkouss, S. Wehner, T. H. Taminiau, and R. Hanson. 2015. "Loophole-Free Bell Inequality Violation Using Electron Spins Separated by 1.3 Kilometres." *Nature* 526(7575):682–686. doi.org/10.1038/Nature15759.

Hernandez, J. M., X. X. Zhang, F. Luis, J. Tejada, J. R. Friedman, M. P. Sarachik, and R. Ziolo. 1997. "Evidence for Resonant Tunneling of Magnetization in Mn_{12} Acetate Complex." *Physical Review B* 55(9):5858–5865. doi.org/10.1103/Physrevb.55.5858.

Higgins, J. S., M. A. Allodi, L. T. Lloyd, J. P. Otto, S. H. Sohail, R. G. Saer, R. E. Wood, S. C. Massey, P. C. Ting, R. E. Blankenship, and G. S. Engel. 2021. "Redox Conditions Correlated with Vibronic Coupling Modulate Quantum Beats in Photosynthetic Pigment-Protein Complexes." *Proceedings of the National Academy of Sciences USA* 118(49). doi.org/10.1073/Pnas.2112817118.

Higgins, J. S., L. T. Lloyd, S. H. Sohail, M. A. Allodi, J. P. Otto, R. G. Saer, R. E. Wood, S. C. Massey, P.-C. Ting, R. E. Blankenship, and G. S. Engel. 2021. "Photosynthesis Tunes Quantum-Mechanical Mixing of Electronic and Vibrational States to Steer Exciton Energy Transfer." *Proceedings of the National Academy of Sciences USA* 118(11):E2018240118. doi.org/10.1073/Pnas.2018240118.

Hill, S. 2013. "Magnetization Tunneling in High-Symmetry Mn_{12} Single-Molecule Magnets." *Polyhedron* 64:128–135. doi.org/10.1016/J.Poly.2013.03.005.

Hill, S., R. S. Edwards, N. Aliaga-Alcalde, and G. Christou. 2003. "Quantum Coherence in an Exchange-Coupled Dimer of Single-Molecule Magnets." *Science* 302(5647):1015–1018. doi.org/10.1126/Science.1090082.

Hill, S., R. S. Edwards, S. I. Jones, N. S. Dalal, and J. M. North. 2003. "Definitive Spectroscopic Determination of the Transverse Interactions Responsible for the Magnetic Quantum Tunneling in Mn_{12}-Acetate." *Physical Review Letters* 90(21):217204. doi.org/10.1103/Physrevlett.90.217204.

Hiscock, H. G., S. Worster, D. R. Kattnig, C. Steers, Y. Jin, D. E. Manolopoulos, H. Mouritsen, and P. J. Hore. 2016. "The Quantum Needle of the Avian Magnetic Compass." *Proceedings of the National Academy of Sciences USA* 113(17): 4634–4639. doi.org/10.1073/Pnas.1600341113.

Hla, S.-W., and K.-H. Rieder. 2003. "STM Control of Chemical Reactions: Single-Molecule Synthesis." *Annual Review of Physical Chemistry* 54(1):307–330. doi.org/10.1146/Annurev.Physchem.54.011002.103852.

Holland, C. M., Y. Lu, and L. W. Cheuk. 2022. "On-Demand Entanglement of Molecules in a Reconfigurable Optical Tweezer Array." *ArXiv preprint*. doi.org/Arxiv:2210.06309.

Hong, C. K., Z. Y. Ou, and L. Mandel. 1987. "Measurement of Subpicosecond Time Intervals Between Two Photons by Interference." *Physical Review Letters* 59(18):2044–2046. doi.org/10.1103/Physrevlett.59.2044.

Hoover, E. E., and J. A. Squier. 2013. "Advances in Multiphoton Microscopy Technology." *Nature Photonics* 7(2):93–101. doi.org/10.1038/Nphoton.2012.361.

Hu, Z., B.-W. Dong, Z. Liu, J.-J. Liu, J. Su, C. Yu, J. Xiong, D.-E. Shi, Y. Wang, B.-W. Wang, A. Ardavan, Z. Shi, S.-D. Jiang, and S. Gao. 2018. "Endohedral Metallofullerene as Molecular High Spin Qubit: Diverse Rabi Cycles in $Gd_2@C_{79}N$." *Journal of the American Chemical Society* 140(3):1123–1130. doi.org/10.1021/Jacs.7b12170.

Hussain, R., G. Allodi, A. Chiesa, E. Garlatti, D. Mitcov, A. Konstantatos, K. S. Pedersen, R. De Renzi, S. Piligkos, and S. Carretta. 2018. "Coherent Manipulation of a Molecular Ln-Based Nuclear Qudit Coupled to an Electron Qubit." *Journal of the American Chemical Society* 140(31):9814–9818. doi.org/10.1021/Jacs.8b05934.

Imtiaz, A., T. M. Wallis, and P. Kabos. 2014. "Near-Field Scanning Microwave Microscopy: An Emerging Research Tool for Nanoscale Metrology." *IEEE Microwave Magazine* 15(1):52–64. doi.org/10.1109/MMM.2013.2288711.

Ishizaki, A., and G. R. Fleming. 2009. "Theoretical Examination of Quantum Coherence in a Photosynthetic System at Physiological Temperature." *Proceedings of the National Academy of Sciences USA* 106(41):17255–17260. doi.org/10.1073/Pnas.0908989106.

Ispasoiu, R. G., and T. Goodson. 2000. "Photon-Number Squeezing by Two-Photon Absorption in an Organic Polymer." *Optics Communications* 178(4):371–376. doi.org/10.1016/S0030-4018(00)00681-7.

Ivanov, M. V., F. H. Bangerter, P. Wójcik, and A. I. Krylov. 2020. "Toward Ultracold Organic Chemistry: Prospects of Laser Cooling Large Organic Molecules." *Physical Review Letters* 11(16):6670–6676. doi.org/10.1021/Acs.Jpclett.0c01960.

Jackson, C. E., C.-Y. Lin, S. H. Johnson, J. Van Tol, and J. M. Zadrozny. 2019. "Nuclear-Spin-Pattern Control of Electron-Spin Dynamics in a Series of V(IV) Complexes." *Chemical Science* 10(36):8447–8454. doi.org/10.1039/C9SC02899D.

Jelezko, F., and J. Wrachtrup. 2006. "Single Defect Centres in Diamond: A Review." *Physica Status Solidi (A)* 203(13): 3207–3225. doi.org/10.1002/Pssa.200671403.

Jenkins, M., T. Hümmer, M. J. Martínez-Pérez, J. García-Ripoll, D. Zueco, and F. Luis. 2013. "Coupling Single-Molecule Magnets to Quantum Circuits." *New Journal of Physics* 15(9):095007. doi.org/10.1088/1367-2630/15/9/095007.

Jeschke, G. 2018. "Dipolar Spectroscopy–Double-Resonance Methods." *eMagRes* 5:1459–1476.

Jo, M.-H., J. E. Grose, K. Baheti, M. M. Deshmukh, J. J. Sokol, E. M. Rumberger, D. N. Hendrickson, J. R. Long, H. Park, and D. C. Ralph. 2006. "Signatures of Molecular Magnetism in Single-Molecule Transport Spectroscopy." *Nano Letters* 6(9):2014–2020. doi.org/10.1021/Nl061212i.

Johnson, S. H., C. E. Jackson, and J. M. Zadrozny. 2020. "Programmable Nuclear-Spin Dynamics in Ti(IV) Coordination Complexes." *Inorganic Chemistry* 59(11):7479–7486. doi.org/10.1021/Acs.Inorgchem.0c00244.

Jones, C. G., M. W. Martynowycz, J. Hattne, T. J. Fulton, B. M. Stoltz, J. A. Rodriguez, H. M. Nelson, and T. Gonen. 2018. "The CryoEM Method MicroED as a Powerful Tool for Small Molecule Structure Determination." *ACS Central Science* 4(11):1587–1592. doi.org/10.1021/Acscentsci.8b00760.

Julien, M. H., Z. H. Jang, A. Lascialfari, F. Borsa, M. Horvatić, A. Caneschi, and D. Gatteschi. 1999. "Proton NMR for Measuring Quantum Level Crossing in the Magnetic Molecular Ring Fe10." *Physical Review Letters* 83(1):227–230. doi.org/10.1103/Physrevlett.83.227.

Kalb, N., A. A. Reiserer, P. C. Humphreys, J. J. W. Bakermans, S. J. Kamerling, N. H. Nickerson, S. C. Benjamin, D. J. Twitchen, M. Markham, and R. Hanson. 2017. "Entanglement Distillation Between Solid-State Quantum Network Nodes." *Science* 356(6341):928–932. doi.org/10.1126/Science.Aan0070.

Kang, G., K. N. Avanaki, M. A. Mosquera, R. K. Burdick, J. P. Villabona-Monsalve, T. Goodson, III, and G. C. Schatz. 2020. "Efficient Modeling of Organic Chromophores for Entangled Two-Photon Absorption." *Journal of the American Chemical Society* 142(23):10446–10458. doi.org/10.1021/Jacs.0c02808.

Kawaguchi, R., K. Hashimoto, T. Kakudate, K. Katoh, M. Yamashita, and T. Komeda. 2023. "Spatially Resolving Electron Spin Resonance of π-Radical in Single-Molecule Magnet." *Nano Letters* 23(1):213–219. doi.org/10.1021/acs.nanolett.2c04049.

Kazmierczak, N. P., R. Mirzoyan, and R. G. Hadt. 2021. "The Impact of Ligand Field Symmetry on Molecular Qubit Coherence." *Journal of the American Chemical Society* 143(42):17305–17315. doi.org/10.1021/Jacs.1c04605.

Kirschvink, J. L., M. M. Walker, and C. E. Diebel. 2001. "Magnetite-Based Magnetoreception." *Current Opinion in Neurobiology* 11(4):462–467. doi.org/10.1016/S0959-4388(00)00235-X.

Kłos, J., and S. Kotochigova. 2020. "Prospects for Laser Cooling of Polyatomic Molecules with Increasing Complexity." *Physical Review Research* 2(1):013384. doi.org/10.1103/Physrevresearch.2.013384.

Köhler, J., J. A. J. M. Disselhorst, M. C. J. M. Donckers, E. J. J. Groenen, J. Schmidt, and W. E. Moerner. 1993. "Magnetic Resonance of a Single Molecular Spin." *Nature* 363(6426):242–244. doi.org/10.1038/363242a0.

Kojima J., and Q-V. Nguyen. 2004. "Entangled Biphoton Virtual-State Spectroscopy of the $A^2\Sigma^+–X^2\Pi$ System of OH." *Chemical Physics Letters* 396(4–6):323–328. doi.org/10.1016/j.cplett.2004.08.051.

Kothe, G., M. Lukaschek, T. Yago, G. Link, K. L. Ivanov, and T.-S. Lin. 2021. "Initializing 214 Pure 14-Qubit Entangled Nuclear Spin States in a Hyperpolarized Molecular Solid." *Journal of Physical Chemistry Letters* 12(14):3647–3654. doi.org/10.1021/Acs.Jpclett.1c00726.

Kovarik, S., R. Robles, R. Schlitz, T. S. Seifert, N. Lorente, P. Gambardella, and S. Stepanow. 2022. "Electron Paramagnetic Resonance of Alkali Metal Atoms and Dimers on Ultrathin MgO." *Nano Letters* 22(10):4176–4181. doi.org/10.1021/Acs.Nanolett.2c00980.

Kozyryev, I., L. Baum, K. Matsuda, H. Boerge, and J. M. Doyle. 2016. "Radiation Pressure Force from Optical Cycling on a Polyatomic Molecule." *Journal of Physics B: Atomic, Molecular, and Optical Physics* 49(13):134002. doi.org/10.1088/0953-4075/49/13/134002.

Kozyryev, I., L. Baum, K. Matsuda, and J. M. Doyle. 2016. "Proposal for Laser Cooling of Complex Polyatomic Molecules." *Chemistry Europe* 17(22):3641–3648. doi.org/10.1002/Cphc.201601051.

Kozyryev, I., Z. Lasner, and J. M. Doyle. 2021. "Enhanced Sensitivity to Ultralight Bosonic Dark Matter in the Spectra of the Linear Radical SrOH." *Physical Review A* 103(4):043313. doi.org/10.1103/Physreva.103.043313.

Kozyryev, I., T. C. Steimle, P. Yu, D.-T. Nguyen, and J. M. Doyle. 2019. "Determination of CaOH and $CaOCH_3$ Vibrational Branching Ratios for Direct Laser Cooling and Trapping." *New Journal of Physics* 21(5):052002. doi.org/10.1088/1367-2630/Ab19d7.

Kragskow, J. G. C., J. Marbey, C. D. Buch, J. Nehrkorn, M. Ozerov, S. Piligkos, S. Hill, and N. F. Chilton. 2022. "Analysis of Vibronic Coupling in a 4f Molecular Magnet with FIRMS." *Nature Communications* 13(1):825. doi.org/10.1038/S41467-022-28352-2.

Krzyaniak, M. D., L. Kobr, B. K. Rugg, B. T. Phelan, E. A. Margulies, J. N. Nelson, R. M. Young, and M. R. Wasielewski. 2015. "Fast Photo-Driven Electron Spin Coherence Transfer: The Effect of Electron-Nuclear Hyperfine Coupling on Coherence Dephasing." *Journal of Materials Chemistry C* 3(30):7962–7967. doi.org/10.1039/C5TC01446H.

Kubo, T., T. Goto, T. Koshiba, K. Takeda, and K. Awaga. 2002. "^{55}Mn NMR in Mn_{12} Acetate: Hyperfine Interaction and Magnetic Relaxation of Cluster." *Physical Review B* 65(22):224425. doi.org/10.1103/Physrevb.65.224425.

Kundu, K., J. Chen, S. Hoffman, J. Marbey, D. Komijani, Y. Duan, A. Gaita-Ariño, J. Stanton, X. Zhang, H.-P. Cheng, and S. Hill. 2023. "Electron-Nuclear Decoupling at a Spin Clock Transition." *Communications Physics* 6(1):38. doi.org/10.1038/s42005-023-01152-w.

Kundu, K., J. R. K. White, S. A. Moehring, J. M. Yu, J. W. Ziller, F. Furche, W. J. Evans, and S. Hill. 2022. "A 9.2-GHz Clock Transition in a Lu(II) Molecular Spin Qubit Arising from a 3,467-MHz Hyperfine Interaction." *Nature Chemistry* 14(4):392–397. doi.org/10.1038/S41557-022-00894-4.

Künstner, S., A. Chu, K. P. Dinse, A. Schnegg, J. E. Mcpeak, B. Naydenov, J. Anders, and K. Lips. 2021. "Rapid-Scan Electron Paramagnetic Resonance Using an EPR-on-a-Chip Sensor." *Magnetic Resonance* 2(2):673–687. doi.org/10.5194/mr-2-673-2021.

Kutas, M., B. Haase, P. Bickert, F. Riexinger, D. Molter, and G. von Freymann. 2020. "Terahertz Quantum Sensing." *Science Advances* 6(11):eaaz8065. doi.org/10.1126/sciadv.aaz8065.

Lecoq, J., N. Orlova, and B. F. Grewe. 2019. "Wide. Fast. Deep: Recent Advances in Multiphoton Microscopy of *In Vivo* Neuronal Activity." *Journal of Neuroscience* 39(46):9042–9052. doi.org/10.1523/Jneurosci.1527-18.2019.

Lee, D.-I., and T. Goodson. 2006. "Entangled Photon Absorption in an Organic Porphyrin Dendrimer." *Journal of Physical Chemistry B* 110(51):25582–25585. doi.org/10.1021/Jp066767g.

Leibfried, D. 2012. "Quantum State Preparation and Control of Single Molecular Ions." *New Journal of Physics* 14(2):023029. doi.org/10.1088/1367-2630/14/2/023029.

Leisegang, M., A. Christ, S. Haldar, S. Heinze, and M. Bode. 2021. "Molecular Chains: Arranging and Programming Logic Gates." *Nano Letters* 21(1):550–555. doi.org/10.1021/Acs.Nanolett.0c03984.

Lelek, M., M. T. Gyparaki, G. Beliu, F. Schueder, J. Griffié, S. Manley, R. Jungmann, M. Sauer, M. Lakadamyali, and C. Zimmer. 2021. "Single-Molecule Localization Microscopy." *Nature Reviews Methods Primers* 1(1):39. doi.org/10.1038/S43586-021-00038-X.

Lemos, G. B., V. Borish, G. D. Cole, S. Ramelow, R. Lapkiewicz, and A. Zeilinger. 2014. "Quantum Imaging with Undetected Photons." *Nature* 512(7515):409–412. doi.org/10.1038/nature13586.

Leuenberger, M. N., and D. Loss. 2001. "Quantum Computing in Molecular Magnets." *Nature* 410(6830):789–793. doi.org/10.1038/35071024.

Li, T., F. Li, C. Altuzarra, A. Classen, and G. S. Agarwal. 2020. "Squeezed Light Induced Two-Photon Absorption Fluorescence of Fluorescein Biomarkers." *Applied Physics Letters* 116(25):254001. doi.org/10.1063/5.0010909.

Li, Y., B. Shen, S. Li, Y. Zhao, J. Qu, and L. Liu. 2021. "Review of Stimulated Raman Scattering Microscopy Techniques and Applications in the Biosciences." *Advanced Biology* 5(1):2000184. doi.org/10.1002/Adbi.202000184.

Lim, H. 2019. "Harmonic Generation Microscopy 2.0: New Tricks Empowering Intravital Imaging for Neuroscience." *Frontiers in Molecular Biosciences* 6:99. doi.org/10.3389/Fmolb.2019.00099.

Lin, Y., D. R. Leibrandt, D. Leibfried, and C.-W. Chou. 2020. "Quantum Entanglement Between an Atom and a Molecule." *Nature* 581(7808):273–277. doi.org/10.1038/S41586-020-2257-1.

Lindner, C., S. Wolf, J. Kiessling, and F. Kühnemann. 2020. "Fourier Transform Infrared Spectroscopy with Visible Light." *Optics Express* 28(4):4426–4432. doi.org/10.1364/oe.382351.

Liu, J., J. Mrozek, W. K. Myers, G. A. Timco, R. E. P. Winpenny, B. Kintzel, W. Plass, and A. Ardavan. 2019. "Electric Field Control of Spins in Molecular Magnets." *Physical Review Letters* 122(3):037202. doi.org/10.1103/Physrevlett.122.037202.

Liu, J., J. Mrozek, A. Ullah, Y. Duan, J. J. Baldoví, E. Coronado, A. Gaita-Ariño, and A. Ardavan. 2021. "Quantum Coherent Spin–Electric Control in a Molecular Nanomagnet at Clock Transitions." *Nature Physics* 17(11):1205–1209. doi.org/10.1038/S41567-021-01355-4.

Liu, K. S., A. Henning, M. W. Heindl, R. D. Allert, J. D. Bartl, I. D. Sharp, R. Rizzato, and D. B. Bucher. 2022. "Surface NMR Using Quantum Sensors in Diamond." *Proceedings of the National Academy of Sciences USA* 119(5):E2111607119. doi.org/10.1073/Pnas.2111607119.

Liu, Y., and K.-K. Ni. 2022. "Bimolecular Chemistry in the Ultracold Regime." *Annual Review of Physical Chemistry* 73(1):73–96. doi.org/10.1146/Annurev-Physchem-090419-043244.

Liu, Y., M.-G. Hu, M. A. Nichols, D. Yang, D. Xie, H. Guo, and K.-K. Ni. 2021. "Precision Test of Statistical Dynamics with State-to-State Ultracold Chemistry." *Nature* 593(7859):379–384. doi.org/10.1038/S41586-021-03459-6.

Lloyd, S. 1993. "A Potentially Realizable Quantum Computer." *Science* 261(5128):1569–1571. doi.org/10.1126/Science.261.5128.1569.

Loh, H., K. C. Cossel, M. C. Grau, K.-K. Ni, E. R. Meyer, J. L. Bohn, J. Ye, and E. A. Cornell. 2013. "Precision Spectroscopy of Polarized Molecules in an Ion Trap." *Science* 342(6163):1220–1222. doi.org/10.1126/Science.1243683.

Loss, D., and D. P. DiVincenzo. 1998. "Quantum Computation with Quantum Dots." *Physical Review A* 57(1):120–126. doi.org/10.1103/Physreva.57.120.

Lotfi, H., M. A. Hassan, M. Kern, and J. Anders. 2022. "A Compact C-Band EPR-on-a-Chip Transceiver in 130-nm SiGe BiCMOS." 17th Conference on Ph.D Research in Microelectronics and Electronics (PRIME), Villasimius (Ca), Italy, June 12–15, 2022.

Loth, S., S. Baumann, C. P. Lutz, D. M. Eigler, and A. J. Heinrich. 2012. "Bistability in Atomic-Scale Antiferromagnets." *Science* 335(6065):196–199. doi.org/10.1126/Science.1214131.

Loth, S., M. Etzkorn, C. P. Lutz, D. M. Eigler, and A. J. Heinrich. 2010. "Measurement of Fast Electron Spin Relaxation Times with Atomic Resolution." *Science* 329(5999):1628–1630. doi.org/10.1126/Science.1191688.

Loudon, R., and P. L. Knight. 1987. "Squeezed Light." *Journal of Modern Optics* 34(6–7):709–759. doi.org/10.1080/09500348714550721.

Lovchinsky, I., J. D. Sanchez-Yamagishi, E. K. Urbach, S. Choi, S. Fang, T. I. Andersen, K. Watanabe, T. Taniguchi, A. Bylinskii, E. Kaxiras, P. Kim, H. Park, and M. D. Lukin. 2017. "Magnetic Resonance Spectroscopy of an Atomically Thin Material Using a Single-Spin Qubit." *Science* 355(6324):503–507. doi.org/10.1126/Science.Aal2538.

Lovchinsky, I., A. O. Sushkov, E. Urbach, N. P. De Leon, S. Choi, K. De Greve, R. Evans, R. Gertner, E. Bersin, C. Müller, L. Mcguinness, F. Jelezko, R. L. Walsworth, H. Park, and M. D. Lukin. 2016. "Nuclear Magnetic Resonance Detection and Spectroscopy of Single Proteins Using Quantum Logic." *Science* 351(6275):836–841. doi.org/10.1126/Science.Aad8022.

Luis, F., P. J. Alonso, O. Roubeau, V. Velasco, D. Zueco, D. Aguilà, J. I. Martínez, L. A. Barrios, and G. Aromí. 2020. "A Dissymmetric [Gd_2] Coordination Molecular Dimer Hosting Six Addressable Spin Qubits." *Communications Chemistry* 3(1):176. doi.org/10.1038/S42004-020-00422-W.

Lunghi, A., and S. Sanvito. 2020. "The Limit of Spin Lifetime in Solid-State Electronic Spins." *Physical Review Letters* 11(15):6273–6278. doi.org/10.1021/Acs.Jpclett.0c01681.

Lutz, P., R. Marx, D. Dengler, A. Kromer, and J. Van Slageren. 2013. "Quantum Coherence in a Triangular Cu_3 Complex." *Molecular Physics* 111(18–19):2897–2902. doi.org/10.1080/00268976.2013.826421.

Macià, F., J. Lawrence, S. Hill, J. M. Hernandez, J. Tejada, P. V. Santos, C. Lampropoulos, and G. Christou. 2008. "Spin Dynamics in Single-Molecule Magnets Combining Surface Acoustic Waves and High-Frequency Electron Paramagnetic Resonance." *Physical Review B* 77(2):020403. doi.org/10.1103/Physrevb.77.020403.

Maeda, K., K. B. Henbest, F. Cintolesi, I. Kuprov, C. T. Rodgers, P. A. Liddell, D. Gust, C. R. Timmel, and P. J. Hore. 2008. "Chemical Compass Model of Avian Magnetoreception." *Nature* 453(7193):387–390. doi.org/10.1038/Nature06834.

Maeda, K., A. J. Robinson, K. B. Henbest, H. J. Hogben, T. Biskup, M. Ahmad, E. Schleicher, S. Weber, C. R. Timmel, and P. J. Hore. 2012. "Magnetically Sensitive Light-Induced Reactions in Cryptochrome are Consistent with Its Proposed Role as a Magnetoreceptor." *Proceedings of the National Academy of Sciences USA* 109(13):4774–4779. doi.org/10.1073/Pnas.1118959109.

Mamin, H. J., M. Kim, M. H. Sherwood, C. T. Rettner, K. Ohno, D. D. Awschalom, and D. Rugar. 2013. "Nanoscale Nuclear Magnetic Resonance with a Nitrogen-Vacancy Spin Sensor." *Science* 339(6119):557–560. doi.org/10.1126/Science.1231540.

Marriott, K. E. R., L. Bhaskaran, C. Wilson, M. Medarde, S. T. Ochsenbein, S. Hill, and M. Murrie. 2015. "Pushing the Limits of Magnetic Anisotropy in Trigonal Bipyramidal Ni(II)." *Chemical Science* 6(12):6823–6828. doi.org/10.1039/C5SC02854J.

Maurer, P. C., G. Kucsko, C. Latta, L. Jiang, N. Y. Yao, S. D. Bennett, F. Pastawski, D. Hunger, N. Chisholm, M. Markham, D. J. Twitchen, J. I. Cirac, and M. D. Lukin. 2012. "Room-Temperature Quantum Bit Memory Exceeding One Second." *Science* 336(6086):1283. doi.org/10.1126/Science.1220513.

Mauser, N., and A. Hartschuh. 2014. "Tip-Enhanced Near-Field Optical Microscopy." *Chemical Society Reviews* 43(4):1248–1262. doi.org/10.1039/C3CS60258C.

McCarron D. 2018. "Laser Cooling and Trapping Molecules." *Journal of Physics B: Atomic, Molecular and Optical Physics* 51(21):212001. doi.org/10.1088/1361-6455/aadfba.

McInnes, E. J. L. 2022. "Molecular Spins Clock in." *Nature Chemistry* 14(4):361–362. doi.org/10.1038/S41557-022-00919-Y.

McKemmish, L. K., R. H. McKenzie, N. S. Hush, and J. R. Reimers. 2011. "Quantum Entanglement Between Electronic and Vibrational Degrees of Freedom in Molecules." *Journal of Chemical Physics* 135(24):244110. doi.org/10.1063/1.3671386.

Micotti, E., Y. Furukawa, K. Kumagai, S. Carretta, A. Lascialfari, F. Borsa, G. A. Timco, and R. E. P. Winpenny. 2006. "Local Spin Moment Distribution in Antiferromagnetic Molecular Rings Probed by NMR." *Physical Review Letters* 97(26):267204. doi.org/10.1103/Physrevlett.97.267204.

Mitra, D., Z. D. Lasner, G.-Z. Zhu, C. E. Dickerson, B. L. Augenbraun, A. D. Bailey, A. N. Alexandrova, W. C. Campbell, J. R. Caram, E. R. Hudson, and J. M. Doyle. 2022. "Pathway Toward Optical Cycling and Laser Cooling of Functionalized Arenes." *Physical Review Letters* 13(30):7029–7035. doi.org/10.1021/Acs.Jpclett.2c01430.

Mitra, D., N. B. Vilas, C. Hallas, L. Anderegg, B. L. Augenbraun, L. Baum, C. Miller, S. Raval, and J. M. Doyle. 2020. "Direct Laser Cooling of a Symmetric Top Molecule." *Science* 369(6509):1366–1369. doi.org/10.1126/Science.Abc5357.

Moffitt, J. R., Y. R. Chemla, S. B. Smith, and C. Bustamante. 2008. "Recent Advances in Optical Tweezers." *Annual Review of Biochemistry* 77:205–228. doi.org/10.1146/Annurev.Biochem.77.043007.090225.

Mola, M., S. Hill, P. Goy, and M. Gross. 2000. "Instrumentation for Millimeter-Wave Magnetoelectrodynamic Investigations of Low-Dimensional Conductors and Superconductors." *Review of Scientific Instruments* 71(1):186–200. doi.org/10.1063/1.1150182.

Moneron, G., and S. W. Hell. 2009. "Two-Photon Excitation STED Microscopy." *Optics Express* 17(17):14567–14573. doi.org/10.1364/OE.17.014567.

Morello, A. 2008. "Quantum Nanomagnets and Nuclear Spins: An Overview." In *Quantum Magnetism*, 125–138. Amsterdam: Springer Netherlands.

Morello, A., and L. J. de Jongh. 2007. "Dynamics and Thermalization of the Nuclear Spin Bath in the Single-Molecule Magnet Mn_{12}–ac: Test for the Theory of Spin Tunneling." *Physical Review B* 76(18):184425. doi.org/10.1103/Physrevb.76.184425.

Morello, A., O. N. Bakharev, H. B. Brom, R. Sessoli, and L. J. De Jongh. 2004. "Nuclear Spin Dynamics in the Quantum Regime of a Single-Molecule Magnet." *Physical Review Letters* 93(19):197202. doi.org/10.1103/Physrevlett.93.197202.

Moreno-Pineda, E., and W. Wernsdorfer. 2021. "Measuring Molecular Magnets for Quantum Technologies." *Nature Reviews Physics* 3(9):645–659. doi.org/10.1038/S42254-021-00340-3.

Morley, G. W., L.-C. Brunel, and J. van Tol. 2008. "A Multifrequency High-Field Pulsed Electron Paramagnetic Resonance/Electron-Nuclear Double Resonance Spectrometer." *Review of Scientific Instruments* 79(6):064703. doi.org/10.1063/1.2937630.

Mukai, Y., R. Okamoto, and S. Takeuchi. 2022. "Quantum Fourier-Transform Infrared Spectroscopy in the Fingerprint Region." *Optics Express* 30(13):22624–22636. doi.org/10.1364/oe.455718.

Mukamel, S., M. Freyberger, W. Schleich, M. Bellini, A. Zavatta, G. Leuchs, C. Silberhorn, R. W. Boyd, L. L. Sánchez-Soto, A. Stefanov, M. Barbieri, A. Paterova, L. Krivitsky, S. Shwartz, K. Tamasaku, K. Dorfman, F. Schlawin, V. Sandoghdar, M. Raymer, A. Marcus, O. Varnavski, T. Goodson, Z.-Y. Zhou, B.-S. Shi, S. Asban, M. Scully, G. Agarwal, T. Peng, A. V. Sokolov, Z.-D. Zhang, M. S. Zubairy, I. A. Vartanyants, E. Del Valle, and F. Laussy. 2020. "Roadmap on Quantum Light Spectroscopy." *Journal of Physics B: Atomic, Molecular, and Optical Physics* 53(7):072002. doi.org/10.1088/1361-6455/Ab69a8.

Müller, C., X. Kong, J. M. Cai, K. Melentijević, A. Stacey, M. Markham, D. Twitchen, J. Isoya, S. Pezzagna, J. Meijer, J. F. Du, M. B. Plenio, B. Naydenov, L. P. Mcguinness, and F. Jelezko. 2014. "Nuclear Magnetic Resonance Spectroscopy with Single Spin Sensitivity." *Nature Communications* 5(1):4703. doi.org/10.1038/Ncomms5703.

Naaman, R., and D. H. Waldeck. 2012. "Chiral-Induced Spin Selectivity Effect." *Journal of Physical Chemistry Letters* 3(16):2178–2187. doi.org/10.1021/Jz300793y.

Nagarajan, K., A. Thomas, and T. W. Ebbesen. 2021. "Chemistry under Vibrational Strong Coupling." *Journal of the American Chemical Society* 143(41):16877–16889. doi.org/10.1021/jacs.1c07420.

Nehrkorn, J., S. M. Greer, B. J. Malbrecht, K. J. Anderton, A. Aliabadi, J. Krzystek, A. Schnegg, K. Holldack, C. Herrmann, T. A. Betley, S. Stoll, and S. Hill. 2021. "Spectroscopic Investigation of a Metal–Metal-Bonded Fe_6 Single-Molecule Magnet with an Isolated S = 19/2 Giant-Spin Ground State." *Inorganic Chemistry* 60(7):4610–4622. doi.org/10.1021/Acs.Inorgchem.0c03595.

Nelson, J. N., M. D. Krzyaniak, N. E. Horwitz, B. K. Rugg, B. T. Phelan, and M. R. Wasielewski. 2017. "Zero Quantum Coherence in a Series of Covalent Spin-Correlated Radical Pairs." *Journal of Physical Chemistry A* 121(11):2241–2252. doi.org/10.1021/Acs.Jpca.7b00587.

Neugebauer, P., D. Bloos, R. Marx, P. Lutz, M. Kern, D. Aguilà, J. Vaverka, O. Laguta, C. Dietrich, R. Clérac, and J. Van Slageren. 2018. "Ultra-Broadband EPR Spectroscopy in Field and Frequency Domains." *Physical Chemistry Chemical Physics* 20(22):15528–15534. doi.org/10.1039/C7CP07443C.

Nguyen, T. N., M. Shiddiq, T. Ghosh, K. A. Abboud, S. Hill, and G. Christou. 2015. "Covalently Linked Dimer of Mn_3 Single-Molecule Magnets and Retention of Its Structure and Quantum Properties in Solution." *Journal of the American Chemical Society* 137(22):7160–7168. doi.org/10.1021/Jacs.5b02677.

Ni, K.-K., S. Ospelkaus, M. H. G. De Miranda, A. Pe'er, B. Neyenhuis, J. J. Zirbel, S. Kotochigova, P. S. Julienne, D. S. Jin, and J. Ye. 2008. "A High Phase-Space-Density Gas of Polar Molecules." *Science* 322(5899):231–235. doi.org/10.1126/Science.1163861.

Ni, K. K., S. Ospelkaus, D. J. Nesbitt, J. Ye, and D. S. Jin. 2009. "A Dipolar Gas of Ultracold Molecules." *Physical Chemistry Chemical Physics* 11(42):9626–9639. doi.org/10.1039/B911779B.

NIST (National Institute of Standards and Technology). 2009. "Physicists Find Way to Control Individual Bits in Quantum Computers." NIST, July. https://www.nist.gov/news-events/news/2009/07/physicists-find-way-control-individual-bits-quantum-computers.

Olshansky, J. H., M. D. Krzyaniak, R. M. Young, and M. R. Wasielewski. 2019. "Photogenerated Spin-Entangled Qubit (Radical) Pairs in DNA Hairpins: Observation of Spin Delocalization and Coherence." *Journal of the American Chemical Society* 141(5):2152–2160. doi.org/10.1021/Jacs.8b13155.

Ospelkaus, S., K.-K. Ni, D. Wang, M. H. G. De Miranda, B. Neyenhuis, G. Quéméner, P. S. Julienne, J. L. Bohn, D. S. Jin, and J. Ye. 2010. "Quantum-State Controlled Chemical Reactions of Ultracold Potassium-Rubidium Molecules." *Science* 327(5967):853–857. doi.org/10.1126/Science.1184121.

Parzuchowski, K. M., A. Mikhaylov, M. D. Mazurek, R. N. Wilson, D. J. Lum, T. Gerrits, C. H. Camp, M. J. Stevens, and R. Jimenez. 2021. "Setting Bounds on Entangled Two-Photon Absorption Cross Sections in Common Fluorophores." *Physical Review Applied* 15(4):044012. doi.org/10.1103/Physrevapplied.15.044012.

Perrin, M. L., E. Burzurí, and H. S. J. van der Zant. 2015. "Single-Molecule Transistors." *Chemical Society Reviews* 44(4): 902–919. doi.org/10.1039/C4CS00231H.

Pietsch, T., S. Egle, M. Keller, H. Fridtjof-Pernau, F. Strigl, and E. Scheer. 2016. "Microwave-Induced Direct Spin-Flip Transitions in Mesoscopic Pd/Co Heterojunctions." *New Journal of Physics* 18(9):093045. doi.org/10.1088/1367-2630/18/9/093045.

Pinto, D., D. Paone, B. Kern, T. Dierker, R. Wieczorek, A. Singha, D. Dasari, A. Finkler, W. Harneit, J. Wrachtrup, and K. Kern. 2020. "Readout and Control of an Endofullerene Electronic Spin." *Nature Communications* 11(1):6405. doi.org/10.1038/S41467-020-20202-3.

Pirandola, S., J. Eisert, C. Weedbrook, A. Furusawa, and S. L. Braunstein. 2015. "Advances in Quantum Teleportation." *Nature Photonics* 9:641–652. doi.org/10.1038/Nphoton.2015.154.

Plasser, F. 2016. "Entanglement Entropy of Electronic Excitations." *Journal of Chemical Physics* 144:194107. doi.org/10.1063/1.4949535.

Potočnik, A., A. Bargerbos, F. A. Y. N. Schröder, S. A. Khan, M. C. Collodo, S. Gasparinetti, Y. Salathé, C. Creatore, C. Eichler, H. E. Türeci, A. W. Chin, and A. Wallraff. 2018. "Studying Light-Harvesting Models with Superconducting Circuits." *Nature Communications* 9(1):904. doi.org/10.1038/s41467-018-03312-x.

Prescimone, A., C. Morien, D. Allan, J. A. Schlueter, S. W. Tozer, J. L. Manson, S. Parsons, E. K. Brechin, and S. Hill. 2012. "Pressure-Driven Orbital Reorientations and Coordination-Sphere Reconstructions in [$CuF_2(H_2O)_2$(pyz)]." *Angewandte Chemie International Edition* 51(30):7490–7494. doi.org/10.1002/Anie.201202367.

Price, H. 2022. "Simulating Four-Dimensional Physics in the Laboratory." *Physics Today* 75(4):38–44. doi.org/10.1063/PT.3.4981.

Purcell, E. M., H. C. Torrey, and R. V. Pound. 1946. "Resonance Absorption by Nuclear Magnetic Moments in a Solid." *Physical Review* 69(1–2):37–38. doi.org/10.1103/Physrev.69.37.

Qian, C., K. Miao, L.-E. Lin, X. Chen, J. Du, and L. Wei. 2021. "Super-Resolution Label-Free Volumetric Vibrational Imaging." *Nature Communications* 12(1):3648. doi.org/10.1038/S41467-021-23951-X.

Ray, K., S. P. Ananthavel, D. H. Waldeck, and R. Naaman. 1999. "Asymmetric Scattering of Polarized Electrons by Organized Organic Films of Chiral Molecules." *Science* 283(5403):814–816. doi.org/10.1126/Science.283.5403.814.

Raymer, M. G., A. H. Marcus, J. R. Widom, and D. L. P. Vitullo. 2013. "Entangled Photon-Pair Two-Dimensional Fluorescence Spectroscopy (EPP-2DFS)." *Journal of Physical Chemistry B* 117(49):15559–15575. doi.org/10.1021/Jp405829n.

Rehberg, M., F. Krombach, U. Pohl, and S. Dietzel. 2011. "Label-Free 3D Visualization of Cellular and Tissue Structures in Intact Muscle with Second and Third Harmonic Generation Microscopy." *PLoS ONE* 6(11):E28237. doi.org/10.1371/Journal.Pone.0028237.

Ribeiro, R. F., L. A. Martínez-Martínez, M. Du, J. Campos-Gonzalez-Angulo, and J. Yuen-Zhou. 2018. "Polariton Chemistry: Controlling Molecular Dynamics with Optical Cavities." *Chemical Science* 9(30):6325–6339. doi.org/10.1039/C8SC01043A.

Riedel, D., I. Söllner, B. J. Shields, S. Starosielec, P. Appel, E. Neu, P. Maletinsky, and R. J. Warburton. 2017. "Deterministic Enhancement of Coherent Photon Generation from a Nitrogen-Vacancy Center in Ultrapure Diamond." *Physical Review X* 7(3):031040. doi.org/10.1103/Physrevx.7.031040.

Rieke, F., and D. A. Baylor. 1998. "Single-Photon Detection by Rod Cells of the Retina." *Reviews of Modern Physics* 70(3):1027–1036. doi.org/10.1103/Revmodphys.70.1027.

Rismani Yazdi, S., R. Nosrati, C. A. Stevens, D. Vogel, P. L. Davies, and C. Escobedo. 2018. "Magnetotaxis Enables Magnetotactic Bacteria to Navigate in Flow." *Small* 14(5):1702982. doi.org/10.1002/Smll.201702982.

Rodgers, L. V. H., L. B. Hughes, M. Xie, P. C. Maurer, S. Kolkowitz, A. C. B. Jayich, and N. P. De Leon. 2021. "Materials Challenges for Quantum Technologies Based on Color Centers in Diamond." *MRS Bulletin* 46(7):623–633. doi.org/10.1557/S43577-021-00137-W.

Röhlsberger, R., J. Evers, and S. Shwartz. 2014. "Quantum and Nonlinear Optics with Hard X-Rays." In *Synchrotron Light Sources and Free-Electron Lasers: Accelerator Physics, Instrumentation and Science Applications*, edited by E. Jaeschke, S. Khan, J. R. Schneider, and J. B. Hastings, 1–28. Cham: Springer International Publishing.

Roslyak, O., C. A. Marx, and S. Mukamel. 2009. "Nonlinear Spectroscopy with Entangled Photons: Manipulating Quantum Pathways of Matter." *Physical Review A* 79(3):033832. doi.org/10.1103/Physreva.79.033832.

Rosner, B. T., and D. W. van der Weide. 2002. "High-Frequency Near-Field Microscopy." *Review of Scientific Instruments* 73(7):2505–2525. doi.org/10.1063/1.1482150.

Rossini, A. J., A. Zagdoun, M. Lelli, A. Lesage, C. Copéret, and L. Emsley. 2013. "Dynamic Nuclear Polarization Surface Enhanced NMR Spectroscopy." *Accounts of Chemical Research* 46(9):1942–1951. doi.org/10.1021/Ar300322x.

Ruamps, R., R. Maurice, L. Batchelor, M. Boggio-Pasqua, R. Guillot, A. L. Barra, J. Liu, E.-E. Bendeif, S. Pillet, S. Hill, T. Mallah, and N. Guihéry. 2013. "Giant Ising-Type Magnetic Anisotropy in Trigonal Bipyramidal Ni(II) Complexes: Experiment and Theory." *Journal of the American Chemical Society* 135(8):3017–3026. doi.org/10.1021/Ja308146e.

Rugar, D., R. Budakian, H. J. Mamin, and B. W. Chui. 2004. "Single Spin Detection by Magnetic Resonance Force Microscopy." *Nature* 430(6997):329–332. doi.org/10.1038/Nature02658.

Rugar, D., H. J. Mamin, M. H. Sherwood, M. Kim, C. T. Rettner, K. Ohno, and D. D. Awschalom. 2015. "Proton Magnetic Resonance Imaging Using a Nitrogen–Vacancy Spin Sensor." *Nature Nanotechnology* 10(2):120–124. doi.org/10.1038/Nnano.2014.288.

Rugg, B. K., M. D. Krzyaniak, B. T. Phelan, M. A. Ratner, R. M. Young, and M. R. Wasielewski. 2019. "Photodriven Quantum Teleportation of an Electron Spin State in a Covalent Donor-Acceptor-Radical System." *Nature Chemistry* 11(11):981–986. doi.org/10.1038/S41557-019-0332-8.

Saeedi, K., S. Simmons, J. Z. Salvail, P. Dluhy, H. Riemann, N. V. Abrosimov, P. Becker, H.-J. Pohl, J. J. L. Morton, and M. L. W. Thewalt. 2013. "Room-Temperature Quantum Bit Storage Exceeding 39 Minutes Using Ionized Donors in Silicon-28." *Science* 342(6160):830–833. doi.org/10.1126/Science.1239584.

Saleh, B. E. A., B. M. Jost, H.-B. Fei, and M. C. Teich. 1998. "Entangled-Photon Virtual-State Spectroscopy." *Physical Review Letters* 80(16):3483–3486. doi.org/10.1103/Physrevlett.80.3483.

Salman, Z., K. H. Chow, R. I. Miller, A. Morello, T. J. Parolin, M. D. Hossain, T. A. Keeler, C. D. P. Levy, W. A. Macfarlane, G. D. Morris, H. Saadaoui, D. Wang, R. Sessoli, G. G. Condorelli, and R. F. Kiefl. 2007. "Local Magnetic Properties of a Monolayer of Mn_{12} Single Molecule Magnets." *Nano Letters* 7(6):1551–1555. doi.org/10.1021/Nl070366a.

Sarovar, M., A. Ishizaki, G. R. Fleming, and K. B. Whaley. 2010. "Quantum Entanglement in Photosynthetic Light-Harvesting Complexes." *Nature Physics* 6(6):462–467. doi.org/10.1038/Nphys1652.

Sato, K., S. Nakazawa, R. Rahimi, T. Ise, S. Nishida, T. Yoshino, N. Mori, K. Toyota, D. Shiomi, Y. Yakiyama, Y. Morita, M. Kitagawa, K. Nakasuji, M. Nakahara, H. Hara, P. Carl, P. Höfer, and T. Takui. 2009. "Molecular Electron-Spin Quantum Computers and Quantum Information Processing: Pulse-Based Electron Magnetic Resonance Spin Technology Applied to Matter Spin-Qubits." *Journal of Materials Chemistry* 19(22):3739–3754. doi.org/10.1039/B819556K.

Sato, K., R. Rahimi, N. Mori, S. Nishida, K. Toyota, D. Shiomi, Y. Morita, A. Ueda, S. Suzuki, K. Furukawa, T. Nakamura, M. Kitagawa, K. Nakasuji, M. Nakahara, H. Hara, P. Carl, P. Höfer, and T. Takui. 2007. "Implementation of Molecular Spin Quantum Computing by Pulsed ENDOR Technique: Direct Observation of Quantum Entanglement and Spinor." *Physica E: Low-Dimensional Systems and Nanostructures* 40(2):363–366. doi.org/10.1016/J.Physe.2007.06.031.

Schlawin, F., K. E. Dorfman, B. P. Fingerhut, and S. Mukamel. 2013. "Suppression of Population Transport and Control of Exciton Distributions by Entangled Photons." *Nature Communications* 4(1):1782. doi.org/10.1038/Ncomms2802.

Schlawin, F., K. E. Dorfman, and S. Mukamel. 2018. "Entangled Two-Photon Absorption Spectroscopy." *Accounts of Chemical Research* 51(9):2207–2214. doi.org/10.1021/Acs.Accounts.8b00173.

Schlegel, C., J. Van Slageren, M. Manoli, E. K. Brechin, and M. Dressel. 2008. "Direct Observation of Quantum Coherence in Single-Molecule Magnets." *Physical Review Letters* 101(14):147203. doi.org/10.1103/Physrevlett.101.147203.

Schnegg, A., J. Behrends, K. Lips, R. Bittl, and K. Holldack. 2009. "Frequency Domain Fourier Transform THz-EPR on Single Molecule Magnets Using Coherent Synchrotron Radiation." *Physical Chemistry Chemical Physics* 11(31):6820–6825. doi.org/10.1039/B905745E.

Scholes, G. D., G. R. Fleming, L. X. Chen, A. Aspuru-Guzik, A. Buchleitner, D. F. Coker, G. S. Engel, R. Van Grondelle, A. Ishizaki, D. M. Jonas, J. S. Lundeen, J. K. Mccusker, S. Mukamel, J. P. Ogilvie, A. Olaya-Castro, M. A. Ratner, F. C. Spano, K. B. Whaley, and X. Zhu. 2017. "Using Coherence to Enhance Function in Chemical and Biophysical Systems." *Nature* 543(7647):647–656. doi.org/10.1038/Nature21425.

Scholl, P., M. Schuler, H. J. Williams, A. A. Eberharter, D. Barredo, K.-N. Schymik, V. Lienhard, L.-P. Henry, T. C. Lang, T. Lahaye, A. M. Läuchli, and A. Browaeys. 2021. "Quantum Simulation of 2D Antiferromagnets with Hundreds of Rydberg Atoms." *Nature* 595:233–238. doi.org/10.1038/s41586-021-03585-1.

Scholten, S. C., A. J. Healey, I. O. Robertson, G. J. Abrahams, D. A. Broadway, and J.-P. Tetienne. 2021. "Widefield Quantum Microscopy with Nitrogen-Vacancy Centers in Diamond: Strengths, Limitations, and Prospects." *Journal of Applied Physics* 130(15):150902. doi.org/10.1063/5.0066733.

Schrödinger, E. 2012. *What Is Life? With Mind and Matter and Autobiographical Sketches, Canto Classics*. Cambridge: Cambridge University Press.

Seo, C., and T.-T. Kim. 2022. "Terahertz Near-Field Spectroscopy for Various Applications." *Journal of the Korean Physical Society* 81:549–561. doi.org/10.1007/S40042-022-00404-2.

Sessoli, R., D. Gatteschi, A. Caneschi, and M. A. Novak. 1993. "Magnetic Bistability in a Metal-Ion Cluster." *Nature* 365(6442):141–143. doi.org/10.1038/365141a0.

Sessoli, R., H. L. Tsai, A. R. Schake, S. Wang, J. B. Vincent, K. Folting, D. Gatteschi, G. Christou, and D. N. Hendrickson. 1993. "High-Spin Molecules: [Mn12O12(O2CR)16(H2O)4]." *Journal of the American Chemical Society* 115(5):1804–1816. doi.org/10.1021/Ja00058a027.

Shapiro, D. A., S. Babin, R. S. Celestre, W. Chao, R. P. Conley, P. Denes, B. Enders, P. Enfedaque, S. James, J. M. Joseph, H. Krishnan, S. Marchesini, K. Muriki, K. Nowrouzi, S. R. Oh, H. Padmore, T. Warwick, L. Yang, V. V. Yashchuk, Y.-S. Yu, and J. Zhao. 2020. "An Ultrahigh-Resolution Soft X-Ray Microscope for Quantitative Analysis of Chemically Heterogeneous Nanomaterials." *Science Advances* 6(51):Eabc4904. doi.org/doi:10.1126/Sciadv.Abc4904.

Shashkova, S., and M. C. Leake. 2017. "Single-Molecule Fluorescence Microscopy Review: Shedding New Light on Old Problems." *Bioscience Reports* 37(4). doi.org/10.1042/Bsr20170031.

Shi, L., A. A. Fung, and A. Zhou. 2021. "Advances in Stimulated Raman Scattering Imaging for Tissues and Animals." *Quantitative Imaging in Medicine and Surgery* 11(3):1078–1101. doi.org/10.21037/Qims-20-712.

Shi, M., P. F. Herskind, M. Drewsen, and I. L. Chuang. 2013. "Microwave Quantum Logic Spectroscopy and Control of Molecular Ions." *New Journal of Physics* 15(11):113019. doi.org/10.1088/1367-2630/15/11/113019.

Shiddiq, M., D. Komijani, Y. Duan, A. Gaita-Ariño, E. Coronado, and S. Hill. 2016. "Enhancing Coherence in Molecular Spin Qubits Via Atomic Clock Transitions." *Nature* 531(7594):348–351. doi.org/10.1038/Nature16984.

Shuman, E. S., J. F. Barry, D. R. Glenn, and D. Demille. 2009. "Radiative Force from Optical Cycling on a Diatomic Molecule." *Physical Review Letters* 103(22):223001. doi.org/10.1103/Physrevlett.103.223001.

Shwartz, S., R. N. Coffee, J. M. Feldkamp, Y. Feng, J. B. Hastings, G. Y. Yin, and S. E. Harris. 2012. "X-Ray Parametric Down-Conversion in the Langevin Regime." *Physical Review Letters* 109(1):013602. doi.org/10.1103/Physrevlett.109.013602.

Simmons, S., R. M. Brown, H. Riemann, N. V. Abrosimov, P. Becker, H.-J. Pohl, M. L. W. Thewalt, K. M. Itoh, and J. J. L. Morton. 2011. "Entanglement in a Solid-State Spin Ensemble." *Nature* 470(7332):69–72. doi.org/10.1038/Nature09696.

Smith, T. A., Z. Wang, and Y. Shih. 2020. "Two-Photon X-Ray Ghost Microscope." *Optics Express* 28(22):32249–32265. doi.org/10.1364/OE.401449.

Sofer, S., O. Sefi, E. Strizhevsky, H. Aknin, S. P. Collins, G. Nisbet, B. Detlefs, Ch. J. Sahle, and S. Shwartz. 2019. "Observation of Strong Nonlinear Interactions in Parametric Down-Conversion of X-Rays into Ultraviolet Radiation." *Nature Communications* 10(1):5673. doi.org/10.1038/S41467-019-13629-W.

Squire, R. H., N. H. March, R. A. Minnick, and R. Turschmann. 2013. "Comparison of Various Types of Coherence and Emergent Coherent Systems." *International Journal of Quantum Chemistry* 113(19):2181–2199. doi.org/10.1002/Qua.24423.

Stamp, P. C. E., and I. S. Tupitsyn. 2004. "Coherence Window in the Dynamics of Quantum Nanomagnets." *Physical Review B* 69(1):014401. doi.org/10.1103/Physrevb.69.014401.

Stasiw, D. E., J. Zhang, G. Wang, R. Dangi, B. W. Stein, D. A. Shultz, M. L. Kirk, L. Wojtas, and R. D. Sommer. 2015. "Determining the Conformational Landscape of ς and π Coupling Using Para-Phenylene and 'Aviram–Ratner' Bridges." *Journal of the American Chemical Society* 137(29):9222–9225. doi.org/10.1021/Jacs.5b04629.

Staudacher, T., F. Shi, S. Pezzagna, J. Meijer, J. Du, C. A. Meriles, F. Reinhard, and J. Wrachtrup. 2013. "Nuclear Magnetic Resonance Spectroscopy on a (5-Nanometer)^3 Sample Volume." *Science* 339(6119):561–563. doi.org/10.1126/Science.1231675.

Stevens, M. A., J. E. Mckay, J. L. S. Robinson, H. El Mkami, G. M. Smith, and D. G. Norman. 2016. "The Use of the Rx Spin Label in Orientation Measurement on Proteins, by EPR." *Physical Chemistry Chemical Physics* 18(8):5799–5806. doi.org/10.1039/C5CP04753F.

Stuhl, B. K., B. C. Sawyer, D. Wang, and J. Ye. 2008. "Magneto-Optical Trap for Polar Molecules." *Physical Review Letters* 101(24):243002. doi.org/10.1103/Physrevlett.101.243002.

Subramanya, M. V. H., J. Marbey, K. Kundu, J. E. Mckay, and S. Hill. 2022. "Broadband Fourier-Transform-Detected EPR at W-Band." *Applied Magnetic Resonance* 54:165–181. doi.org/10.1007/S00723-022-01499-3.

Sun, K., M. F. Gelin, and Y. Zhao. 2022. "Engineering Cavity Singlet Fission in Rubrene." *Physical Review Letters* 13(18):4090–4097. doi.org/10.1021/Acs.Jpclett.2c00801.

Sushkov, A. O., I. Lovchinsky, N. Chisholm, R. L. Walsworth, H. Park, and M. D. Lukin. 2014. "Magnetic Resonance Detection of Individual Proton Spins Using Quantum Reporters." *Physical Review Letters* 113(19):197601. doi.org/10.1103/Physrevlett.113.197601.

Suturina, E. A., J. Nehrkorn, J. M. Zadrozny, J. Liu, M. Atanasov, T. Weyhermüller, D. Maganas, S. Hill, A. Schnegg, E. Bill, J. R. Long, and F. Neese. 2017. "Magneto-Structural Correlations in Pseudotetrahedral Forms of the $[Co(SPh)_4]^{2-}$ Complex Probed by Magnetometry, MCD Spectroscopy, Advanced EPR Techniques, and Ab Initio Electronic Structure Calculations." *Inorganic Chemistry* 56(5):3102–3118. doi.org/10.1021/Acs.Inorgchem.7b00097.

Svidzinsky, A., G. Agarwal, A. Classen, A. V. Sokolov, A. Zheltikov, M. S. Zubairy, and M. O. Scully. 2021. "Enhancing Stimulated Raman Excitation and Two-Photon Absorption Using Entangled States of Light." *Physical Review Research* 3(4):043029. doi.org/10.1103/Physrevresearch.3.043029.

Szoke, S., M. He, B. P. Hickam, and S. K. Cushing. 2021. "Designing High-Power, Octave Spanning Entangled Photon Sources for Quantum Spectroscopy." *Journal of Chemical Physics* 154(24):244201. doi.org/10.1063/5.0053688.

Szoke, S., H. Liu, B. P. Hickam, M. He, and S. K. Cushing. 2020. "Entangled Light–Matter Interactions and Spectroscopy." *Journal of Materials Chemistry C* 8(31):10732–10741. doi.org/10.1039/D0tc02300k.

Tabakaev, D., A. Djorović, L. La Volpe, G. Gaulier, S. Ghosh, L. Bonacina, J. P. Wolf, H. Zbinden, and R. T. Thew. 2022. "Spatial Properties of Entangled Two-Photon Absorption." *Physical Review Letters* 129(18):183601. doi.org/10.1103/Physrevlett.129.183601.

Tabakaev, D., M. Montagnese, G. Haack, L. Bonacina, J. P. Wolf, H. Zbinden, and R. T. Thew. 2021. "Energy-Time-Entangled Two-Photon Molecular Absorption." *Physical Review A* 103(3):033701. doi.org/10.1103/Physreva.103.033701.

Takahashi, S., and S. Hill. 2005. "Rotating Cavity for High-Field Angle-Dependent Microwave Spectroscopy of Low-Dimensional Conductors and Magnets." *Review of Scientific Instruments* 76(2):023114. doi.org/10.1063/1.1852859.

Takahashi, S., R. Hanson, J. Van Tol, M. S. Sherwin, and D. D. Awschalom. 2008. "Quenching Spin Decoherence in Diamond Through Spin Bath Polarization." *Physical Review Letters* 101(4):047601. doi.org/10.1103/Physrevlett.101.047601.

Takahashi, S., I. S. Tupitsyn, J. Van Tol, C. C. Beedle, D. N. Hendrickson, and P. C. E. Stamp. 2011. "Decoherence in Crystals of Quantum Molecular Magnets." *Nature* 476(7358):76–79. doi.org/10.1038/Nature10314.

Takahashi, S., J. Van Tol, C. C. Beedle, D. N. Hendrickson, L.-C. Brunel, and M. S. Sherwin. 2009. "Coherent Manipulation and Decoherence of S=10 Single-Molecule Magnets." *Physical Review Letters* 102(8):087603. doi.org/10.1103/Physrevlett.102.087603.

Tegmark, M. 2000. "Importance of Quantum Decoherence in Brain Processes." *Physical Review E* 61(4):4194–4206. doi.org/10.1103/Physreve.61.4194.

Thiele, S., F. Balestro, R. Ballou, S. Klyatskaya, M. Ruben, and W. Wernsdorfer. 2014. "Electrically Driven Nuclear Spin Resonance in Single-Molecule Magnets." *Science* 344(6188):1135–1138. doi.org/10.1126/Science.1249802.

Thirunavukkuarasu, K., S. M. Winter, C. C. Beedle, A. E. Kovalev, R. T. Oakley, and S. Hill. 2015. "Pressure Dependence of the Exchange Anisotropy in an Organic Ferromagnet." *Physical Review B* 91(1):014412. doi.org/10.1103/Physrevb.91.014412.

Thomann, H., T. V. Morgan, H. Jin, S. J. N. Burgmayer, R. E. Bare, and E. I. Stiefel. 1987. "Protein Nitrogen Coordination to the Iron-Molybdenum Center of Nitrogenase from Clostridium Pasteurianum." *Journal of the American Chemical Society* 109(25):7913–7914. doi.org/10.1021/Ja00259a067.

Togan, E., Y. Chu, A. S. Trifonov, L. Jiang, J. Maze, L. Childress, M. V. G. Dutt, A. S. Sørensen, P. R. Hemmer, A. S. Zibrov, and M. D. Lukin. 2010. "Quantum Entanglement Between an Optical Photon and a Solid-State Spin Qubit." *Nature* 466(7307):730–734. doi.org/10.1038/Nature09256.

Truppe, S., H. J. Williams, M. Hambach, L. Caldwell, N. J. Fitch, E. A. Hinds, B. E. Sauer, and M. R. Tarbutt. 2017. "Molecules Cooled Below the Doppler Limit." *Nature Physics* 13(12):1173–1176. doi.org/10.1038/Nphys4241.

Turin, L. 1996. "A Spectroscopic Mechanism for Primary Olfactory Reception." *Chemical Senses* 21(6):773–791. doi.org/10.1093/Chemse/21.6.773.

Uber, J. S., M. Estrader, J. Garcia, P. Lloyd-Williams, A. Sadurní, D. Dengler, J. Van Slageren, N. F. Chilton, O. Roubeau, S. J. Teat, J. Ribas-Ariño, and G. Aromí. 2017. "Molecules Designed to Contain Two Weakly Coupled Spins with a Photoswitchable Spacer." *Chemistry: A European Journal* 23(55):13648–13659. doi.org/10.1002/Chem.201702171.

Ullah, A., Z. Hu, J. Cerdá, J. Aragó, and A. Gaita-Ariño. 2022. "Electrical Two-Qubit Gates within a Pair of Clock-Qubit Magnetic Molecules." *npj Quantum Information* 8(1). https://doi.org/10.1038/s41534-022-00647-8.

Upton, L., M. Harpham, O. Suzer, M. Richter, S. Mukamel, and T. Goodson, III. 2013. "Optically Excited Entangled States in Organic Molecules Illuminate the Dark." *Physical Review Letters* 4(12):2046–2052. doi.org/10.1021/Jz400851d.

Van Doorslaer, S. 2017. "Hyperfine Spectroscopy: ESEEM." *eMagRes* 6:51–70.

van Slageren, J. 2019. "Spin–Electric Coupling." *Nature Materials* 18(4):300–301. doi.org/10.1038/S41563-019-0314-7.

van Slageren, J., S. Vongtragool, B. Gorshunov, A. A. Mukhin, N. Karl, J. Krzystek, J. Telser, A. Müller, C. Sangregorio, D. Gatteschi, and M. Dressel. 2003. "Frequency-Domain Magnetic Resonance Spectroscopy of Molecular Magnetic Materials." *Physical Chemistry Chemical Physics* 5(18):3837–3843. doi.org/10.1039/B305328H.

van Tol, J., L.-C. Brunel, and R. J. Wylde. 2005. "A Quasioptical Transient Electron Spin Resonance Spectrometer Operating at 120 and 240 GHz." *Review of Scientific Instruments* 76(7):074101. doi.org/10.1063/1.1942533.

Varnavski, O., and T. Goodson, III. 2020. "Two-Photon Fluorescence Microscopy at Extremely Low Excitation Intensity: The Power of Quantum Correlations." *Journal of the American Chemical Society* 142(30):12966–12975. doi.org/10.1021/Jacs.0c01153.

Varnavski, O., C. Gunthardt, A. Rehman, G. D. Luker, and T. Goodson, III. 2022. "Quantum Light-Enhanced Two-Photon Imaging of Breast Cancer Cells." *Physical Review Letters* 13(12):2772–2781. doi.org/10.1021/Acs.Jpclett.2c00695.

Varnavski, O., B. Pinsky, and T. Goodson, III. 2017. "Entangled Photon Excited Fluorescence in Organic Materials: An Ultrafast Coincidence Detector." *Physical Review Letters* 8(2):388–393. doi.org/10.1021/Acs.Jpclett.6b02378.

Vilas, N. B., C. Hallas, L. Anderegg, P. Robichaud, A. Winnicki, D. Mitra, and J. M. Doyle. 2022. "Magneto-Optical Trapping and Sub-Doppler Cooling of a Polyatomic Molecule." *Nature* 606(7912):70–74. doi.org/10.1038/S41586-022-04620-5.

Villabona-Monsalve, J. P., O. Varnavski, B. A. Palfey, and T. Goodson, III. 2018. "Two-Photon Excitation of Flavins and Flavoproteins with Classical and Quantum Light." *Journal of the American Chemical Society* 140(44):14562–14566. doi.org/10.1021/Jacs.8b08515.

Vincent, R., S. Klyatskaya, M. Ruben, W. Wernsdorfer, and F. Balestro. 2012. "Electronic Read-Out of a Single Nuclear Spin Using a Molecular Spin Transistor." *Nature* 488(7411):357–360. doi.org/10.1038/Nature11341.

Volkovich, S., and S. Shwartz. 2020. "Subattosecond X-Ray Hong-Ou-Mandel Metrology." *Optics Letters* 45(10):2728–2731. doi.org/10.1364/Ol.382044.

Wang, H., and W. Xiong. 2021. "Vibrational Sum-Frequency Generation Hyperspectral Microscopy for Molecular Self-Assembled Systems." *Annual Review of Physical Chemistry* 72(1):279–306. doi.org/10.1146/Annurev-Physchem-090519-050510.

Wang, X., J. E. Mckay, B. Lama, J. Van Tol, T. Li, K. Kirkpatrick, Z. Gan, S. Hill, J. R. Long, and H. C. Dorn. 2018. "Gadolinium Based Endohedral Metallofullerene $Gd_2@C_{79}N$ as a Relaxation Boosting Agent for Dissolution DNP at High Fields." *Chemical Communications* 54(19):2425–2428. doi.org/10.1039/C7CC09765D.

Wang, Z., S. Datta, C. Papatriantafyllopoulou, G. Christou, N. S. Dalal, J. Van Tol, and S. Hill. 2011. "Spin Decoherence in an Iron-Based Magnetic Cluster." *Polyhedron* 30(18):3193–3196. doi.org/10.1016/J.Poly.2011.04.009.

Warner, M., S. Din, I. S. Tupitsyn, G. W. Morley, A. M. Stoneham, J. A. Gardener, Z. Wu, A. J. Fisher, S. Heutz, C. W. M. Kay, and G. Aeppli. 2013. "Potential for Spin-Based Information Processing in a Thin-Film Molecular Semiconductor." *Nature* 503(7477):504–508. doi.org/10.1038/Nature12597.

Wasielewski, M. R., M. D. E. Forbes, N. L. Frank, K. Kowalski, G. D. Scholes, J. Yuen-Zhou, M. A. Baldo, D. E. Freedman, R. H. Goldsmith, T. Goodson, M. L. Kirk, J. K. Mccusker, J. P. Ogilvie, D. A. Shultz, S. Stoll, and K. Birgitta Whaley. 2020. "Exploiting Chemistry and Molecular Systems for Quantum Information Science." *Nature Reviews Chemistry* 4(9):490–504. doi.org/10.1038/S41570-020-0200-5.

Weiden, N., H. Käss, and K. P. Dinse. 1999. "Pulse Electron Paramagnetic Resonance (EPR) and Electron–Nuclear Double Resonance (ENDOR) Investigation of $N@C_{60}$ in Polycrystalline C_{60}." *Journal of Physical Chemistry B* 103(45): 9826–9830. doi.org/10.1021/Jp9914471.

Weijers, H. W., W. D. Markiewicz, A. V. Gavrilin, A. J. Voran, Y. L. Viouchkov, S. R. Gundlach, P. D. Noyes, D. V. Abraimov, H. Bai, S. T. Hannahs, and T. P. Murphy. 2016. "Progress in the Development and Construction of a 32-T Superconducting Magnet." *IEEE Transactions on Applied Superconductivity* 26(4):1–7. doi.org/10.1109/TASC.2016.2517022.

Williams, H. J., L. Caldwell, N. J. Fitch, S. Truppe, J. Rodewald, E. A. Hinds, B. E. Sauer, and M. R. Tarbutt. 2018. "Magnetic Trapping and Coherent Control of Laser-Cooled Molecules." *Physical Review Letters* 120(6):163201. doi.org/10.1103/PhysRevLett.120.163201.

Willke, P., T. Bilgeri, X. Zhang, Y. Wang, C. Wolf, H. Aubin, A. Heinrich, and T. Choi. 2021. "Coherent Spin Control of Single Molecules on a Surface." *ACS Nano* 15(11):17959–17965. doi.org/10.1021/Acsnano.1c06394.

Wilson, A., J. Lawrence, E. C. Yang, M. Nakano, D. N. Hendrickson, and S. Hill. 2006. "Magnetization Tunneling in High-Symmetry Single-Molecule Magnets: Limitations of the Giant Spin Approximation." *Physical Review B* 74(14):140403. doi.org/10.1103/Physrevb.74.140403.

Wiltschko, R., and W. Wiltschko. 2012. "Magnetoreception." In *Sensing in Nature*, edited by C. López-Larrea, 126–141. New York: Springer US.

Wiltschko, W., and R. Wiltschko. 2005. "Magnetic Orientation and Magnetoreception in Birds and Other Animals." *Journal of Comparative Physiology A* 191(8):675–693. doi.org/10.1007/S00359-005-0627-7.

Wolf, F., Y. Wan, J. C. Heip, F. Gebert, C. Shi, and P. O. Schmidt. 2016. "Non-Destructive State Detection for Quantum Logic Spectroscopy of Molecular Ions." *Nature* 530(7591):457–460. doi.org/10.1038/Nature16513.

Wolfowicz, G., and J. J. L. Morton. 2016. "Pulse Techniques for Quantum Information Processing." *eMagRes* 5:1515–1528.

Wong, L. J., and I. Kaminer. 2021. "Prospects in X-Ray Science Emerging from Quantum Optics and Nanomaterials." *Applied Physics Letters* 119(13):130502. doi.org/10.1063/5.0060552.

Wrachtrup, J., C. Von Borczyskowski, J. Bernard, M. Orrit, and R. Brown. 1993. "Optical Detection of Magnetic Resonance in a Single Molecule." *Nature* 363(6426):244–245. doi.org/10.1038/363244a0.

Wright, K., K. M. Beck, S. Debnath, J. M. Amini, Y. Nam, N. Grzesiak, J. S. Chen, N. C. Pisenti, M. Chmielewski, C. Collins, K. M. Hudek, J. Mizrahi, J. D. Wong-Campos, S. Allen, J. Apisdorf, P. Solomon, M. Williams, A. M. Ducore, A. Blinov, S. M. Kreikemeier, V. Chaplin, M. Keesan, C. Monroe, and J. Kim. 2019. "Benchmarking an 11-Qubit Quantum Computer." *Nature Communications* 10(1):5464. doi.org/10.1038/S41467-019-13534-2.

Yan, R., Y. Zhang, Y. Li, L. Xia, Y. Guo, and Q. Zhou. 2020. "Structural Basis for the Recognition of SARS-CoV-2 by Full-Length Human ACE2." *Science* 367(6485):1444–1448. doi.org/10.1126/science.abb2762.

Yang, K., W. Paul, S.-H. Phark, P. Willke, Y. Bae, T. Choi, T. Esat, A. Ardavan, A. J. Heinrich, and C. P. Lutz. 2019. "Coherent Spin Manipulation of Individual Atoms on a Surface." *Science* 366(6464):509–512. doi.org/10.1126/Science.Aay6779.

Ye, L., and S. Mukamel. 2020. "Interferometric Two-Photon-Absorption Spectroscopy with Three Entangled Photons." *Applied Physics Letters* 116(17):174003. doi.org/10.1063/5.0004617.

Young, L., K. Ueda, M. Gühr, P. H. Bucksbaum, M. Simon, S. Mukamel, N. Rohringer, K. C. Prince, C. Masciovecchio, M. Meyer, A. Rudenko, D. Rolles, C. Bostedt, M. Fuchs, D. A. Reis, R. Santra, H. Kapteyn, M. Murnane, H. Ibrahim, F. Légaré, M. Vrakking, M. Isinger, D. Kroon, M. Gisselbrecht, A. L'Huillier, H. J. Wörner, and S. R. Leone. 2018. "Roadmap of Ultrafast X-Ray Atomic and Molecular Physics." *Journal of Physics B: Atomic, Molecular, and Optical Physics* 51(3):032003. doi.org/10.1088/1361-6455/Aa9735.

Yu, C.-J., M. J. Graham, J. M. Zadrozny, J. Niklas, M. D. Krzyaniak, M. R. Wasielewski, O. G. Poluektov, and D. E. Freedman. 2016. "Long Coherence Times in Nuclear Spin-Free Vanadyl Qubits." *Journal of the American Chemical Society* 138(44):14678–14685. doi.org/10.1021/Jacs.6b08467.

Yu, C.-J., S. von Kugelgen, D. W. Laorenza, and D. E. Freedman. 2021. "A Molecular Approach to Quantum Sensing." *ACS Central Science* 7(5):712–723. doi.org/10.1021/Acscentsci.0c00737.

Zadrozny, J. M., A. T. Gallagher, T. D. Harris, and D. E. Freedman. 2017. "A Porous Array of Clock Qubits." *Journal of the American Chemical Society* 139(20):7089–7094. doi.org/10.1021/Jacs.7b03123.

Zadrozny, J. M., J. Liu, N. A. Piro, C. J. Chang, S. Hill, and J. R. Long. 2012. "Slow Magnetic Relaxation in a Pseudotetrahedral Cobalt(II) Complex with Easy-Plane Anisotropy." *Chemical Communications* 48(33):3927–3929. doi.org/10.1039/C2CC16430B.

Zadrozny, J. M., J. Niklas, O. G. Poluektov, and D. E. Freedman. 2015. "Millisecond Coherence Time in a Tunable Molecular Electronic Spin Qubit." *ACS Central Science* 1(9):488–492. doi.org/10.1021/Acscentsci.5b00338.

Zadrozny, J. M., D. J. Xiao, J. R. Long, M. Atanasov, F. Neese, F. Grandjean, and G. J. Long. 2013. "Mössbauer Spectroscopy as a Probe of Magnetization Dynamics in the Linear Iron(I) and Iron(II) Complexes $[Fe(C(SiMe_3)_3)_2]^{1-/0}$." *Inorganic Chemistry* 52(22):13123–13131. doi.org/10.1021/Ic402013n.

Zhang, X., C. Wolf, Y. Wang, H. Aubin, T. Bilgeri, P. Willke, A. J. Heinrich, and T. Choi. 2022. "Electron Spin Resonance of Single Iron Phthalocyanine Molecules and Role of Their Non-Localized Spins in Magnetic Interactions." *Nature Chemistry* 14(1):59–65. doi.org/10.1038/S41557-021-00827-7.

Zhu, G.-Z., D. Mitra, B. L. Augenbraun, C. E. Dickerson, M. J. Frim, G. Lao, Z. D. Lasner, A. N. Alexandrova, W. C. Campbell, J. R. Caram, J. M. Doyle, and E. R. Hudson. 2022. "Functionalizing Aromatic Compounds with Optical Cycling Centres." *Nature Chemistry* 14(9):995–999. doi.org/10.1038/S41557-022-00998-X.

Zoltowski, B. D., Y. Chelliah, A. Wickramaratne, L. Jarocha, N. Karki, W. Xu, H. Mouritsen, P. J. Hore, R. E. Hibbs, C. B. Green, and J. S. Takahashi. 2019. "Chemical and Structural Analysis of a Photoactive Vertebrate Cryptochrome from Pigeon." *Proceedings of the National Academy of Sciences USA* 116(39):19449–19457. doi.org/10.1073/pnas.1907875116.

4

Experimental and Computational Approaches for Scaling Qubit Design and Function

Key Takeaways

- In many qubit platforms, knowledge is lacking of both what the origins of decoherence are and/or how they can be controlled through chemical design.
- Large-scale, reproducible synthesis of current molecular qubits (e.g., molecular color centers) is a major hurdle to the realization of QIS platforms.
- Error mitigation has rarely been leveraged to improve the information processing capabilities of chemical systems due to the lack of high-accuracy molecular noise models.
- The computational modeling of molecules and materials for QIS can be accelerated via embedding and sampling approaches.
- High-throughput evaluations for identifying and characterizing quantum materials (solid-state materials and molecules) are underutilized.
- In fields such as materials science, catalysis, and structural biology, international collaboration through shared databases has led to scientific success. These databases share common features such as open-access, user-friendly interfaces and centralized information.
- Not all problems in quantum chemistry are suitable for quantum computers because many problems can be solved to sufficiently high levels of precision using classical computers to resolve the relevant effects. At the same time, quantum computers have the potential to simulate the dynamics of any chemical process efficiently and prepare ground states of some chemical systems believed to be classically intractable.
- The chemistry problems that are most difficult for classical computers, but seem to provide an advantage for quantum computers, are so-called "strongly correlated" systems that tend to involve the breaking of multiple chemical bonds, transition metals, free radicals, or some unusual states of the electrons leading to high entanglement. The dynamics of electrons also tend to generate large amounts of entanglement that may require a quantum computer to simulate. Simulating more accurate model chemistries is an important benefit of leveraging the potential high-accuracy calculations offered by quantum computers.

- The computational modeling of large, heterogenous systems can be accelerated via embedding and sampling approaches on classical systems. Currently, the limits of a suite of embedding and other reduced-scaling techniques are underdeveloped. Improving these tools will ultimately lead to a deeper understanding of QIS challenges.
- Computational or theoretical tools for modeling the entanglement and coherence within quantum information processors remain costly. The high cost to use classical computational tools for modeling quantum phenomena is inhibiting key research activities such as direct simulation of large quantum information processors.
- Based on current knowledge of quantum algorithms, noisy intermediate-scale quantum computers, quantum annealers, and analog quantum computers are unlikely to solve problems near the classically intractable regime without error correction.
- Fault-tolerant quantum computing (FTQC) shows promise for solving problems in chemistry, especially the modern electronic structure problem. FTQC is the goal that the field of quantum computing is pushing toward, as it offers the most impactful future.

4.1 EXPLOITING THE ADVANTAGES OF BOTTOM-UP CHEMICAL SYNTHESIS FOR CONSTRUCTING QUANTUM ARCHITECTURES

As discussed in Chapter 2, molecules offer an unmatched combination of atomic-scale tunability, portability, and scalability (Atzori and Sessoli 2019; Gaita-Ariño et al. 2019; Graham, Zadrozyn, et al. 2017; Jackson et al. 2021; Ye, Seo, and Galli 2019; C.-J. Yu et al.). With molecular chemistry, one can exploit the advantages used in medicinal chemistry for centuries—each molecule within a tablet of aspirin is chemically identical down to the sub-angstrom level and has the potential to maintain its fundamental characteristics in a wide range of chemical environments. This high level of structural precision combined with tunability makes molecules perfect for designer applications within QIS. Molecules can be tuned to have extrinsic compatibility with their environments for quantum sensing applications (Atzori and Sessoli 2019; Lovchinsky et al. 2016; C.-J. Yu et al. 2021). For example, they can be designed to be water soluble for integration into biological media or engineered to be compatible with surfaces. Their nanometer-scale size provides access to tiny sensing locations, as well as proximal placement within hybrid systems and spatial differentiation of sensor response. Furthermore, because molecules can be functionalized, they can be specifically tethered to an analyte, thereby enabling sub-diffraction imaging by initially identifying the molecule location and subsequently executing a quantum imaging experiment. Finally, molecular structures offer a unique means to address the scalability challenge of QIS: synthesizing large-scale, reproducible quantum states for applications from computing to communications (Figure 4-1). The aggregate of these attributes primes molecules to be key elements in the future quantum internet.

4.2 DEVELOPING TECHNIQUES FOR SYNTHESIZING MOLECULAR QUBITS THAT RETAIN THEIR DESIRABLE QUANTUM PROPERTIES IN DIFFERENT HOST CHEMICAL ENVIRONMENTS

For physical qubits to be useful, several challenges related to the underlying atomic-scale properties of the constituent materials must be addressed. First, it must be possible to fabricate large quantities of precise qubits reliably so that their functionality is known and largely retained upon synthesis or fabrication without the need for extensive characterization. Each of these qubits then needs to be entangled with other qubits and scaled up to construct larger, multiqubit systems. How can larger-scale quantum architectures be synthesized with sufficient control over disorder and decoherence? As pointed out in previous chapters of this report, loss of quantum coherence occurs during interactions between a qubit and its environment; decoherence processes need to be understood, controlled,

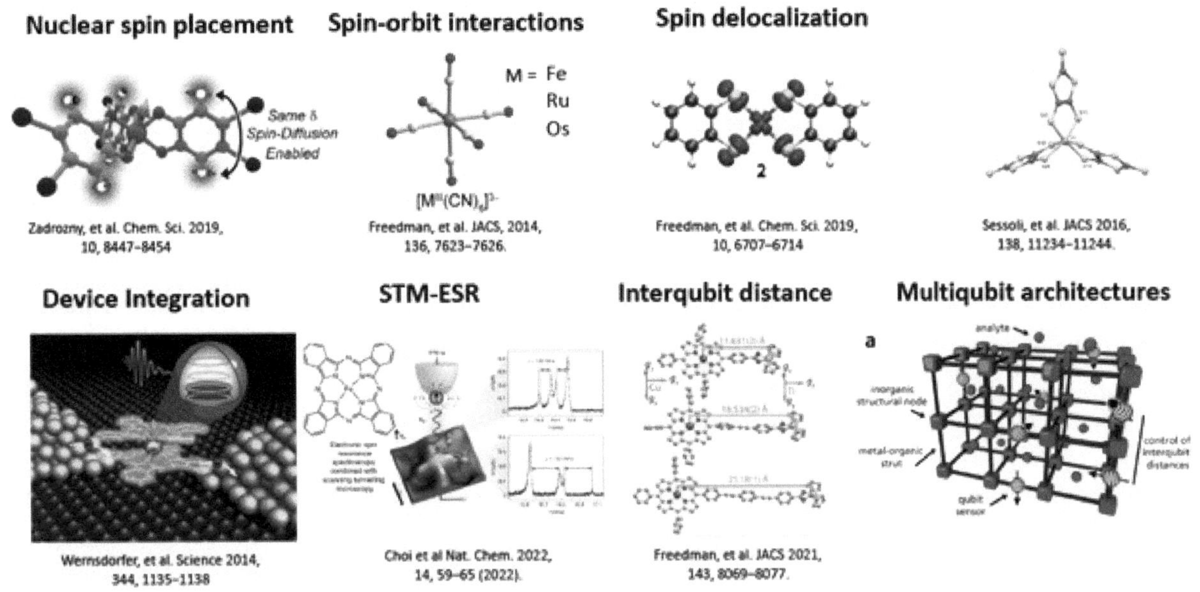

FIGURE 4-1 The many roles that chemistry can assume in the synthesis and design of molecular qubits and QIS materials. Molecules demonstrate a high modularity for QIS, enabling both intrinsic and extrinsic modulation. Previous results demonstrate that structural control (e.g., through nuclear and nuclear spin placement) and electronic control (e.g., through spin–orbit coupling of molecules) can be combined to control the quantum properties of QIS systems. Molecules can also be integrated into larger systems, break junctions, or, in a recent experiment, a single-molecule scanning tunneling microscope–electron spin resonance (STM-ESR). Chemical approaches may also be employed to build from one molecule into larger systems to form multiqubit arrays.
SOURCE: Atzori et al. 2016; Fataftah et al. 2019; Graham et al. 2014.; Jackson et al. 2019; von Kugelgen et al. 2021; Thiele et al. 2014; Zadrozny et al. 2017; Zhang et al. 2022.

and mitigated during each of these steps. Additionally, to build qubit architectures comprising dissimilar materials/chemistries (e.g., in hybrid quantum devices), heterogeneous interfaces between molecules and host materials need to be studied carefully and controlled. While current physical qubit proposals cover a spectrum of physical systems including molecular and solid-state spin centers (Miao et al. 2020), trapped ions (Bruzewicz et al. 2019; Georgescu 2020), superconducting (Devoret, Wallraff, and Martinis 2004), and topological systems (Freedman et al. 2003), all have decoherence and scaling issues that critically affect their potential as scalable, fault-tolerant systems. Exquisite control over constituent chemical environments at an atomic scale is central to understanding and controlling coherence and scaling in any qubit system. Chemists are needed at each step of this process:

1. Synthesis or fabrication of high-quality, chemically precise molecules and materials that have near-optimal or tunable quantum properties across the range of quantum architectures;
2. Development of a fundamental understanding of decoherence processes that arise from integrating molecular qubits into larger coherent quantum architectures;
3. Design and synthesis of chemical environments that are robust to decoherence that can decouple the fundamental degrees of freedom of qubits from unwanted environment-induced noise;
4. Bottom-up and top-down fabrication of heterogeneous material platforms that can transduce quantum information coherently from one chemical environment to another by leveraging surface chemistry and single-digit nanofabrication; and
5. Design and fabrication of routes to couple physical qubit systems, especially molecules, that retain their quantum coherence but that can be patterned or arranged in repeatable, identical arrays.

Finally, while each of these steps is crucial for short- and long-term challenges in QIS, the ultimate goal is the fundamental design of precise, coupled, coherent physical qubit systems with on-demand properties for a specific quantum application. This can only be achieved through a concerted effort in chemical design, incorporating state-of-the-art synthesis, fabrication, characterization, and theory. Next, an overview of architecture-specific challenges to scaling is provided.

4.2.1 Investigating and Controlling the Interactions among Qubits and between Qubits and Their Environments

4.2.1a Chemical Design of Spin in Molecules

Several transition metal and f-block elements can serve as nuclear or electronic spin qubits. As described in Chapter 2, many promising examples of such systems, including nanomaterial–molecular (Rabl et al. 2010), nuclear-electronic spin (Abobeih et al. 2018), and light–matter hybrid (Ebbesen 2016) systems, have been published previously (Thiele et al. 2014). These studies exemplify how nuclear spins that can be used as quantum resources can be controlled through precise molecular design and synthesis (Sørensen et al. 2017). In fact, the ability to control the molecular environment around a metal atom can help to create a qubit with impressive insensitivity to the local environment. Systems in which the crystal field is tuned to create an avoided crossing "clock transition" with a large energy difference demonstrate the power of molecular design to create a robust qubit (Ishikawa, Sugita, and Wernsdorfer 2005; Kundu et al. 2022; McInnes 2022; Shiddiq et al. 2016). Scaling up systems to create arrays of molecular qubits requires a controlled synthesis with controllable inter-qubit distances. Examples of these types of controllable systems are metal–organic frameworks (MOFs) with constructions that are conducive to particular applications (Gaita-Ariño et al. 2019; Yamabayashi et al. 2018; Yu et al. 2020). More generally, interfacing tunable chemical ligands in scalable organic frameworks is already the focus of several classes of organic frameworks (MOFs along with hydrogen-bonded organic frameworks and covalent organic frameworks) and can potentially form highly ordered arrays.

4.2.1b Chemical Design of Solid-State Spin Defects

Defects in semiconductors have become powerful hosts for spin-based qubits in the solid state within materials including diamond, silicon carbide, and silicon (Awschalom et al. 2018). By exploiting defects (Figure 4-2) generated via electron or ion radiation as well as naturally occurring defects in these materials, researchers have

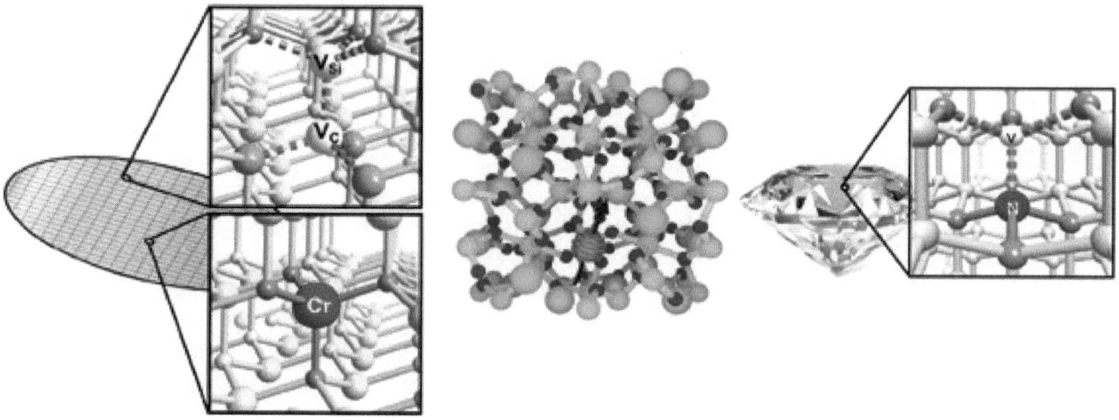

FIGURE 4-2 Vacancy-based spin qubits whose properties can be finely tuned via chemistry. (left) Chromium impurity in 4H-SiC. (middle, right) Nitrogen-vacancy center in diamond.
SOURCE: Daniel Laorenza, from the Freedman Laboratory.

demonstrated precise quantum control of individual electron and nuclear spins, robust entanglement, electron–nuclear quantum registers, and increased control of the spin–optical interface at the level of single photons (Anderson et al. 2022). Moreover, driven by predictive theoretical work, several defect-based states have been engineered for practical applications including quantum memories and optical emission (Singh et al. 2022) in the telecommunications regime. Impressive quantum sensing of electric, magnetic, and strain fields has been shown along with dramatic improvements in small-volume and single-spin nuclear magnetic resonance (Morello et al. 2010). However, these material systems face considerable challenges for QIS and engineering, including the creation of scalable and precise quantum states and the mitigation of the impact of strain (Hruszkewycz et al. 2018; Rose et al. 2018), isotopic variations (Balasubramanian et al. 2009; Bourassa et al. 2020), decoherence-driven charge fluctuations from interfaces (Bluvstein et al. 2019; Sangtawesin et al. 2019), and unintentional dopants. Chemistry can play a key role in resolving many of these challenges, especially by exploiting the broad atomic-scale synthesis capabilities for a target qubit, the host matrix, and their mutual interactions. For example, an electron spin qubit interaction with a selected proximal nuclear spin (Graham, Krzyaniak, et al. 2017; Graham, Yu, et al. 2017; Yu et al. 2016) may be optimized through ligand design between weak and strong coupling regimes. In addition, spin-lattice relaxation (Amdur et al. 2022) and the operating temperature of a spin qubit may be engineered by modifying the chemical structure of the ligand and thereby the degree of vibronic coupling at the qubit frequency. Chemistry also provides an opportunity to enhance qubit coherence by controlling the symmetry of the qubit itself or its packing within its host environment (Bayliss et al. 2022; Headley et al. 2016).

The selection of appropriate hosts and defects, and their mutual interactions with one another, remains both a challenge and an opportunity for scientific research. In particular, opportunities that build upon the success of density functional theory (DFT) and coupled cluster expansion techniques in predicting coherent properties of solid-state spin defects (Seo et al. 2017; Weber et al. 2010) may enable the discovery of robust spin qubit candidates. There have been considerable advances in understanding the mechanisms of spin decoherence in semiconductors and solid-state nanostructures (Kanai et al. 2022). State-of-the-art synchrotron spectroscopies, coherent Bragg diffraction imaging, and high-resolution magnetic resonance techniques have proven to be powerful tools to improve the quality of host materials. In addition, improvements in first-principles approaches to understanding and predicting properties (e.g., DFT) have led to a deeper physical understanding of quantum decoherence and successful predictions for mitigating many sources of noise, from dynamical decoupling to unique pulse sequences. Furthermore, there have been successful efforts aimed at harnessing the capabilities of today's electronic technologies for quantum-state control, including electrical control and readout of single quantum states (Sangtawesin et al. 2019), the creation of decoherence-free subspaces to locally enhance coherence (Miao et al. 2020), and the integration of photonics to both enhance the efficiency of quantum emitters and entangle nearby qubits (Dibos et al. 2018; Lukin, Guidry, and Vučković 2020; Wan et al. 2020). Recently, rare-earth ions in oxide semiconductor hosts—including silicon-compatible materials—have emerged with encouraging properties as single quantum memories, with impressively long coherence times and single-shot photonic readout.

4.2.1c Chemical Design of Superconducting Qubits

Chemistry also offers a potential path toward improving the coherence times of superconducting qubits. While currently one of the most scalable platforms (Arute et al. 2019), superconducting qubits are still plagued with hardware coherence issues that prohibit the realization of fault-tolerant systems (Sheridan et al. 2021; Siddiqi 2021). In superconducting systems (comprising qubits and a resonator), the central hurdle is mitigating decoherence processes that significantly suppress coherence times to well below their theoretical limits. Decoherence processes in superconducting systems currently are dominated by materials-based decoherence through the inevitable inhomogeneities present in multicomponent materials platforms such as defects and interfaces (Krantz et al. 2019). Therefore, reducing materials-based losses in superconducting qubits requires extensive knowledge of the chemical and structural makeup of such complex chemical systems, and predictive routes to their control and mitigation via physical and chemical means. The former requires advanced spectroscopies and microscopies, in tandem with theory, to probe and uncover the structure–property relationships of these heterogeneous nanoscale systems (both their bulk and interfacial properties), often at the forefront of nanoscale characterization. For example, many superconducting qubit systems are made up of amorphous superconductors, for which conventional diffraction and spectroscopies that rely on reciprocal space cannot be used; novel techniques such as fluctuation electron microscopy (Kennedy et al. 2020) can be used to probe these noncrystalline systems to elucidate information on the bonding environment. Current characterization needs also include nondestructive interfacial probes that can give information on the structure–property relationships of buried interfaces that are often nanometers in extent.

Defect-based decoherence in superconducting qubits is the known Achilles' heel for superconducting qubits—these comprise primarily parasitic "two-level systems" (TLSs) that are commonly formed at the interfaces between different superconducting components (Martinis et al. 2005). As shown in Figure 4-3, in both Al- and Nb-based superconducting qubits (the most commonly employed materials platforms to date), native oxides form at the interfaces of these elemental systems, resulting in trapped structural configurations that make up the TLS. These metastable chemical configurations can tunnel between equivalent configurations, resulting in an oscillating dipole that electromagnetically couples to the qubit—resulting in decoherence (Phillips 1987). Chemistry can play a unique role in understanding and designing a more favorable chemical environment to reduce, or even eliminate, the chemical motifs associated with TLS noise and other defect-based decoherence. Understanding, reducing, and/or entirely circumventing these parasitic TLSs is a central challenge for developing superconducting circuits. For instance, TLS noise is directly related to the local bonding environment in amorphous suboxides—these suboxides can be passivated with appropriate selective chemical treatment of lossy surfaces (Altoé et al. 2022), and/or modified through doping and/or alloying to reduce the prevalence of the most deleterious TLS by controlling local chemical motifs/coordination (Hamdan, Trinastic, and Cheng 2014). Oxygen off-stoichiometry has been identified as a key cause of parasitic magnetic-based decoherence in Nb oxide thin films (Sheridan et al. 2021), providing a chemical indicator of decoherence in these systems. Another emerging area is the exploration of novel superconducting materials and chemistries as qubits and resonators that are more robust to deleterious TLS formation and/or are comprised of stacked van der Waals bonded two-dimensional (2D) materials, which can circumvent lossy interfaces between dissimilar three-dimensional elements (Altoé et al. 2022).

4.2.1d Chemical Design of Trapped Molecule/Ion Qubits

Ions and molecules that are electromagnetically confined by electromagnetic fields in ultrahigh vacuum are "trapped" ions and molecules. Upon laser-cooling to their motional ground states, they can be used as multiqubit systems (Bollinger et al. 1991). Decoherence for trapped ions is related to both the electric field noise of the laser/trap system and materials-based losses from the electrode surfaces, which are shown to reduce multiqubit

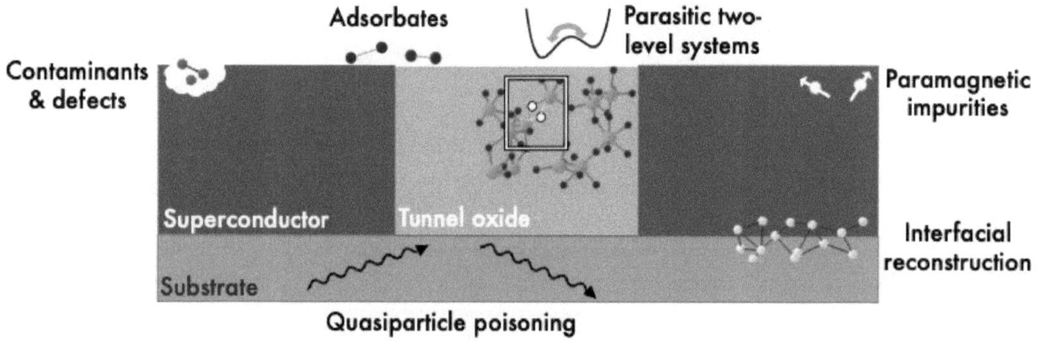

FIGURE 4-3 Superconducting qubits. Summary of the most deleterious defects in superconducting qubits depicted here for a Josephson tunnel junction comprising a superconducting metal (e.g., Nb or Al) sandwiching an oxide tunnel junction (e.g., AlOx, NbOx) which is grown on a semiconducting substrate (e.g., silicon). Atomic tunneling systems known as parasitic two-level systems are found in the amorphous oxide. Quasiparticle poisoning occurs from the presence of athermal phonons. Paramagnetic impurities can be induced through off-stoichiometry or by common absorbates such as H_2 and O_2. Finally, conventional fabrication residuals and contaminants are found throughout the structures, particularly at interfaces and surfaces. SOURCE: Sinéad M. Griffin.

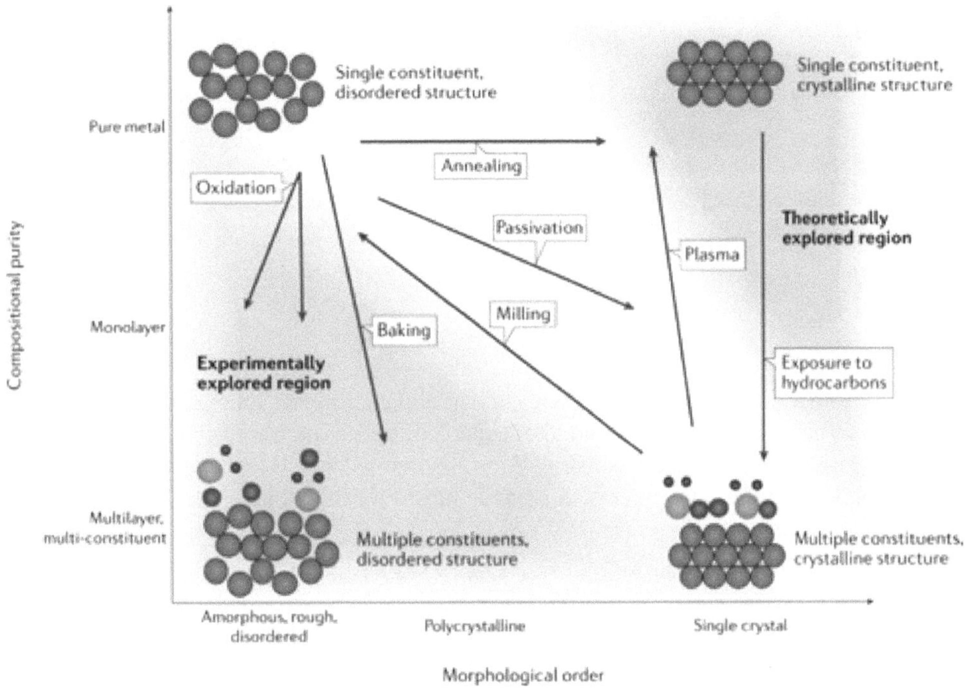

FIGURE 4-4 The many roles that chemistry can assume in the synthesis and design of trapped ions/molecular qubits. Chemical and structural complexity in trapped ion/molecular systems including theoretical and experimental parameter space. The axes represent the amount of compositional and morphological order. The assumed paths taken by a surface in this space under various treatments and procedures are depicted by the arrows. Cartoon surface cross sections in the four quadrants of the diagram represent potential atomic and molecular structures and compositions at these extreme points.
SOURCE: Brown et al. 2021.

gate fidelities (Brown et al. 2021). Understanding the microscopic origins of these decoherence channels is an ongoing challenge requiring exquisite knowledge and control of the electrode composition—possible sources include vibrations (Wineland et al. 1998), excitations, surface diffusion of chemisorbed atoms/molecules (Kim et al. 2017), and fluctuating "patch charges" on inhomogeneous electrode surfaces (Hite et al. 2013). Current and future challenges for trapped ion/molecule scaling will require surface chemistry approaches (Figure 4-4) to understand lossy electrode surfaces and to design robust, noise-tolerant components and surface treatments that can be cycled for several operations.

4.2.1e Chemical Aspects for the Design of Topological Qubits

While still in their infancy with regard to their demonstrated potential for QIS systems, topological qubits present an exciting possibility for robust, scalable qubits (Kitaev 2003). Owing to its fundamental nature (e.g., the topology of the material), the topological qubit should be inherently robust to certain external perturbations such as defects, strain, and interfaces—in stark contrast to other qubit platforms—and does not require complex error correction. However, while these perturbations often do not destroy the topology of the material, they can make identifying reciprocal-space features extremely difficult by inducing mid-gap states. This, in particular, is the case for both identifying and manipulating Majorana zero modes—the fundamental topological qubit—in physical systems (Nayak et al. 2008). While several routes to achieving Majorana zero modes have been theoretically proposed, no experiments to date have shown definitive evidence of their appearance and braiding (Kayyalha et al. 2020; P. Yu et al. 2021), which motivates materials and chemical approaches to achieving these highly sought-after emergent particles (Figure 4-5; Flensberg, von Oppen, and Stern 2021). Current chemical and materials approaches

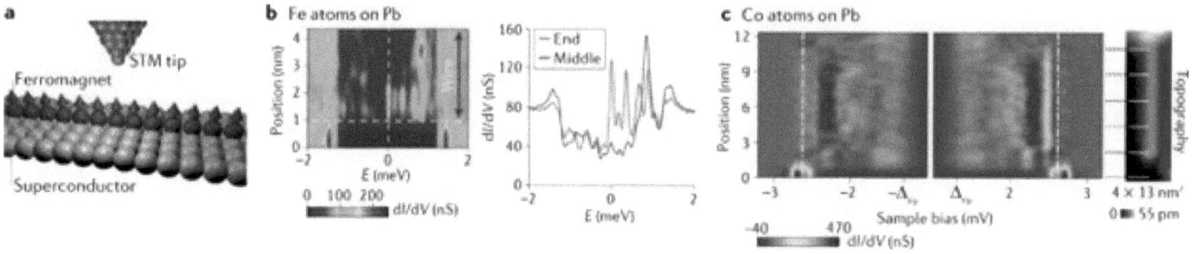

FIGURE 4-5 Chemical tuning of topological qubits.
SOURCE: Flensberg, von Oppen, and Stern 2021.

to combining strong spin–orbit coupling, superconductivity, and appropriate time-reversal symmetry breaking to achieve Majorana zero modes include semiconductor/superconducting interfaces and molecular/superconducting interfaces (see Figure 4-6). Huge surface chemistry efforts are required to make clean, precisely controlled surfaces with deterministic molecular placement and/or thin nanowires to pass the topological protocol.

4.2.1f Designing Hybrid Quantum Architectures That Mutually Enhance Each Other's Quantum Properties

Hybrid quantum computers synergistically combine the strengths of classical computing with the opportunities of quantum computing. For instance, a classical optimization algorithm can be used to guide the quantum circuit parameters, while the quantum components solve subtasks more efficiently than would be possible classically.

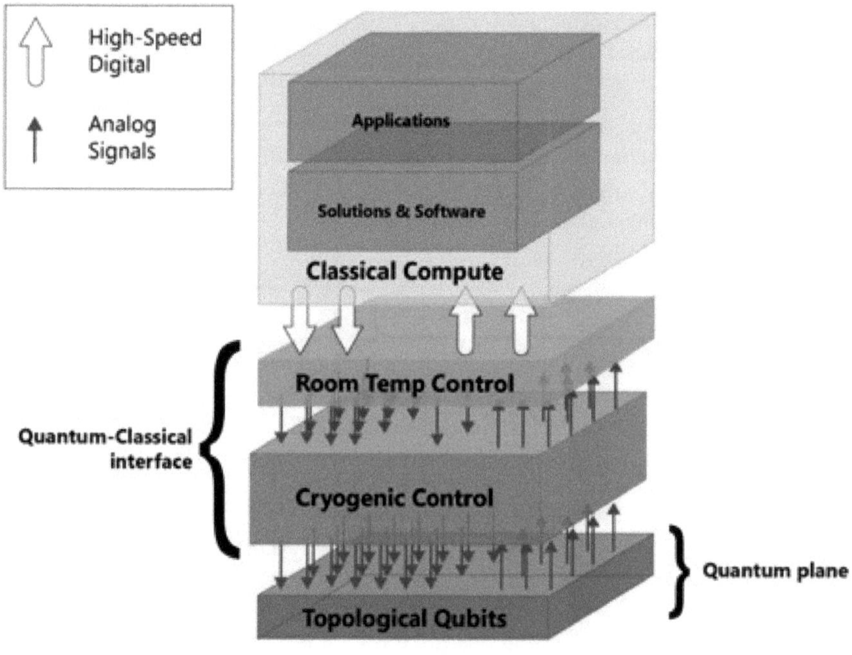

FIGURE 4-6 Practical quantum computing encompasses hybrid quantum–classical approaches that combine components of quantum computing systems with classical computing, with current progress and challenges for hybrid algorithms. For the physical manifestation of hybrid quantum–classical computers, several dissimilar physical components will need to be interfaced and connected that preserve quantum and classical information efficiently. Moreover, for systems combining more than one kind of physical qubit that require the transduction of quantum information between dissimilar qubits, chemical approaches are needed to maximally couple and transduce quantum information between these disparate modes.
SOURCE: Nayak 2021.

Further information is detailed below about hybrid quantum–classical algorithms, such as the variational quantum eigensolver (VQE), which have been applied to a range of problems in quantum chemistry. However, for their practical implementation, hybrid quantum computers must additionally combine the hardware requirements of quantum computers (i.e., scalability, coherence, and low error rates) with traditional classical computing components. This will require efficient and low-loss integration of qubit systems with conventional semiconductor technologies such as complementary metal-oxide semiconductor circuits.

4.2.2 Developing Noise Models and Quantum Error-Mitigation Techniques for Individual Qubits, Systems of Qubits, and Quantum Architectures That Can Be Experimentally Validated

No physical system is able to store and operate on information perfectly without errors. Indeed, one of the biggest challenges for quantum information protocols is protection from noise. Error-mitigation techniques that reduce noise thus have to be applied to qubits just as they are routinely applied to classical bits. In contrast to error-correction strategies, error-mitigation strategies involve developing noise models and do not necessitate making and manipulating copies of qubit information, which often makes them more practical in the context of chemistry. Such techniques could accelerate the adoption of moderately imperfect qubits and may enable the use of larger arrays of qubits at the same error rates as smaller arrays without correction, paving the way to larger, multiqubit technologies (Kandala et al. 2019).

But, to apply error-mitigation techniques, detailed models of qubit errors that depend upon the details of the specific qubit technology must be developed. If error models are known, they can be used to analytically reduce errors. Often the origin of this noise stems from molecular processes, meaning that first principles of computational modeling can play a key role in determining the source of the noise and analyzing its form to perform corrections. For example, substantial research has been conducted on quantitatively developing error models for superconducting qubits (Burnett et al. 2019) and ion traps (Foulon et al. 2022) based on the theory of TLSs. Research points to chemical defects at the surface of the oxide and other materials that make up superconducting qubits (Altoé et al. 2022) and the presence of adsorbates on the surface of the electrodes of a trapped-ion quantum computer as leading to anomalous 1/f noise that causes related qubits to decohere rapidly (Bruno et al. 2015; Foulon et al. 2022; Noel et al. 2019; Sedlacek et al. 2018; Wang et al. 2015). First-principles computational simulations of the chemistry at these surfaces enable one to predict how this noise varies based on defect and adsorbate concentrations, adsorbate type, temperature, and frequency, thus yielding models that can be applied directly to measurements to reduce noise and related errors (Aliferis and Preskill 2008; Georgopoulos, Emary, and Zuliani 2021; Tuckett, Bartlett, and Flammia 2018).

Similar models could be developed for molecular qubits. Simple proof-of-principle quantum algorithms for quantum error mitigation have been demonstrated in nuclear spin-based qubits (qudits), where qudits are d-dimensional ($d > 2$) quantum systems. Molecular strategies for this typically involve single-ion magnets, in which the electron and nuclear spin states are weakly anisotropic or exchange-coupled, leading to multiple spin states that are low in energy. For suitably engineered systems, unequal energy spacings allow addressing via microwave-resonant pulses (Aguilà, Roubeau, and Aromí 2021; Chicco et al. 2021; Gimeno et al. 2021; Hussain et al. 2018; Jenkins et al. 2017; Luis et al. 2011; Moreno-Pineda et al. 2017, 2018). Molecular electron–nuclear spin-based qudits (Ferrando-Soria et al. 2016; Godfrin et al. 2017; Luis et al. 2020) have been utilized for the implementation of quantum mitigation codes, where the molecular systems function as noisy intermediate-scale quantum (NISQ) units (Chiesa et al. 2020, 2021; Macaluso et al. 2020) and the electronic structure is "protected from decoherence." Taking advantage of nuclear spin structures with greater than two quantum levels ($d > 2$) in qudits leads to the possibility for long coherence times, due to isolation of the system from the environment, but, consequently, long manipulation times. Strategies to shorten the manipulation times for gate operations involve taking advantage of electron–nuclear coupling (hyperfine interactions) to perform operations on nuclear spin states at rates much shorter than the decoherence times (Castro et al. 2022; Chizzini et al. 2022; Hussain et al. 2018). A combination of further experiments and theory will be necessary to realize these nascent error-mitigation strategies (Castro et al. 2022; Chizzini et al. 2022; Hussain et al. 2018). One should distinguish important problems related to calculating excited-state energies from more formidable challenges of calculating entanglement or coherence in molecules with similar approaches. While the committee realizes that both are very important to chemical understanding, the calculations and approaches should specify the problems to be solved and the potential outcomes initially.

4.2.3 Understanding and Advancing the Limits of Classical Electronic Structure Algorithms and Modeling Approaches That Can Guide the Design of Molecular or Solid-State Qubits and Scalable Quantum Architectures

A promising route to investigating and ultimately designing scalable quantum information platforms is via both classical and quantum computational simulation techniques. Over the past few decades, bolstered by exponential increases in computational power, the simulation of quantum systems on classical computers, including molecules and materials, has evolved from the modeling of approximate coarse-grained systems to provide basic physical insights to the fully *ab initio* simulation of real and complex materials that can provide predictions that can be compared directly with experiments. Accordingly, many classical computational methods now exist, including DFT, coupled cluster theory (CC; Bartlett and Musiał 2007), and quantum Monte Carlo (QMC) methods (Foulkes et al. 2001), that can now model relatively large quantum systems with both speed and accuracy (Box 4-1). Where systems are too large or complex to be modeled directly using these methods, reasonably accurate and efficient quantum embedding schemes (Huang, Pavone, and Carter 2011; Sun and Chan 2016) that model critical aspects of systems using highly accurate quantum mechanics or less accurate quantum or classical methods have been developed. Indeed, simulation is now viewed as the third pillar of scientific discovery; just as it has been leveraged to provide key insights into the design of catalysts and the identification of potential pharmaceuticals, it now also has the potential to provide invaluable predictions regarding how qubits decohere and how they can be scaled into larger, multiqubit platforms (Lordi and Nichol 2021; Philbin and Narang 2021). For instance, quantum simulations performed on classical computers have recently been leveraged to characterize T-center qubits in Si (Dhaliah et al. 2022) and neutral group IV spin qubit vacancies in diamond (Ma et al. 2020). Nonetheless, several key challenges curtail the straightforward simulation of systems employed for QIS applications and quantum sensors on classical resources.

The rapid advancement of supercomputing power, in tandem with the development of increasingly sophisticated and predictive computational modeling, has made computational and theoretical modeling indispensable for QIS. Such *in silico* investigations have the advantage of systematically isolating and identifying qubit interactions toward control and design—for instance, accurate modeling of qubit–environment interactions such as qubit–phonon interactions.

4.2.3a Challenges for Modeling Quantum Systems with Classical Resources

As famously noted by Richard P. Feynman in his *Lectures on Computation* (Feynman, Hey, and Allen 1998), accurately modeling quantum systems on classical computers is a fundamentally expensive task because the number of states a quantum system can potentially occupy grows exponentially with the size of a system. While not all states are equally important and often approximations based on the locality of processes can be made, this scaling inherently limits the ultimate size of quantum systems that can be modeled using quantum resources. As a rule of thumb subject to change in the coming years, tens of isolated atoms can be modeled with high accuracy, while thousands of isolated atoms can be modeled with reasonable accuracy. Incorporating the effects of the environment, as a way to model decoherence, inevitably reduces these numbers even further (Friesner 2005). Moreover, many systems being considered for QIS are definitively strongly correlated and entangled, which means that high-accuracy methods must be employed to uncover useful design principles. Developing new techniques that can exploit physical insights to cheaply yet accurately model quantum information systems thus remains both a challenge and an opportunity for the community. Potential ways of addressing this challenge include the use of physically inspired embedding techniques, new ways of more directly and efficiently computing entanglement and modeling decoherence, the curation and more active use of QIS materials databases, and the employment of increasingly available quantum computing resources.

4.2.3b Modeling Large-Scale Systems Requires Advanced Sampling and Embedding

As previously described, many quantum sensors and quantum information platforms are highly complex, often hybrid materials that integrate qubits into an optimized host environment. The inherently large size of these

BOX 4-1
Modern Electronic Structure Algorithms

Predicting the properties and behavior of molecules and materials of use in QIS draws heavily upon an extensive yet growing suite of simulation methods termed electronic structure methods (Helgaker, Jørgensen, and Olsen 2000; Shavitt and Bartlett 2009; Szabo and Ostlund 1989). What unites this wide range of methods is that they all solve the electronic Schrödinger equation to determine the ground, excited, and/or thermal configurations of electrons within quantum systems, which in turn determine the geometry, conductivity, and magnetism, among other properties, of molecules and materials. A key challenge for all of these methods is that to precisely describe a given quantum system, they must enumerate all of the many possible electronic states of that system—which can very easily surpass the trillions in number. Thus, exact full configuration interaction (FCI) (also called exact diagonalization) calculations can only be performed for relatively small molecules or systems. As shown in Figure 4-1-1, to be able to describe QIS systems of physically relevant dimensions, different methods such as density functional theory (DFT) (Jones 2015; Koch and Holthausen 2000), one of the modern workhorses of materials discovery and design; coupled cluster theory (CC) (Helgaker, Jørgensen, and Olsen 2000; Shavitt and Bartlett 2009), the gold standard for highly accurate quantum chemistry calculations; and empirical methods based on experimental or other parameters (Thiel 2014) must be employed, which sacrifice differing levels of accuracy for speed. As no method has yet been—or may likely ever be—developed that does not suffer from this trade-off, modern-day practitioners must therefore continually ask what level of accuracy they reasonably need to solve their physical problem of interest with the computing resources they have. For QIS applications, a combination of methods is often employed to model large QIS systems with varying levels of accuracy. QIS systems also typically necessitate time-dependent treatments (involving time-dependent solutions of the Schrödinger equation), as well as treatments of phonons and photons, which play a key role in decoherence and measurements, respectively. A central challenge for the accurate modeling of QIS systems moving forward will be the integration of all of these critical elements into relatively inexpensive theoretical frameworks and related software that can be made widely accessible for computational scientists and experimentalists alike.

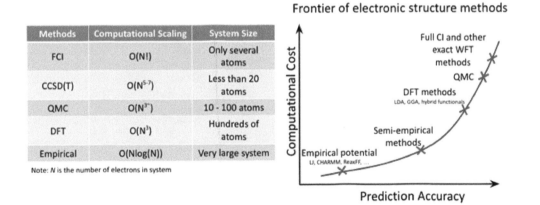

Methods	Computational Scaling	System Size
FCI	O(N!)	Only several atoms
CCSD(T)	$O(N^{5-7})$	Less than 20 atoms
QMC	$O(N^{3*})$	10 - 100 atoms
DFT	$O(N^{3})$	Hundreds of atoms
Empirical	O(Nlog(N))	Very large system

Note: *N* is the number of electrons in system

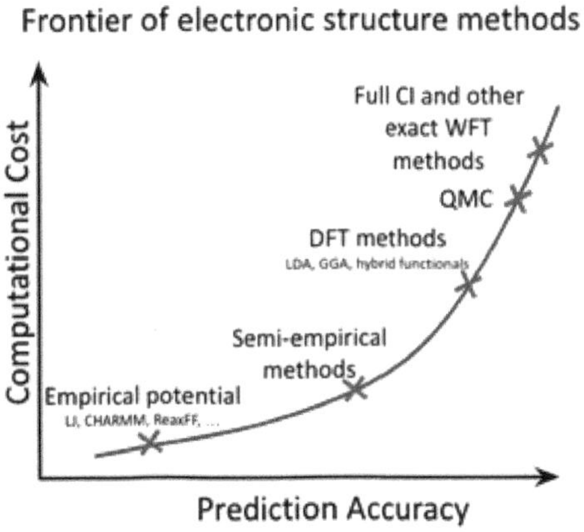

FIGURE 4-1-1 (left) Key classes of electronic structure methods for QIS and other applications and their estimated scalings in conventional implementations. (right) Typically, the computational cost of these methods increases with their accuracy, leaving practitioners having to find reasonable compromises between speed and accuracy for their intended applications.
SOURCE: Friesner 2005.

systems combined with the typical desire to understand the decoherence processes involving many electronic and phononic degrees of freedom at play in these materials makes directly determining their electronic structure a formidable challenge (Philbin and Narang 2021). While algorithmic advances, including linear scaling and tensor contraction techniques (Hohenstein, Parrish, and Martínez 2012), continue to be made that accelerate the direct modeling of these systems, scaling challenges are often mitigated by different forms of divide-and-conquer techniques: methods that divide the system into smaller portions, which can be modeled with highly accurate theories that can either be stitched back together to recreate the electronic structure of the entire system (Pruitt et al. 2014) or integrated into lower-accuracy descriptions of larger portions of the system.

One of the most productive divide-and-conquer strategies for qubits is embedding (Huang, Pavone, and Carter 2011; Sun and Chan 2016). In embedding theories, a smaller portion of a system that necessitates high-accuracy quantum simulations using techniques such as multireference perturbation theories (Shavitt and Bartlett 2009), CC (Shavitt and Bartlett 2009), QMC (Foulkes et al. 2001), or configuration interaction (Szabo and Ostlund 1989) is modeled surrounded by a larger portion of the system that is treated using lower-accuracy quantum or classical methods. In the context of quantum information systems, the atoms, vacancies, or complexes that constitute the system's physical qubits are typically modeled with high accuracy, while their hosts are modeled with lower accuracy.

In regard to the various theoretical mechanisms for coupling defect center–based spin qubits to other qubit platforms, potential interface coupling mechanisms such as superconducting qubits, other defect centers, and photons were discussed Wang and colleagues (2021) and are graphically illustrated in Figure 4-7. For the case of gigahertz platforms, dipole-, phonon-, and magnon-mediated mechanisms have been demonstrated. In the case of a dipole-mediated process, challenges remain in the selectivity of the interface mechanism due to the requirement for strict distance and orientation dependence between qubits. While the cavity phonon–mediated dipole coupling may overcome these limitations, other limiting factors present further challenges for the cavity-phonon mechanism. For example, in this mechanism, dipole selection rules limit the use of certain qubit states. Also, further

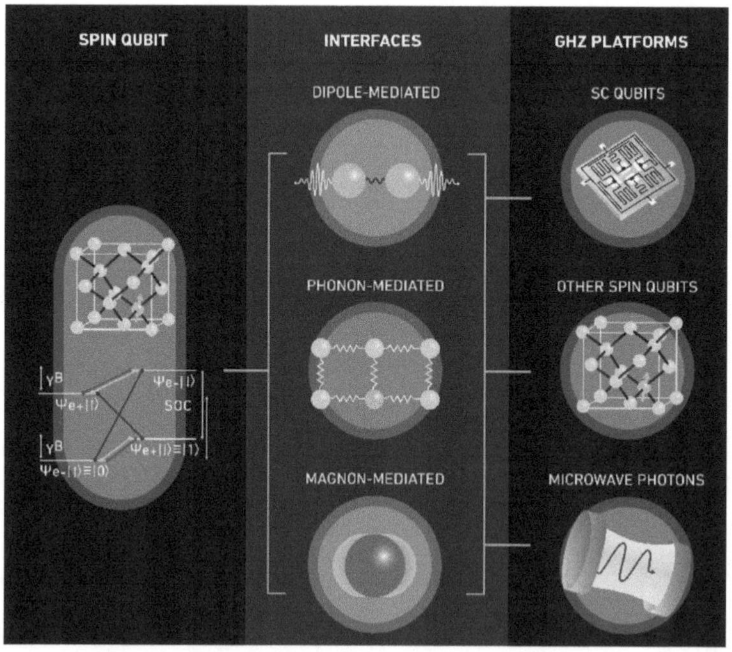

FIGURE 4-7 Spin qubits are capable of coupling to key gigahertz quantum systems, such as superconducting (SC) qubits, other spin qubits, and microwave photons via dipole-mediated (red), phonon-mediated (green), and magnon-mediated (blue) interactions. Such complimentary functionalities allow quantum information processing, storage, and transmission properties to be incorporated into one composite platform.
SOURCE: Wang, Haas, and Narang 2021.

investigation could be undertaken to determine the usefulness of this mechanism, as there may be limitations in the ability to tune two qubits, on demand into resonance for fast entanglement. Another solid-state interface coupling mechanism is that of phonon-mediated quantum interfaces. This approach seems to have the potential to couple qubits over arbitrary distances as long as the qubit states exhibit strain susceptibility. Wang and colleagues (2021) suggest further investigation and development at the nanoscale regarding the systems engineering of the cavity, the materials used, as well as a deeper understanding of the phonon-mediated mechanism. Finally, a more recent approach utilizes a magnon-mediated interaction to effectively couple defect center–based spin qubits with other qubits. This approach offers an enhanced coupling rate in comparison to phonon-mediated interactions. Wang and colleagues (2021) stressed that nanomagnonic systems offer a unique opportunity for nanotechnologists and may have a significant impact on quantum device innovation and translation.

An alternative divide-and-conquer approach to modeling, particularly disordered quantum systems, is through sampling. Disorder is a leading cause of decoherence in many materials and thin films of use in QIS applications (Foulon et al. 2022; Harrelson et al. 2021; Ray et al. 2019). For example, in recent work, researchers studied decoherence in superconducting Nb thin films by sampling over forms of disorder (Harrelson et al. 2021; Sheridan et al. 2021). In sampling techniques, different portions of a typically heterogeneous material are randomly selected to represent the whole, are modeled using high-accuracy methods, and then are averaged to recreate the electronic structure of the larger system. The sampling can be performed according to a uniform distribution or, if information is known about the system, to a more complicated distribution through importance sampling. Alternatively, samples can be generated from snapshots of molecular dynamics simulations or replica exchange Monte Carlo simulations (Frenkel and Smit 2002). Even more recently, experimental data have been used to train machine learning models to learn forms of disorder present over length scales not easily attainable within simulations (Kilgour and Simine 2022). Given sufficient experimental data on surfaces, this presents a promising path forward while emphasizing the need for more detailed experimental characterization of quantum materials for quantum information.

These strategies are but a few of many possible strategies for capturing the complexity of quantum information systems without invoking descriptions that may be more expensive or detailed than necessary. Given these techniques' inherent approximations, continued research could determine their accuracy, further improve their efficiency, and understand for which materials and properties they perform best. Properly modeling decoherence will moreover necessitate increased exploration into the most efficient ways to integrate descriptions of phonons and electromagnetism into the *ab initio* modeling of molecules and materials (Philbin and Narang 2021). The committee concludes that the computational modeling of large, heterogenous systems can be accelerated via embedding and sampling approaches on classical systems. Currently, the suite of embedding and other reduced-scaling techniques is underdeveloped. Improving these tools will ultimately lead to a deeper understanding of QIS challenges.

4.2.3c Theoretical and Computational Approaches for Modeling Entanglement

Even if scaling hurdles to simulating large quantum systems on a classical computer can be surmounted, quantifying entanglement and coherence in these systems remains a costly proposition because calculating these quantities requires an accurate accounting of multielectron interactions—often in the presence of phonons and other non-electronic interactions, and usually within a time-dependent framework.

In recent years, several new modeling techniques have emerged for computing entanglement (typically in terms of nth-order Rényi entanglement entropies) in interacting systems, including density matrix renormalization group (DMRG), tensor network, and QMC techniques. Nonetheless, these techniques scale steeply as high-polynomial powers with the size and number of electrons in the system. Even within these frameworks, the computation of the entanglement itself necessitates more time-consuming algorithms than the computation of more traditional quantities like energies and charge densities. To mitigate these steep scalings, attention has turned toward using embedding algorithms, which partition a system into a region that is treated with near-exact levels of theory and a surrounding region that is treated with less accuracy. In the context of quantum information, high-accuracy regions may, for example, be placed around transition metal or lanthanide atoms that host different spin states, while surrounding organic groups may be treated with lower levels of theory. Such embedding methods, including DFT-based embeddings, dynamical mean field theory, and density matrix embedding theory, enable only the most physically/chemically important regions of a system to be treated, thus substantially reducing the expense of quantum calculations at the cost of modest approximations. In their original forms, these methods were overwhelmingly developed to treat electrons alone. In part inspired by QIS needs, more recent investigations have begun to incorporate phonons/vibrations (Sandhoefer and Chan 2016) into these treatments because of the important role that they assume in decoherence (Sheng et al. 2022). In sum, the committee concludes that computational or theoretical tools for modeling the entanglement and coherence within quantum information processors remain costly. The high cost to use classical computational tools for modeling quantum phenomena is inhibiting key research activities such as direct simulation of large quantum information processors.

4.2.3d Validating the Modeling of Quantum Information Systems

The synergy between experiment and theory can help uncover the molecular and electronic structure that can further develop and improve qubits. As simulations become capable of accurately determining essential properties such as T_1 and T_2 relaxation times (Ariciu et al. 2019) and the optical response in optically addressable qubits (Goh, Pandharkar, and Gagliardi 2022), models will continue to play an important role in understanding and predicting the feasibility of using a system in a QIS application. Validating computational models is often hindered by the lack of experiments on systematically altered systems. Graham, Yu, and colleagues (2017) demonstrated the power of systematic modifications to vanadyl complexes with increasing distance between vanadium(IV) and a propyl moiety that demonstrates the effect on decoherence as a function of nuclear and electronic spin distance. The growing infrastructure of automated and robotic systems (Seifrid et al. 2022) that synthesize and analyze holds great promise in generating both systematically modified systems and data that can hasten validations.

4.2.3e Leveraging and Developing Machine Learning, Chemical Informatics, Chemical Databases, Molecular Simulations, and Quantum Computing Algorithms to Inform and Facilitate Qubit Design

Materials-informatics approaches are used broadly across the fields of chemistry, materials science, and physics to accelerate the discovery of novel materials and molecules. The explosion of computing power over the past two decades along with improvements in theoretical and computational approaches enable predictive calculations of many compounds, which can be collected into databases. The Materials Genome Initiative kickstarted the foundation of several materials databases with properties predicted from first-principles calculations with a range of applications and properties targeted. For instance, the Materials Project was originally developed for the discovery of battery materials but has expanded to have success across a range of applications including novel photovoltaics, multiferroics, and topological materials. It has recently been extended into the molecular space to include an MOF database (Rosen et al. 2021, 2022). Equivalent databases are starting to become available for molecular compounds (Organic Materials Database) and more complex systems (Automated Interactive Infrastructure and Database for Computational Science and Materials Cloud Two-dimensional Crystals Database).

The use of such materials and cheminformatics approaches for QIS applications is in its infancy yet presents an enormous opportunity to accelerate both the discovery of novel qubits and the understanding of their control in a systematic fashion. High-throughput searches have predicted new topological materials (Frey et al. 2020), and "low-throughput" calculations have been used to identify new spin defects in hexagonal boron nitride (Bhang et al. 2021). First-principles approaches such as these were used to search large combinations of possible defects in Si for new spin defect qubit candidates, with the resulting 400 combinations available in the Quantum Defects Genome database (Xiong et al. 2023). The first-principles approach includes careful benchmarking of the theoretical approaches with experimental measurements, which will improve the predictive power of such approaches. These tools can enable discovery through the exploration of compounds that are synthesizable and have yet to be tested for potential use in QIS technologies (Lilienfeld, Müller, and Tkatchenko 2020). Cheminformatics and machine learning have also been used to determine synthesis routes and to improve synthesis conditions through, for example, the design of experiments methodology (Weissman and Anderson 2015). Keeping the databases fully open and available would advance the field quickly.

4.2.4 Open, High-Quality QIS Software and Data Repositories

As demonstrated in this chapter, theory and computer models play an important role in the advancement of chemistry in the QIS field. Open-source and commercially available electronic structure codes have played an integral role in studying and understanding spectroscopic results in recent studies. The number of codes and applications is extensive, and attempts have been made to track more established software packages. Two examples are referenced here; however, other lists with additional detail may be available (Giannozzi et al. 2009; Scientific Library n.d.). Open-source code enables researchers to replicate and analyze each other's predictions and build upon past theoretical achievements and code bases instead of having to perennially reinvent the wheel. This accelerates the overall pace of scientific discovery. Moreover, open-source code democratizes discovery by granting researchers who do not have the staff, bandwidth, or expertise to maintain their own codes—including experimentalists who may wish to validate their own results—to still make valuable contributions to the field. Alternatively, commercial software is often more user-friendly, well documented, and regularly updated. Thus, commercial software can be the more favorable option.

However, often free code is not well documented, is difficult to install and use, and ultimately remains inaccessible to many. Software development faces a challenge in striking a balance between ease of human readability and performance-driven computer readability. A set of minimal requirements for open-source code would increase the longevity and transferability of software developed for research and minimize the replication of software tools.

These challenges are not unique to computational chemistry codes, and the computational science community is continually developing tools (e.g., container technology such as Docker, Singularity, and others; see Docker n.d.) to improve the experience of using code across operating systems and environments. While these tools help the user, programmers also need to be considered; this is particularly important in academic groups, in which users often program or reprogram their own software to suit their specific needs. In an information-gathering meeting, the committee heard from Dr. Arman Zaribafiyan (see Appendix B), from Good Chemistry, who highlighted that although the tasks of software engineers and scientists in industry are often clearly delineated, integrated collaboration between the two disciplines is often necessary to solve grand challenges like those posed by the field of QIS. QIS-specific software packages have yet to nucleate, but ensuring that those that eventually rise to the fore remain open will be important for enabling rapid advances.

Equally valuable to the community are the products not only of software but of experiments—that is, data. As described earlier, other communities, including the materials community, have started to develop databases that can be leveraged to guide discovery (Box 4-2). As QIS-specific databases begin to emerge, ensuring that they remain open for widespread exploration will be important. In the case of QIS applications, both theoretical and experimental data, including calculations of decoherence times and entanglement, spectra, and structures, are invaluable not only for validating one another but for facilitating mutual method development. Some of these QIS-relevant data can be harvested from other existing molecular and materials databases, and such efforts to aggregate data have to be undertaken and supported. However, the amount of QIS-specific data is only expected to grow, and the community will need to establish well-structured databases with clear guidelines in the near future. FAIR—Findability, Accessibility, Interoperability, and Reusability—standards (GO FAIR 2017) that emphasize proper labeling for and reusability of data should serve as baselines and be applied wherever possible to ensure professional data management and stewardship.

However, certain challenges associated with intellectual property and proprietary/commercial data need to be carefully balanced to maximize their utility across different sectors while considering the needs of the data sponsors and originators. While public funders' data should be accessible and follow the FAIR guidelines, data originators may be at a disadvantage if immediate sharing is required. To address this issue, first the community can draw inspiration from other communities with rich data infrastructures. For example, the National Aeronautics and Space Administration allows a data embargo period of one year for scientists who proposed and led the data collection to have time to analyze and publish. Second, it is essential to incorporate robust attribution frameworks and guidelines into all databases so that those who contributed to the data and infrastructure are adequately acknowledged and cited. Finally, incentivizing industry to share their potentially valuable data openly continues to be a challenge that can be overcome through collaboration between open repositories and industry.

BOX 4-2
The Role of Cheminformatics and Databases in Modern Chemical Discovery

While cheminformatics has long assumed a role in chemistry and pharmaceutical discovery, in particular, advances in machine learning techniques over the past decade that make them both more accurate and more efficient have made learning from chemical data an even more important mode of scientific inquiry. Indeed, given the computational expense associated with predicting QIS phenomena from first principles (i.e., knowledge of just the atoms involved), cheminformatics has become a comparatively inexpensive alternative for predicting chemical properties that can be used to guide experiments. Pivotal successful examples of using cheminformatics and machine learning techniques to guide discovery include the Materials Project and the Open Catalyst Project, which are now heavily used to inform materials and catalytic discovery. In cheminformatics, neural networks or other models (e.g., regression models) learn functional relationships between data points and their related properties, such as their energies or magnetic moments. These models can then be used to predict the properties of new molecular or material inputs. Key to the success of these endeavors is thus the aggregation of large amounts of both simulation and experimental data, which can be used as training sets for machine learning and other models. While most chemistry experiments are conducted on tens of samples, cheminformatics techniques often require more than thousands of data points to make meaningful predictions. High-throughput techniques are therefore necessary for running enough experiments to provide sufficient data for discovery, and expedient computational methods are essential for filling in remaining data points. As data can only be learned from if they are properly named and formatted for transparency, it will be crucial that all data, experimental data included, follow agreed-upon standards for formatting. Databases containing data for QIS applications are currently lacking but will need to be developed, populated, and made open source to enable QIS-motivated data-driven discovery in the future.

4.3 MODELING CHEMISTRY PROBLEMS USING QUANTUM COMPUTERS

Up to this point, the discussions have focused on the current challenges of and progress made with classical computational tools to study complex quantum systems (e.g., qubits) and large molecules. Even with the advancement made, classical computation still faces severe limitations (e.g., computing power and cost) in providing accurate theoretical models for understanding chemical systems employed in QIS applications. Currently, quantum computers face similar challenges to their classical counterparts. Small data input/output, large compute, and super-quadratic speedup of quantum algorithms are three main criteria to determine whether a quantum application can reach a reasonable crossover time and crossover size compared to its classical rival. Although still in the early stages of development, a mature quantum computer may eventually offer an alternative approach for providing enhanced computational tools that surpass the limitations of classical systems. The remainder of this chapter discusses the current challenges and future opportunities in using quantum computers to study chemistry problems that are, at this time, considered extremely difficult to simulate using classical algorithms.

4.3.1 Identifying Important Open Chemistry Problems for Quantum Computers That Are Unresolved Due to Classically Intractable Electronic Structure

While advancements are being made in developing the hardware for quantum computers, many challenges associated with *adopting* this technology remain (Argüello-Luengo et al. 2019) to gain truly valuable, and classically intractable, insights about chemistry. The first challenge, as demonstrated in this chapter, is that classical computers actually do work quite well for resolving many chemical problems in part because much of chemistry occurs around low-energy electronic states. Such states have limited and structured entanglement, enabling classically efficient methods like DFT to work well often despite making fundamentally uncontrolled approximations. Classically computable methods are especially viable when the calculations do not require high accuracy and when systems are not "strongly correlated." A strong correlation (high entanglement) of electrons tends to occur near bond-breaking transitions, in free radicals, and around transition metal centers. Much work remains in understanding how and for what problems quantum computers might provide otherwise inaccessible new chemical insights. First, even if one has an oracle for efficient and completely accurate energy calculations, this does not immediately resolve all chemically relevant questions. An insightful chemist is needed to interrogate the quantum computer thoughtfully to derive a chemical understanding. Second, quantum computers might scale polynomially, but that does not mean the calculations are free (or even as efficient as DFT); indeed, the simulation of chemical Hamiltonians still takes considerable quantum computing resources. Thus, currently, we remain limited in how many electrons can be explicitly treated on a quantum computer; we will need improved methods of embedding the most strongly correlated subsystems inside quantum computations, with the environment treated at lower levels of theory on classical computers.

Finally, as classical methods continue to improve, we face the prospect that the "truly hard" chemistry problems that require a quantum computer will become increasingly esoteric. The problem does not lie in the class of nondeterministic polynomial time (NP) but rather the challenge lies in the fact that the proposed problems cannot be verified efficiently by a classical witness. A generalization of this challenge is Quantum Merlin Arthur (QMA)-Hard, which is the quantum version of the class NP; thus, any problem in NP could be solved in polynomial time. This generalization implies that some quantum chemistry problems exist, although of unknown measure, that would be inefficient to solve even on a quantum computer. However, it is still unclear if natural molecular or material Hamiltonians exist that are expected to be found in their ground states in nature that would qualify as QMA-Hard. If that is the case, questions arise about how those systems get into their ground states since any quantum dynamics can be efficiently simulated by a quantum computer (Schuch and Verstraete 2009). Furthermore, see Goings and colleagues (2022) for a case study proposing a target that would be appropriate for a quantum computer and benchmarking the best classical methods against the projected quantum costs. The following section discusses how various chemistry research problems could be vetted to determine their viability for quantum computing applications.

4.3.1a Evaluation of Chemistry Use Cases with Quantum Computers

Below is a set of criteria that could be used to evaluate research proposals that identify chemistry problems that could be addressed using a quantum computer:

1. The challenges the investigator is aiming to study need either to have significant scientific merit or industrial relevance or to address a fundamental chemistry question.
2. An explanation for why these quantities cannot be obtained using classical methods needs to be provided.
3. A compilation is involved of the proposed quantum algorithm to an error-corrected instruction set that is compatible with a scalable, fault-tolerant error-correcting code such as the surface code.
4. Estimation of computing time and resources of the best classical methods are included in the proposal. Any research will benchmark these more precisely.
5. The proposed work aims to answer the following question: what quantities are being measured and to what precision?

These investigations will serve as case studies that clearly target quantum computing efforts and their role in facilitating our understanding of chemistry. Given the complexity involved between ongoing developments of the quantum computer's architecture and the need to have appropriate chemical problems to study with these new technologies, the committee has provided Recommendation 4-2, summarized at the end of the chapter.

4.3.2 Studying How Quantum Computing Algorithms for Dynamics Can Be Used to Accelerate Chemistry and Spectroscopy

The most promising problems to solve on quantum computers are electronic structures, which can have exponential computational complexity and for which there are several known quantum algorithms to provide exponential speedup. Electronic structure is the foundation for understanding chemical properties and reactivities. For example, electronic structure calculations have been used often to help assign peaks for complicated experimental spectroscopy; the molecular orbitals from electronic structure calculations are often used to predict and interpret the reactivity of chemical species.

Although challenges are evident, one of the most significant promises of quantum computing is that it can provide exponential speedup for a certain class of problems. The electronic structure problem is the problem of solving for the ground state of electrons interacting with each other through the Coulomb potential and with the point charges arising from the nuclei. Such ground states characterize the electronic wave functions of molecular systems and materials.

Subspace full configuration interaction (FCI) is not commonly used in routine chemistry applications due to the very restricted size (small molecules) of the affordable one-particle basis and because reliable relative energies may not require ultimate accuracy of total energies (Liu et al. 2022). In the spirit of FCI, many lower-scaling algorithms have been developed for chemistry applications. These traditional algorithms represent the state of the art for solving the electronic Schrödinger equation with high accuracy (Eriksen et al. 2020; Motta et al. 2017; Williams et al. 2020). The core idea is still to solve the eigenvalue problem either by using predefined restrictions of the many-electron basis (coupled cluster with a full treatment of singles and doubles (CCSD(T)), or Davidson correction (MRCI+Q)) or through an iterative construction of the basis-set expansion (DMRG, FCIQMC). The restriction to a selected finite set of active orbitals in all FCI-type approaches generates an artificial distinction of electronic correlations into two groups: (1) static electronic correlations, which are typically characterized by orbitals that occur in determinants with large weight in the wave function expansion; and (2) dynamical electronic correlations, which refers to orbitals present in determinants with small to vanishing weights. This artificial split into static and dynamical electronic correlations can be overcome if a routine numerical approach is used to obtain results of FCI quality on a one-particle basis of one to a few thousand orbitals. There are several FCI-type approaches to address static and dynamical electronic correlations. Even though it is difficult to rigorously assess the error in the energy after a fixed number of optimization cycles with predefined parameters, these traditional methods likely will remain the reference methods for most quantum chemistry calculations, at least in the near term.

Certain critical chemical problems and processes have electronic structures that are too complicated to be handled by any classical methods. FeMo cofactor (FeMoCo) of the nitrogenase enzyme responsible for nitrogen fixation has been discussed frequently and likely represents the most complicated electronic structure (Lee et al. 2021; Li et al. 2019; Reiher et al. 2017). Metalloporphyrin, which often plays an important role in biological processes, is another example of a complex electronic structure (but simpler and more typical than FeMoCo) that supports the need for quantum computers. Even though the active space size in both examples is similar, due to the nature of the electronic structure, the estimated run time on quantum computers is different. Thus, the high-accuracy treatment of strong correlation offered by quantum computers is relevant for understanding mechanisms in inorganic catalysis, for example. Solving the modern electronic structure problem will provide deeper insight into the properties and behaviors of other strongly correlated systems, such as transition metal and heavy metal complexes and biradicals. As demonstrated in Chapter 3 these classes of molecules also have potential utility in a variety of quantum applications. Therefore, gaining more accurate details about the nature of their chemical processes like bond breaking, excited-state transitions, and magnetic properties will be critical for streamlining the development of functional materials. However, to reach high levels of accurate predictions, progress first needs to be made toward solving the modern electronic structure problem.

The committee was presented with expert information regarding the possibility of utilizing quantum computers with new strategies for accurate calculations of highly correlated systems. The committee found that chemistry could play a role in the development of new approaches and algorithms specific to the use of a quantum computer in chemistry. One such idea, of many, involves a method of moments coupled cluster (MMCC) formulism. Some success has been reported on utilizing this method to effectively correct the energies of approximate coupled cluster formulations using moments that are not used to determine approximate cluster amplitudes (Peng and Kowalski 2022). Reports have suggested that particular quantum algorithms for computing MMCC ground-state energies offer a clear advantage over classical computing approaches. While great work remains to be performed to illustrate the robustness of such an approach, this example demonstrates a quantum computing approach that may provide the utility to calculate a strongly correlated system, which was previously thought to be impossible with classical approaches. This and other approaches on the horizon are important in the development of our understanding of the true advantage of the quantum computing approach. Researchers in this area, in all settings, should be encouraged in their use of new approaches with quantum computers in chemistry that illustrate a strong advantage (McArdle et al. 2020).

4.3.2a Developing More Efficient Methods of Encoding Chemical Systems on Quantum Computers

Calculating total electronic energies with known accuracy is the basis for any theoretical description of molecular systems. As encoding the exact determinant space for large systems having many orbitals (>18) is very challenging on classical hardware, a key goal of all these novel methods is to approximate the full determinant space. Quantum computing holds promise to accomplish this goal, provided that a sufficiently large quantum computer can be built. The most studied approach to solving the electronic structure problem is quantum phase estimation (QPE), which is a fully quantum algorithm that can solve exact solutions of the chemical wave functions and energies. Bayesian phase difference estimation (Sugisaki et al. 2022) and other methods (Arute et al. 2019) are also used; however, the remaining discussion will be focused on QPE. QPE provides an appealing quantum speedup compared to classical rivalries like FCI. For classical computers, as explained in Box 4-1, FCI is a classical algorithm that solves the electronic structure problem; however, it uses exponential scaling.

In QPE, one chooses a trial state, a target error in the eigenvalue estimate, and a desired success probability. The algorithm then returns an estimate of a randomly selected eigenstate. Importantly, the eigenvalue is sampled with some probability. If ground-state energies are desired, then the trial state should be chosen to make the overlap with the true ground state reasonably large. Many ways exist to prepare initial states including using matrix product states, adiabatic-state preparation, and unitary coupled clusters. Different research groups are developing procedures to increase the efficiency of preparing the initial states (Lee et al. 2023). Importantly, if an initial state is difficult to establish, as is the case in strongly correlated systems, then the initial-state overlap will be weak.

There are concerns that sometimes the dependence on the initial-state overlap can lead to poor scaling in the quantum algorithm, especially when the initial overlap is weak (Aspuru-Guzik et al. 2005; Lee et al. 2023). Furthermore, this algorithm measures the eigenvalue of unitary matrices. Quantum circuits can be used to synthesize unitary matrices that share an eigenbasis with the molecular Hamiltonian—for example, via realizing time evolution. Quantum walks are also a popular alternative (Babbush et al. 2018). Developments are under way to increase the efficiencies of these quantum algorithm approaches and lower their costs. Finally, a major setback of using QPE, at the moment, is that it requires quantum hardware that may not be completely accessible in the near future.

4.3.2b VQE to Solve for Electronic Structure

An alternative to a fully quantum algorithm like QPE is VQE. VQE is considered a meta-algorithm, or algorithmic framework, for a quantum–classical hybrid approach where the quantum computer does some work and the classical computer does some work (Chen et al. 2021). VQE is attractive for studying quantum chemistry problems using a popular quantum computer known as the NISQ (McClean et al. 2016).

Running quantum algorithms on real hardware is essential for understanding their strengths and limitations, especially in the NISQ. Simulations assume a perfect system and do not effectively take into account the decoherence errors, loss of accuracy, and time-outs that can occur on real systems (Yamamoto et al. 2022). Quantum computers have evolved from exploratory physics experiments into cloud-accessible hardware as early stage commercial products that are available to a broad audience of chemists and those in other disciplines (Endo, Benjamin, and Li 2018). NISQ-era computers can experience a wide range of complex errors. Many of these errors result from noise in or miscalibrations of the lowest-level components in the system, quantum gate operations, and decoherence in the qubits themselves. The Quantum Economic Development Consortium sponsored a quantum benchmarking project that resulted in an open-source suite of bench tools to run on various quantum hardware systems (Lubinski et al. 2023). This work could be extended to benchmark real chemical structure calculations and molecular energy measurement algorithms on real hardware (Gulka et al. 2021).

VQE is attractive for NISQ because it does not require long coherent circuits like QPE (O'Malley et al. 2016). However, the number of circuit repetitions required is dramatically more for VQE, which can eventually lead to other challenges such as reaching the limits of NISQ supporting capabilities (Peruzzo et al. 2014; Wecker, Hastings, and Troyer 2015).

The compatibility between VQE and NISQ is exemplified in the simulation of electronic structures, where VQE can be modified or constrained to find the electronic state with a specific number of electrons, electron spin, or other properties (Ryabinkin, Genin, and Izmaylov 2019). Another significant simulation problem addressed by VQE is the dynamic correlated states in a quantum system. The state characterizes the dynamics of quantum particles through the (correlated) system and provides insight into its resulting optical, magnetic, and transport properties. The state also allows access to many static observables, notably the total energy of the system. The ability to understand the dynamic correlated-state system can apply to the understanding of high-temperature semiconductors although the observation will be limited by circuit depth (Chen et al. 2021).

VQE uses the output of a parameterized quantum circuit as an "ansatz" (i.e., a mathematical assumption about the form of an unknown function that is made to find a solution to an equation or other problem) or a trial for the true ground state. The energy of the ground state is calculated using the quantum computer. Then, the classical computer optimizes the parameters of the quantum circuit to lower the energy of the ansatz (Grimsley et al. 2019; Peruzzo et al. 2014).

Furthermore, the committee received expert information from Professor Giulia Galli (see Appendix B), who has made progress in the use of quantum computers for chemical calculations. Galli's work involves using quantum calculations to predict and design novel molecular and nano-qubits. The long-term goal of the research is to create new simulations on quantum computers for greater accuracy and speed. As illustrated in Figure 4-8, many of the chemistry problems explored using VQE on NISQ are on very simple chemical systems, like diatomic molecules (H_2, LiH, NaH, and others), and still require significant error correction to the raw data (see graph in Figure 4-8).

VQE has also been implemented to simulate two intermediate-scale chemistry problems: the binding energy of hydrogen chains, as large as H_{12}, and the isomerization mechanism of diazene (Arute et al. 2020). The simulation was performed using a circuit consisting of 12 qubits and up to 72 qubit gates. However, the authors reiterate the 50-qubit barrier (known as the classically intractable regime), where the key building blocks of the proposed VQE algorithm are potentially scalable to larger systems that cannot be simulated classically (Arute et al. 2020; O'Brien et al. 2022). Many NISQ VQE chemistry experiments have only been 12 qubits, and data suggest that scaling all up to 50 or more qubits to surpass the classically intractable regime will be challenging.

4.3.2c Exploring How Quantum Machine Learning Can Be Used to Accelerate Chemical Research by Processing Quantum Data from Entangled Sensor Arrays or Quantum Simulations of Chemistry

Another widely anticipated application area for quantum computing is machine learning. Quantum machine learning (QML) (Biamonte et al. 2017) consists of two distinct branches: (1) using quantum computers to accelerate classical machine learning as performed by classical computers and (2) using the quantum advantage of quantum computers to construct fully nonclassical quantum neural networks. While quantum computers hold promise for accelerating classical machine learning algorithms, here we instead focus on the potential for quantum computers to analyze data more rapidly. Perhaps the most pressing challenge in the field of QML is to answer the following question: on what types of data sets do we expect QML to provide a quantum advantage? Currently, the compelling answer to this question is that QML is particularly promising for modeling "quantum data" (Huang et al. 2021). Quantum data are data that consist of quantum states rather than classical information. Examples of quantum data include wave function output by quantum simulations as well as data that are obtained from quantum sensors and then transduced to quantum computers. Section 4.2.3a discusses using classical computers to simulate "quantum data"; however, this technique is limited in providing "enough" data due to computational power and cost.

Recent work has shown that when learning from quantum data is applied, it is possible to achieve an exponential advantage in learning. Quantum data can be generated from performing quantum simulations of chemical systems, and learning is applied on top of those data (Huang et al. 2022). This exponential advantage arises because exponentially fewer data are needed if the data are "quantum data" and entangled. Furthermore, because this is a query advantage (i.e., needing less data) as opposed to a conventional quantum speedup (i.e., an algorithm with faster run time), it is possible to reach a quantum advantage with few qubits, especially in the context where data are limited. For example, this advantage could help in understanding topological order in systems, which is a

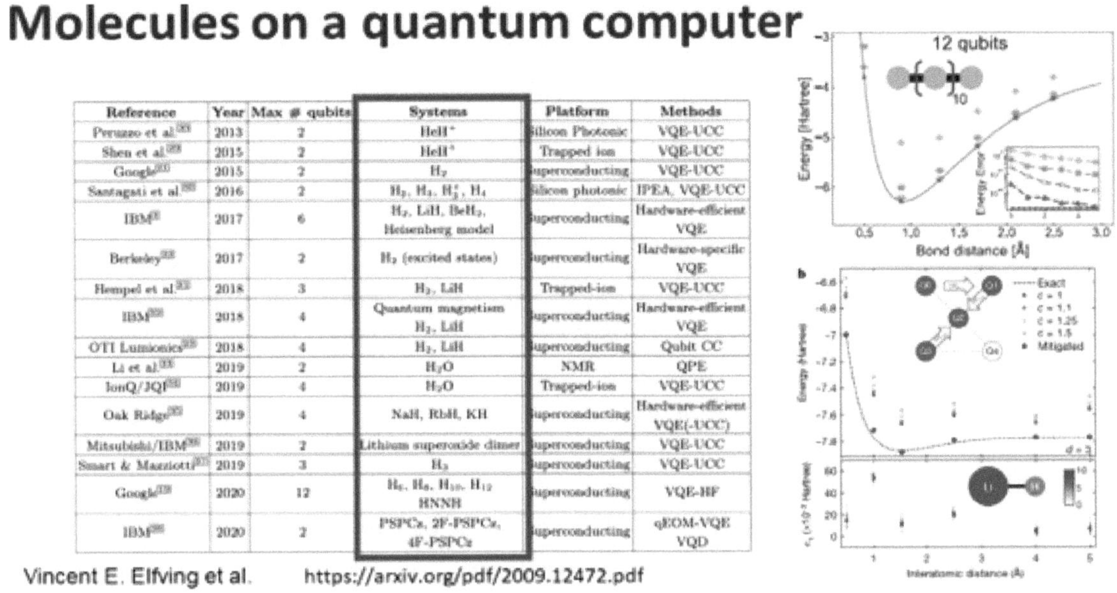

Reference	Year	Max # qubits	Systems	Platform	Methods
Peruzzo et al.	2013	2	HeH^+	ilicon Photonic	VQE-UCC
Shen et al.	2015	2	HeH^+	Trapped ion	VQE-UCC
Google	2015	2	H_2	uperconducting	VQE-UCC
Santagati et al.	2016	2	H_2, H_3, H_3^+, H_4	ilicon photonic	IPEA, VQE-UCC
IBM	2017	6	H_2, LiH, BeH_2, Heisenberg model	uperconducting	Hardware-efficient VQE
Berkeley	2017	2	H_2 (excited states)	uperconducting	Hardware-specific VQE
Hempel et al.	2018	3	H_2, LiH	Trapped-ion	VQE-UCC
IBM	2018	4	Quantum magnetism H_2, LiH	uperconducting	Hardware-efficient VQE
OTI Lumionics	2018	4	H_2, LiH	uperconducting	Qubit CC
Li et al.	2019	2	H_2O	NMR	QPE
IonQ/JQI	2019	4	H_2O	Trapped-ion	VQE-UCC
Oak Ridge	2019	4	NaH, RbH, KH	uperconducting	Hardware-efficient VQE(-UCC)
Mitsubishi/IBM	2019	2	Lithium superoxide dimer	uperconducting	VQE-UCC
Smart & Mazziotti	2019	3	H_3	uperconducting	VQE-UCC
Google	2020	12	H_6, H_8, H_{10}, H_{12} HNNH	uperconducting	VQE-HF
IBM	2020	2	PSPCz, 2F-PSPCz, 4F-PSPCz	uperconducting	qEOM-VQE VQD

FIGURE 4-8 (a) Experiments with quantum computing hardware applied to simulating molecular and material chemistry (this selection covers many but not all results). Years and top-to-bottom order, in order of appearance on prepublication service arXiv. Maximum qubit number denotes the number of qubits actually used in the simulation, potentially on sub-lattices of larger chips. All systems were discretized into near-minimal basis sets (i.e., STO-3G or similar) or utilized severe approximations to reduce the number of qubits in other ways. The "superconducting" platform denotes any variant of superconducting platform where microwave pulses are used to control qubits defined by flux or charge quanta on superconducting islands. VQE-UCC stands for any strategy combining the variational quantum eigensolver (VQE) algorithm with the chemistry-inspired unitary coupled cluster ansatz approach. VQE-HF performs the Hartree-Fock procedure on-chip using VQE. IPEA and QPE are forms of quantum phase estimation implementations. The quantum equation-of-motion VQE (qEOM-VQE) and variational quantum deflation (VQD) methods are used to compute excited-state energies. *Recently, IonQ conducted a 12-qubit experiment on the trapped-ion platform using VQE-UCC (Zhao et al. 2022). **Google conducted a chemistry experiment on 16 qubits in 2022 using fermionic quantum Monte Carlo methods (Huggins et al. 2022). (b) Binding curve simulations H_{12} with error mitigation. Comparison of Sycamore's raw performance (yellow) with postselection (green), purification (blue), and error-mitigated combined with variational relaxation (red). All points except for red were calculated by using the optimal basis rotation angles computed from a classical simulation; thus, the variational optimization shown is only used to correct systematic errors in the circuit realization. (c) Hardware-efficient VQE applied to the electronic structure problem of LiH using trial state preparation circuit depths $d = 3$. The experimental results from the extrapolation of energies obtained from the final stretch factors, compared to the exact energy, for a range of interatomic distances.
SOURCES: Arute et al. 2020; Elfving et al. 2020; Kandala et al. 2019.

global property of the wave function that cannot be observed by local measurements. Another potential impact of this area on chemistry is that quantum sensors (either molecular sensors or sensors using a different technology), where the state of the sensor can be transduced to a quantum computer, might allow new tools for investigations within a chemical laboratory. This direction has many synergies with molecular sensors and molecular qubits in the context of quantum signal transduction, and both technologies would likely need to be further developed to make this a reality.

4.3.3 Developing More Efficient Quantum Algorithms for Fault-Tolerant Quantum Computers to Simulate Molecular Systems

Quantum simulations, like classical theory, are facing accuracy issues in predicting large molecules. Currently, the most studied model for such systems is medium-sized inorganic catalyst molecules, which are a subject of homogeneous catalysis research. The challenge is to find an application that is insurmountable by current quantum computers. And, the system is large enough to be predicted accurately with a quantum computer without the need for gross oversimplifications. Because the architecture of quantum computers is still premature, the technology is understandably plagued with defects that ultimately cause data to be highly erroneous. Having an understanding of the basic architecture of quantum computers will help the chemist determine which types of chemistry problems will be suitable for a quantum computer versus a classical computer. Box 4-3 elaborates further on quantum digital gates.

Two major architectures are competing for popular acceptance: fault-tolerant quantum computing (FTQC) and, as described earlier, NISQ. NISQ exists now, and is small scale and non–error corrected. NISQ may use digital or analog paradigms, and current devices are based on both approaches. For FTQC, the digital (gate) approach to error correction for fault tolerance is better developed as a concept, hence the belief that it will be the most likely useful, far-future solution. Whether it would be possible to realize a fault-tolerant analog quantum computer remains unknown.

The ultimate goal for quantum computing systems is to design platforms capable of eliminating the amount and intensity of noise generated in the data, which is directly related to the overall quality of analysis. Two

BOX 4-3
What Is a Quantum Logic Gate?

The standard model of quantum computing, also known as the digital or gate model, involves algorithms that are instantiated as quantum circuits consisting of a series of quantum logic gates (Barends et al. 2014; Paz-Silva and Lidar 2013; Van Meter and Itoh 2005). A quantum logic gate is the quantum equivalent of logic gates used in classical computation. In quantum computation, these gates are realized by time evolution under Hamiltonian elements acting on one or two qubits at a time. All reversible classical logic gates can be executed on a quantum computer. But quantum logic gates also include some new operations such as gates that put quantum bits in a superposition of two logical states and gates that entangle qubits. Whenever a quantum logic gate is executed natively in hardware, it has a finite error probability that is referred to as the gate fidelity, f. Thus, if a quantum circuit involves L gates, the total success probability for the quantum computer to output a sample from the intended state is roughly $O(f^L)$. Thus, the probability that the quantum circuit occurs without an error is exponentially decreasing in the number of logic gates if the logic gates have a finite error probability that is not decreasing in the number of gates.

The "digital gate model," or standard model of quantum computing, is the only model of quantum computing that is known to be compatible with fault-tolerant quantum computing (FTQC). Within noisy intermediate-scale quantum (NISQ), a few competing models target special purpose applications, including quantum annealing (Johnson et al. 2011) and analog quantum simulators (Buluta and Nori 2009). Quantum annealing is a paradigm that focuses mostly on solving optimization problems. However, despite almost two decades of research, no compelling evidence suggests that existing quantum annealers actually solve real-world optimization problems with either a considerable scaling advantage or faster wall clock time on finite instances of real applications. Quantum simulators are analog devices that are designed to realize specific Hamiltonians of physical interest. For example, one can simulate certain parameter regimes of Hubbard models using quantum simulators (Salfi et al. 2016). While quantum simulators have scaled to realize fairly large and possibly classically intractable quantum systems (Daley et al. 2022), they have limited control over the precision of experiments and the parameter regimes that can be studied. The Hamiltonians of interest in chemistry typically involve a Coulomb operator, which would require a high degree of control and connectivity to realize in an analog quantum simulator. Recently, some preliminary descriptions have shown how chemistry could be simulated in this analog fashion (Argüello-Luengo et al. 2019). However, to date, the vast majority of work on quantum computing for chemistry (more than 95 percent of all papers on the topic) focuses on the gate model—either NISQ or FTQC.

main approaches have been executed in the development process to address this problem: error correction and error mitigation.

Unlike with classical computers, one cannot simply copy quantum information and use a "repetition code" to correct errors as a consequence of the quantum no-cloning theorem (Wootters and Zurek 1982). Instead, a theory of quantum error correction is now very developed that is based on topology (Kitaev 2006). Topology is a global property of objects. For example, one cannot tell if a donut and a coffee cup are topologically the same or different just by examining a small piece of one of those objects; one must examine the entire object. Hence, topological information is nonlocal. By contrast, errors that occur in individual two-qubit gates or errors that occur from the system interacting with the environment (in most cases) are local. Thus, quantum error correction is based on the idea that if quantum information is encoded in global topological properties of quantum states, it will be robust to errors. A well-developed theory of FTQC now exists.

With FTQC codes, as one increases the number of physical qubits used to store a single "logical qubit," the error rate decreases exponentially. The most popular error-correcting code that can be made fully fault tolerant (meaning that one can exponentially suppress errors in all aspects of the code, including logical operations, measurements, decoding, and encoding, so long as error rates are "low enough") is the surface code. The surface code is a topological code (a 2D variant of the Toric code) composed to work for qubits on a planar lattice (Fowler et al. 2012). The surface code is the most popular code because it has the lowest "threshold" of any known code involving only 2D connectivity. 2D connectivity is attractive because devices with higher connectivity pose extreme engineering challenges. The surface code, or close variants like the honeycomb code (Haah and Hastings 2022), is the plan of record for most efforts to build an error-corrected quantum computer, including the leading ion trap, photonics, and superconducting qubit efforts.

The best two-qubit gate fidelities for quantum hardware today that have scaled to more than a few qubits have roughly $f = 0.99$ (Egan et al. 2021; Google AI n.d.). This value is near the error threshold for the surface code. At those error rates, one would require at least several thousand physical qubits for each logical qubit. Thus, a quantum computer with a few hundred logical qubits would require nearly one million physical qubits. The overheads decrease somewhat if the error rates come down further. For example, with $1e^{-4}$ or $1e^{-5}$ error rates, the total number of physical qubits for a processor with a few hundred logical qubits might be reduced to closer to 100,000. But it is considered unlikely that error rates as low as $1e^{-5}$ will be reached with our current paradigms.

One possibility for getting physical error rates that low is topological quantum computing (Nayak et al. 2008), but so far not even a single topological qubit has been produced. If we were to use a platform with $f = 0.999$ gate fidelities as a NISQ platform (this would be an extremely accurate NISQ device), then already for $L = 1e^5$ gates (a relatively "small" quantum circuit) the total success probability would be on the order of $4e^{-5}$. This means that the quantum computer would need to be run $1/(4e^{-5})$ times, or about 20,000 times, just to see a single realization that is not errant. How plausible that is depends on the hardware. Superconducting qubits have gate times on the order of tens of nanoseconds, so this might take only tens of seconds. But ion traps, for example, are several orders of magnitude slower. And it can be very difficult to determine when an error has occurred. Strategies for detecting when errors have occurred so that one can postselect on error-free outcomes, but which do not actually remove the exponential scaling associated with these errors, are referred to as "error mitigation" (which is quite a different topic entirely from "error correction," despite similar sounding names; see Section 2.2.3 for further details) (Cai et al. 2022).

Most of the time, FTQC algorithms are studied formally. For example, scientists aim to quantify (either asymptotically or in terms of finite resources like constant factors) the number of logical or physical qubits and quantum gates required to perform a precise task. An example of this task is simulating a wave function, starting in its initial state under the molecular Hamiltonian for time (t) and having it evolve within a controlled error (ε). The parameters considered in this system would include the number of particles, number of basis functions, size of particles, and other variables that could affect the behavior of the system.

Work to improve the cost of these algorithms has been ongoing. Su and colleagues (2021) and Lee and colleagues (2021) attempt to show how the cost of quantum algorithms for chemistry in both plane wave and molecular orbital basis sets has improved over the years. Note that while these studies contain very detailed analyses of the costings, the costings are sometimes difficult to compare.

4.3.4 Closing the Loop: Modeling Quantum Information Systems on Quantum Computers

Modeling even the smallest of quantum systems requires classical resources that scale—at best, polynomially, and at worst, exponentially—with system size. On one hand, capturing the size and complexity of realistic quantum information processors and sensors represents a stark challenge for modern classical simulation. On the other hand, this challenge presents a grand and fascinating opportunity for using quantum computers to model future versions of themselves—and other quantum information devices.

Although modern quantum computers are inherently noisy, they have shown substantial speedups in determining the properties and dynamics of a wide array of quantum systems, including popular tight-binding models, small molecules, and simple materials. These speedups can also be extended to the modeling of quantum devices. For instance, by leveraging embedding theories to reduce the size of the quantum problem to be solved, Huang, Govoni, and Galli (2022) and Ma, Govoni, and Galli (2020) recently used quantum computers to model spin defects in semiconductors, including the negatively charged nitrogen-vacancy center, the neutral silicon-vacancy center, and the Cr impurity in 4H-SiC (Vorwerk et al. 2022), which are some of the most popular forms of qubits eyed for future quantum technologies. Quantum simulations using the QPE and VQE algorithms were able to determine the ground and select excited states of the nitrogen-vacancy center within 0.2 eV (Ma, Govoni, and Galli 2020). This demonstration highlights the promise of using quantum computers to inform the design of future quantum computers in a continuous self-improvement loop.

Despite these demonstrations, it remains challenging to fully realize the intricacy of quantum information processors on modern quantum computers. These processors can involve numerous degrees of freedom, which can be difficult to map to most modern machines' finite numbers of qubits. Even if this mapping can be performed, the solutions to many quantum problems necessitate levels of accuracy that are difficult for quantum computers to achieve given modern levels of noise and tomography costs. More importantly, most quantum sensors and other devices are inherently open systems that exchange energy and other quantities with their environments, necessitating that some aspect of their environments also be represented on limited computing resources. This leaves room for exciting developments that will overcome these challenges and will benefit both current technologies and the future quantum devices they can synergistically advance.

4.4 SUMMARY OF RESEARCH PRIORITIES AND RECOMMENDATIONS

The following fundamental research priorities have been identified by the committee and extensively discussed in Chapter 4 as those that the Department of Energy and the National Science Foundation should prioritize within the target research area of "experimental and computational approaches for scaling qubit design and function."

Research Priorities:

- **Exploit the advantages of bottom-up chemical synthesis for constructing quantum architectures.**
- **Develop techniques for synthesizing molecular qubits that retain their desirable quantum properties in different host chemical environments.**
- **Investigate and control the interactions among qubits and between qubits and their environments.**
- **Design hybrid quantum architectures that mutually enhance each other's quantum properties.**
- **Develop noise models and quantum error-mitigation techniques for individual qubits, systems of qubits, and quantum architectures that can be experimentally validated.**
- **Understand and advance the limits of classical electronic structure algorithms and modeling approaches that can guide the design of molecular or solid-state qubits and scalable quantum architectures.**
- **Leverage and develop machine learning, chemical informatics, chemical databases, and molecular simulations to inform and facilitate qubit design.**
- **Identify important open chemistry problems including those with applications to QIS that are unresolved due to classically intractable electronic structure.**

- **Develop more efficient methods of encoding chemical systems on quantum computers (e.g., better basis sets, quantization, fermion mappings, embedding theories).**
- **Develop more efficient quantum algorithms for fault-tolerant quantum computers to simulate molecular systems, including those with QIS relevance.**
- **Study how quantum computing algorithms for dynamics can be used to accelerate chemistry and spectroscopy.**
- **Explore how quantum machine learning can be used to accelerate chemical research by processing quantum data from entangled sensor arrays or quantum simulations of chemistry.**

The committee makes Recommendation 4-1 to address issues concerning open-access databases and the importance of sharing data across relevant research entities to move the field of QIS and chemistry forward. The related discussion included ways to improve the organization and types of relevant experimental and theoretical data that would be useful for the research community. The committee puts forth Recommendation 4-2 to ensure that efforts in QIS research involving quantum-accelerated calculations keep chemistry-related problems at the forefront of the research questions to solve.

Recommendation 4-1. The Department of Energy and the National Science Foundation should establish open-access, centralized databases that include quantum information science (QIS)–relevant data to enhance predictions and expedite new discoveries. These databases should contain (1) structure–property relationships, (2) results of electronic structure calculations, (3) spectroscopic data, (4) experimental characterization of quantum devices, and (5) other data to inform QIS investigations. These data should be obtained from QIS studies contributed by scientists and engineers across industry, academia, and government. These agencies should also create a centralized database to house a body of experimental work that demonstrates discrete quantum use cases for chemistry.

Recommendation 4-2. The Department of Energy, the National Science Foundation, and other funding agencies, both public and private, should develop initiatives to support multidisciplinary research in quantum information science to address how quantum-accelerated calculations could solve chemistry problems. In connection with these initiatives, the research community should establish a set of standards for how to evaluate quantum advantage in specific chemistry use cases.

REFERENCES

Abobeih, M. H., J. Cramer, M. A. Bakker, N. Kalb, M. Markham, D. J. Twitchen, and T. H. Taminiau. 2018. "One-Second Coherence for a Single Electron Spin Coupled to a Multi-Qubit Nuclear-Spin Environment." *Nature Communications* 9(1):2552. doi.org/10.1038/S41467-018-04916-Z.

Aguilà, D., O. Roubeau, and G. Aromí. 2021. "Designed Polynuclear Lanthanide Complexes for Quantum Information Processing." *Dalton Transactions* 50(35):12045–12057. doi.org/10.1039/D1DT01862K.

Aliferis, P., and J. Preskill. 2008. "Fault-Tolerant Quantum Computation Against Biased Noise." *Physical Review A* 78(5):052331. doi.org/10.1103/Physreva.78.052331.

Altoé, M. V. P., A. Banerjee, C. Berk, A. Hajr, A. Schwartzberg, C. Song, M. Alghadeer, S. Aloni, M. J. Elowson, J. M. Kreikebaum, E. K. Wong, S. M. Griffin, S. Rao, A. Weber-Bargioni, A. M. Minor, D. I. Santiago, S. Cabrini, I. Siddiqi, and D. F. Ogletree. 2022. "Localization and Mitigation of Loss in Niobium Superconducting Circuits." *PRX Quantum* 3(2):020312. doi.org/10.1103/Prxquantum.3.020312.

Amdur, M. J., K. R. Mullin, M. J. Waters, D. Puggioni, M. K. Wojnar, M. Gu, L. Sun, P. H. Oyala, J. M. Rondinelli, and D. E. Freedman. 2022. "Chemical Control of Spin–Lattice Relaxation to Discover a Room Temperature Molecular Qubit." *Chemical Science* 13(23):7034–7045. doi.org/10.1039/D1SC06130E.

Anderson, C. P., E. O. Glen, C. Zeledon, A. Bourassa, Y. Jin, Y. Zhu, C. Vorwerk, A. L. Crook, H. Abe, J. Ul-Hassan, T. Ohshima, N. T. Son, G. Galli, and D. D. Awschalom. 2022. "Five-Second Coherence of a Single Spin with Single-Shot Readout in Silicon Carbide." *Science Advances* 8(5):Eabm5912. doi.org/10.1126/Sciadv.Abm5912.

Argüello-Luengo, J., A. González-Tudela, T. Shi, P. Zoller, and J. I. Cirac. 2019. "Analogue Quantum Chemistry Simulation." *Nature* 574(7777):215–218. https://doi.org/10.1038/s41586-019-1614-4.

Ariciu, A.-M., D. H. Woen, D. N. Huh, L. E. Nodaraki, A. K. Kostopoulos, C. A. P. Goodwin, N. F. Chilton, E. J. L. McInnes, R. E. P. Winpenny, W. J. Evans, and F. Tuna. 2019. "Engineering Electronic Structure to Prolong Relaxation Times in Molecular Qubits by Minimising Orbital Angular Momentum." *Nature Communications* 10(1):3330. doi.org/10.1038/S41467-019-11309-3.

Arute, F., K. Arya, R. Babbush, D. Bacon, J. C. Bardin, R. Barends, R. Biswas, S. Boixo, F. G. S. L. Brandao, D. A. Buell, B. Burkett, Y. Chen, Z. Chen, B. Chiaro, R. Collins, W. Courtney, A. Dunsworth, E. Farhi, B. Foxen, A. Fowler, C. Gidney, M. Giustina, R. Graff, K. Guerin, S. Habegger, M. P. Harrigan, M. J. Hartmann, A. Ho, M. Hoffmann, T. Huang, T. S. Humble, S. V. Isakov, E. Jeffrey, Z. Jiang, D. Kafri, K. Kechedzhi, J. Kelly, P. V. Klimov, S. Knysh, A. Korotkov, F. Kostritsa, D. Landhuis, M. Lindmark, E. Lucero, D. Lyakh, S. Mandrà, J. R. McClean, M. Mcewen, A. Megrant, X. Mi, K. Michielsen, M. Mohseni, J. Mutus, O. Naaman, M. Neeley, C. Neill, M. Y. Niu, E. Ostby, A. Petukhov, J. C. Platt, C. Quintana, E. G. Rieffel, P. Roushan, N. C. Rubin, D. Sank, K. J. Satzinger, V. Smelyanskiy, K. J. Sung, M. D. Trevithick, A. Vainsencher, B. Villalonga, T. White, Z. J. Yao, P. Yeh, A. Zalcman, H. Neven, and J. M. Martinis. 2019. "Quantum Supremacy Using a Programmable Superconducting Processor." *Nature* 574(7779):505–510. doi.org/10.1038/S41586-019-1666-5.

Arute, F., K. Arya, R. Babbush, D. Bacon, J. C. Bardin, R. Barends, S. Boixo, M. Broughton, B. B. Buckley, D. A. Buell, B. Burkett, N. Bushnell, Y. Chen, Z. Chen, B. Chiaro, R. Collins, W. Courtney, S. Demura, A. Dunsworth, E. Farhi, A. Fowler, B. Foxen, C. Gidney, M. Giustina, R. Graff, S. Habegger, M. P. Harrigan, A. Ho, S. Hong, T. Huang, W. J. Huggins, L. Ioffe, S. V. Isakov, E. Jeffrey, Z. Jiang, C. Jones, D. Kafri, K. Kechedzhi, J. Kelly, S. Kim, P. V. Klimov, A. Korotkov, F. Kostritsa, D. Landhuis, P. Laptev, M. Lindmark, E. Lucero, O. Martin, J. M. Martinis, J. R. McClean, M. Mcewen, A. Megrant, X. Mi, M. Mohseni, W. Mruczkiewicz, J. Mutus, O. Naaman, M. Neeley, C. Neill, H. Neven, M. Y. Niu, T. E. O'Brien, E. Ostby, A. Petukhov, H. Putterman, C. Quintana, P. Roushan, N. C. Rubin, D. Sank, K. J. Satzinger, V. Smelyanskiy, D. Strain, K. J. Sung, M. Szalay, T. Y. Takeshita, A. Vainsencher, T. White, N. Wiebe, Z. J. Yao, P. Yeh, and A. Zalcman. 2020. "Hartree-Fock on a Superconducting Qubit Quantum Computer." *Science* 369(6507): 1084–1089. doi.org/10.1126/Science.Abb9811.

Aspuru-Guzik, A., A. D. Dutoi, P. J. Love, and M. Head-Gordon. 2005. "Simulated Quantum Computation of Molecular Energies." *Science* 309(5741):1704–1707. doi.org/10.1126/Science.1113479.

Atzori, M., and R. Sessoli. 2019. "The Second Quantum Revolution: Role and Challenges of Molecular Chemistry." *Journal of the American Chemical Society* 141(29):11339–11352. doi.org/10.1021/Jacs.9b00984.

Atzori, M., E. Morra, L. Tesi, A. Albino, M. Chiesa, L. Sorace, and R. Sessoli. 2016. "Quantum Coherence Times Enhancement in Vanadium(IV)-Based Potential Molecular Qubits: The Key Role of the Vanadyl Moiety." *Journal of the American Chemical Society* 138 (35): 11234–44.

Awschalom, D. D., R. Hanson, J. Wrachtrup, and B. B. Zhou. 2018. "Quantum Technologies with Optically Interfaced Solid-State Spins." *Nature Photonics* 12(9):516–527. doi.org/10.1038/S41566-018-0232-2.

Babbush, R., C. Gidney, D. W. Berry, N. Wiebe, J. McClean, A. Paler, A. Fowler, and H. Neven. 2018. "Encoding Electronic Spectra in Quantum Circuits with Linear T Complexity." *Physical Review X* 8(4):041015. doi.org/10.1103/Physrevx.8.041015.

Balasubramanian, G., P. Neumann, D. Twitchen, M. Markham, R. Kolesov, N. Mizuochi, J. Isoya, J. Achard, J. Beck, J. Tissler, V. Jacques, P. R. Hemmer, F. Jelezko, and J. Wrachtrup. 2009. "Ultralong Spin Coherence Time in Isotopically Engineered Diamond." *Nature Materials* 8(5):383–387. doi.org/10.1038/nmat2420.

Barends, R., J. Kelly, A. Megrant, A. Veitia, D. Sank, E. Jeffrey, T. C. White, J. Mutus, A. G. Fowler, B. Campbell, Y. Chen, Z. Chen, B. Chiaro, A. Dunsworth, C. Neill, P. O'Malley, P. Roushan, A. Vainsencher, J. Wenner, A. N. Korotkov, A. N. Cleland, and J. M. Martinis. 2014. "Superconducting Quantum Circuits at the Surface Code Threshold for Fault Tolerance." *Nature* 508(7497):500–503. doi.org/10.1038/nature13171.

Bartlett, R. J., and M. Musiał. 2007. "Coupled-Cluster Theory in Quantum Chemistry." *Reviews of Modern Physics* 79(1): 291–352. doi.org/10.1103/Revmodphys.79.291.

Bayliss, S. L., P. Deb, D. W. Laorenza, M. Onizhuk, G. Galli, D. E. Freedman, and D. D. Awschalom. 2022. "Enhancing Spin Coherence in Optically Addressable Molecular Qubits through Host-Matrix Control." *Physical Review X* 12(3):031028. doi.org/10.1103/PhysRevX.12.031028.

Bhang J., H. Ma, D. Yim, G. Galli, and H. Seo. 2021. "First-Principles Predictions of Out-Of-Plane Group IV and V Dimers as High-Symmetry, High-Spin Defects in Hexagonal Boron Nitride." *ACS Applied Materials and Interfaces* 13(38): 45768–45777. doi.org/10.1021/acsami.1c16988.

Biamonte, J., P. Wittek, N. Pancotti, P. Rebentrost, N. Wiebe, and S. Lloyd. 2017. "Quantum Machine Learning." *Nature* 549(7671):195–202. doi.org/10.1038/Nature23474.

Bluvstein, D., Z. Zhang, C. A. Mclellan, N. R. Williams, and A. C. Bleszynski Jayich. 2019. "Extending the Quantum Coherence of a Near-Surface Qubit by Coherently Driving the Paramagnetic Surface Environment." *Physical Review Letters* 123(14):146804. doi.org/10.1103/Physrevlett.123.146804.

Bollinger, J. J., D. J. Heizen, W. M. Itano, S. L. Gilbert, and D. J. Wineland. 1991. "A 303-MHz Frequency Standard Based on Trapped Be^+ Ions." *IEEE Transactions on Instrumentation and Measurement* 40(2):126–128. doi.org/10.1109/TIM.1990.1032897.

Bourassa, A., C. P. Anderson, K. C. Miao, M. Onizhuk, H. Ma, A. L. Crook, H. Abe, J. Ul-Hassan, T. Ohshima, N. T. Son, G. Galli, and D. D. Awschalom. 2020. "Entanglement and Control of Single Nuclear Spins in Isotopically Engineered Silicon Carbide." *Nature Materials* 19(12):1319–1325. doi.org/10.1038/s41563-020-00802-6.

Brown, K. R., J. Chiaverini, J. M. Sage, and H. Häffner. 2021. "Materials Challenges for Trapped-Ion Quantum Computers." *Nature Reviews Materials* 6(10):892–905. doi.org/10.1038/S41578-021-00292-1.

Bruno, A., G. de Lange, S. Asaad, K. L. van der Enden, N. K. Langford, and L. DiCarlo. 2015. "Reducing Intrinsic Loss in Superconducting Resonators by Surface Treatment and Deep Etching of Silicon Substrates." *Applied Physics Letters* 106(18):182601. https://doi.org/10.1063/1.4919761.

Bruzewicz, C. D., J. Chiaverini, R. Mcconnell, and J. M. Sage. 2019. "Trapped-Ion Quantum Computing: Progress and Challenges." *Applied Physics Reviews* 6(2):021314. doi.org/10.1063/1.5088164.

Buluta, I., and F. Nori. 2009. "Quantum Simulators." *Science* 326(5949):108–111. doi.org/10.1126/science.1177838.

Burnett, J.J., A. Bengtsson, M. Scigliuzzo, D. Niepce, M. Kudra, P. Delsing, and J. Bylander. 2019. "Decoherence Benchmarking of Superconducting Qubits." *npj Quantum Information* 5:54. doi.org/10.1038/s41534-019-0168-5.

Cai, Z., R. Babbush, S. C. Benjamin, S. Endo, W. J. Huggins, Y. Li, J. R. McClean, and T. E. O'Brien. 2022. "Quantum Error Mitigation." *ArXiv preprint*. arXiv:2210.00921.

Castro, A., A. García Carrizo, S. Roca, D. Zueco, and F. Luis. 2022. "Optimal Control of Molecular Spin Qudits." *Physical Review Applied* 17(6):064028. doi.org/10.1103/Physrevapplied.17.064028.

Chen, H., M. Nusspickel, J. Tilly, and G. H. Booth. 2021. "Variational Quantum Eigensolver for Dynamic Correlation Functions." *Physical Review A* 104(3). https://doi.org/10.1103/physreva.104.032405.

Chicco, S., A. Chiesa, G. Allodi, E. Garlatti, M. Atzori, L. Sorace, R. De Renzi, R. Sessoli, and S. Carretta. 2021. "Controlled Coherent Dynamics of [VO(TPP)], a Prototype Molecular Nuclear Qudit with an Electronic Ancilla." *Chemical Science* 12(36):12046–12055. doi.org/10.1039/D1SC01358K.

Chiesa, A., E. Macaluso, F. Petiziol, S. Wimberger, P. Santini, and S. Carretta. 2020. "Molecular Nanomagnets as Qubits with Embedded Quantum-Error Correction." *Physical Review Letters* 11(20):8610–8615. doi.org/10.1021/Acs.Jpclett.0c02213.

Chiesa, A., F. Petiziol, E. Macaluso, S. Wimberger, P. Santini, and S. Carretta. 2021. "Embedded Quantum-Error Correction and Controlled-Phase Gate for Molecular Spin Qubits." *AIP Advances* 11(2):025134. doi.org/10.1063/9.0000166.

Chizzini, M., L. Crippa, L. Zaccardi, E. Macaluso, S. Carretta, A. Chiesa, and P. Santini. 2022. "Quantum Error Correction with Molecular Spin Qudits." *Physical Chemistry Chemical Physics* 24(34):20030–20039. doi.org/10.1039/D2CP01228F.

Daley, A. J., I. Bloch, C. Kokail, S. Flannigan, N. Pearson, M. Troyer, and P. Zoller. 2022. "Practical Quantum Advantage in Quantum Simulation." *Nature* 607(7920):667–676. doi.org/10.1038/s41586-022-04940-6.

Devoret, M. H., A. Wallraff, and J. M. Martinis. 2004. "Superconducting Qubits: A Short Review." *ArXiv*. arxiv.org/abs/cond-mat/0411174.

Dhaliah, D., Y. Xiong, A. Sipahigil, S. M. Griffin, and G. Hautier. 2022. "First-Principles Study of the T Center in Silicon." *Physical Review Materials* 6(5):L053201. doi.org/10.1103/Physrevmaterials.6.L053201.

Dibos, A. M., M. Raha, C. M. Phenicie, and J. D. Thompson. 2018. "Atomic Source of Single Photons in the Telecom Band." *Physical Review Letters* 120(24):243601. doi.org/10.1103/Physrevlett.120.243601.

Docker. n.d. "Telepresence." 9 May 2023, www.docker.com.

Ebbesen, T. W. 2016. "Hybrid Light–Matter States in a Molecular and Material Science Perspective." *Accounts of Chemical Research* 49(11):2403–2412. doi.org/10.1021/Acs.Accounts.6b00295.

Egan, L., D. M. Debroy, C. Noel, A. Risinger, D. Zhu, D. Biswas, M. Newman, M. Li, K. R. Brown, M. Cetina, and C. Monroe. 2021. "Fault-Tolerant Control of an Error-Corrected Qubit." *Nature* 598(7880):281–286. doi.org/10.1038/S41586-021-03928-Y.

Elfving, V. E., B. W. Broer, M. Webber, J. Gavartin, M. D. Halls, K. P. Lorton, and A. Bochevarov. 2020. "How will Quantum Computers Provide an Industrially Relevant Computational Advantage in Quantum Chemistry?" *ArXiv preprint*. arXiv:2009.12472.

Endo, S., S. C. Benjamin, and Y. Li. 2018. "Practical Quantum Error Mitigation for Near-Future Applications." *Physical Review X* 8(3):031027. doi:10.1103/PhysRevX.8.031027.

Eriksen, J. J., T. A. Anderson, J. E. Deustua, K. Ghanem, D. Hait, M. R. Hoffmann, S. Lee, D. S. Levine, I. Magoulas, J. Shen, N. M. Tubman, K. B. Whaley, E. Xu, Y. Yao, N. Zhang, A. Alavi, G. K.-L. Chan, M. Head-Gordon, W. Liu, P. Piecuch, S. Sharma, S. L. Ten-No, C. J. Umrigar, and J. Gauss. 2020. "The Ground State Electronic Energy of Benzene." *Physical Review Letters* 11(20):8922–8929. doi.org/10.1021/Acs.Jpclett.0c02621.

Fataftah, M., M. D. Krzyaniak, B. Vlaisavljevich, M. R. Wasielewski, J. M. Zadrozny, and D. E. Freedman. 2019. "Metal–Ligand Covalency Enables Room Temperature Molecular Qubit Candidates." *Chemical Science* 10(27):6707–14. https://doi.org/10.1039/C9SC00074G.

Ferrando-Soria, J., E. Moreno Pineda, A. Chiesa, A. Fernandez, S. A. Magee, S. Carretta, P. Santini, I. J. Vitorica-Yrezabal, F. Tuna, G. A. Timco, E. J. L. McInnes, and R. E. P. Winpenny. 2016. "A Modular Design of Molecular Qubits to Implement Universal Quantum Gates." *Nature Communications* 7(1):11377. doi.org/10.1038/Ncomms11377.

Feynman, R. P., J. Hey, and R. W. Allen. 1998. *Feynman Lectures on Computation*. Boston: Addison-Wesley Longman.

Flensberg, K., F. von Oppen, and A. Stern. 2021. "Engineered Platforms for Topological Superconductivity and Majorana Zero Modes." *Nature Reviews Materials* 6(10):944–958. doi.org/10.1038/S41578-021-00336-6.

Foulkes, W. M. C., L. Mitas, R. J. Needs, and G. Rajagopal. 2001. "Quantum Monte Carlo Simulations of Solids." *Reviews of Modern Physics* 73(1):33–83. doi.org/10.1103/Revmodphys.73.33.

Foulon, B. L., K. G. Ray, C.-E. Kim, Y. Liu, B. M. Rubenstein, and V. Lordi. 2022. "1/Ω Electric-Field Noise in Surface Ion Traps from Correlated Adsorbate Dynamics." *Physical Review A* 105(1):013107. doi.org/10.1103/Physreva.105.013107.

Fowler, A. G., M. Mariantoni, J. M. Martinis, and A. N. Cleland. 2012. "Surface Codes: Towards Practical Large-Scale Quantum Computation." *Physical Review A* 86(3):032324. https://doi.org/10.1103/PhysRevA.86.032324.

Freedman, M., A. Kitaev, M. Larsen, and Z. Wang. 2003. "Topological Quantum Computation." *Bulletin of the American Mathematical Society* 40(1):31–38.

Frenkel, D., and B. Smit. 2002. *Understanding Molecular Simulation: From Algorithms to Applications, Understanding Molecular Simulation (Second Edition)*. San Diego: Academic Press.

Frey, N. C., M. K. Horton, J. M. Munro, S. M. Griffin, K. A. Persson, and V. B. Shenoy. 2020. "High-Throughput Search for Magnetic and Topological Order in Transition Metal Oxides." *Science Advances* 6(50):Eabd1076. doi.org/10.1126/Sciadv.Abd1076.

Friesner, R. A. 2005. "*Ab Initio* Quantum Chemistry: Methodology and Applications." *Proceedings of the National Academy of Sciences USA* 102(19):6648–6653. doi.org/10.1073/Pnas.0408036102.

Gaita-Ariño, A., F. Luis, S. Hill, and E. Coronado. 2019. "Molecular Spins for Quantum Computation." *Nature Chemistry* 11(4):301–309. doi.org/10.1038/S41557-019-0232-Y.

Georgescu, I. 2020. "Trapped Ion Quantum Computing Turns 25." *Nature Reviews Physics* 2(6):278–278. doi.org/10.1038/S42254-020-0189-1.

Georgopoulos, K., C. Emary, and P. Zuliani. 2021. "Modeling and Simulating the Noisy Behavior of Near-Term Quantum Computers." *Physical Review A* 104(6):062432. doi.org/10.1103/Physreva.104.062432.

Giannozzi, P., S. Baroni, N. Bonini, M. Calandra, R. Car, C. Cavazzoni, D. Ceresoli, G. L. Chiarotti, M. Cococcioni, I. Dabo, A. Dal Corso, S. De Gironcoli, S. Fabris, G. Fratesi, R. Gebauer, U. Gerstmann, C. Gougoussis, A. Kokalj, M. Lazzeri, L. Martin-Samos, N. Marzari, F. Mauri, R. Mazzarello, S. Paolini, A. Pasquarello, L. Paulatto, C. Sbraccia, S. Scandolo, G. Sclauzero, A. P. Seitsonen, A. Smogunov, P. Umari, and R. M. Wentzcovitch. 2009. "QUANTUM ESPRESSO: A Modular and Open-Source Software Project for Quantum Simulations of Materials." *Journal of Physics: Condensed Matter* 21(39):395502. doi.org/10.1088/0953-8984/21/39/395502.

Gimeno, I., A. Urtizberea, J. Román-Roche, D. Zueco, A. Camón, P. J. Alonso, O. Roubeau, and F. Luis. 2021. "Broad-Band Spectroscopy of a Vanadyl Porphyrin: A Model Electronuclear Spin Qudit." *Chemical Science* 12(15):5621–5630. doi.org/10.1039/D1SC00564B.

GO FAIR. 2017. "FAIR Principles." https://www.go-fair.org/fair-principles/.

Godfrin, C., A. Ferhat, R. Ballou, S. Klyatskaya, M. Ruben, W. Wernsdorfer, and F. Balestro. 2017. "Operating Quantum States in Single Magnetic Molecules: Implementation of Grover's Quantum Algorithm." *Physical Review Letters* 119(18):187702. doi.org/10.1103/Physrevlett.119.187702.

Goh, T., R. Pandharkar, and L. Gagliardi. 2022. "Multireference Study of Optically Addressable Vanadium-Based Molecular Qubit Candidates." *Journal of Physical Chemistry A* 126(36):6329–6335. doi.org/10.1021/acs.jpca.2c04730.

Goings, J. J., A. White, J. Lee, C. S. Tautermann, M. Degroote, C. Gidney, T. Shiozaki, R. Babbush, and N. C. Rubin. 2022. "Reliably Assessing the Electronic Structure of Cytochrome P450 on Today's Classical Computers and Tomorrow's Quantum Computers." *Proceedings of the National Academy of Sciences USA* 119(38):E2203533119. doi.org/10.1073/Pnas.2203533119.

Google AI. n.d. "Advancing AI for Everyone." 9 May 2023. https://ai.google/.

Graham, M. J., M. D. Krzyaniak, M. R. Wasielewski, and D. E. Freedman. 2017. "Probing Nuclear Spin Effects on Electronic Spin Coherence Via EPR Measurements of Vanadium(IV) Complexes." *Inorganic Chemistry* 56(14):8106–8113. doi.org/10.1021/Acs.Inorgchem.7b00794.

Graham, M. J., C.-J. Yu, M. D. Krzyaniak, M. R. Wasielewski, and D. E. Freedman. 2017. "Synthetic Approach to Determine the Effect of Nuclear Spin Distance on Electronic Spin Decoherence." *Journal of the American Chemical Society* 139(8):3196–3201. doi.org/10.1021/Jacs.6b13030.

Graham, M. J., J. M. Zadrozny, M. S. Fataftah, and D. E. Freedman. 2017. "Forging Solid-State Qubit Design Principles in a Molecular Furnace." *Chemistry of Materials* 29(5):1885–1897. doi.org/10.1021/Acs.Chemmater.6b05433.

Graham, M. J., J. M. Zadrozny, M. Shiddiq, J. S. Anderson, M. S. Fataftah, S. Hill, and D. E. Freedman. 2014. "Influence of Electronic Spin and Spin–Orbit Coupling on Decoherence in Mononuclear Transition Metal Complexes." *Journal of the American Chemical Society* 136(21):7623–7626. doi.org/10.1021/ja5037397.

Grimsley, H. R., S. E. Economou, E. Barnes, and N. J. Mayhall. 2019. "An Adaptive Variational Algorithm for Exact Molecular Simulations on a Quantum Computer." *Nature Communications* 10(1):3007. doi.org/10.1038/S41467-019-10988-2.

Gulka, M., D. Wirtitsch, V. Ivády, J. Vodnik, J. Hruby, G. Magchiels, E. Bourgeois, A. Gali, M. Trupke, and M. Nesladek. 2021. "Room-Temperature Control and Electrical Readout of Individual Nitrogen-Vacancy Nuclear Spins." *Nature Communications* 12(1):4421. doi: 10.1038/s41467-021-24494-x.

Haah, J., and M. B. Hastings. 2022. "Boundaries for the Honeycomb Code." *Quantum* 6:693. doi.org/10.22331/q-2022-04-21-693.

Hamdan, R., J. P. Trinastic, and H. P. Cheng. 2014. "Molecular Dynamics Study of the Mechanical Loss in Amorphous Pure and Doped Silica." *Journal of Chemical Physics* 141(5):054501. doi.org/10.1063/1.4890958.

Harrelson, T. F., E. Sheridan, E. Kennedy, J. Vinson, A. T. N'Diaye, M. V. P. Altoé, A. Schwartzberg, I. Siddiqi, D. F. Ogletree, M. C. Scott, and S. M. Griffin. 2021. "Elucidating the Local Atomic and Electronic Structure of Amorphous Oxidized Superconducting Niobium Films." *Applied Physics Letters* 119(24):244004. doi.org/10.1063/5.0069549.

Headley, M. B., A. Bins, A. Nip, E. W. Roberts, M. R. Looney, A. Gerard, and M. F. Krummel. 2016. "Visualization of Immediate Immune Responses to Pioneer Metastatic Cells in the Lung." *Nature* 531(7595):513–517. doi.org/10.1038/nature16985.

Helgaker, T., P. Jørgensen, and J. Olsen. 2000. *Molecular Electronic-Structure Theory: Helgaker/Molecular Electronic-Structure Theory*. New York: John Wiley and Sons.

Hite, D. A., Y. Colombe, A. C. Wilson, D. T. C. Allcock, D. Leibfried, D. J. Wineland, and D. P. Pappas. 2013. "Surface Science for Improved Ion Traps." *MRS Bulletin* 38(10):826–833. doi.org/10.1557/Mrs.2013.207.

Hohenstein, E. G., R. M. Parrish, and T. J. Martínez. 2012. "Tensor Hypercontraction Density Fitting. I. Quartic Scaling Second- and Third-Order Møller-Plesset Perturbation Theory." *Journal of Chemical Physics* 137(4):044103. doi.org/10.1063/1.4732310.

Hruszkewycz, S. O., S. Maddali, C. P. Anderson, W. Cha, K. C. Miao, M. J. Highland, A. Ulvestad, D. D. Awschalom, and F. J. Heremans. 2018. "Strain Annealing of SiC Nanoparticles Revealed Through Bragg Coherent Diffraction Imaging for Quantum Technologies." *Physical Review Materials* 2(8):086001. doi.org/10.1103/Physrevmaterials.2.086001.

Huang, B., M. Govoni, and G. Galli. 2022. "Simulating the Electronic Structure of Spin Defects on Quantum Computers." *PRX Quantum* 3(1):010339. doi.org/10.1103/Prxquantum.3.010339.

Huang, C., M. Pavone, and E. A. Carter. 2011. "Quantum Mechanical Embedding Theory Based on a Unique Embedding Potential." *Journal of Chemical Physics* 134(15):154110. doi.org/10.1063/1.3577516.

Huang, H.-Y., M. Broughton, J. Cotler, S. Chen, J. Li, M. Mohseni, H. Neven, R. Babbush, R. Kueng, J. Preskill, and J. R. McClean. 2022. "Quantum Advantage in Learning from Experiments." *Science* 376(6598):1182–1186. doi.org/10.1126/Science.Abn7293.

Huang, H.-Y., M. Broughton, M. Mohseni, R. Babbush, S. Boixo, H. Neven, and J. R. McClean. 2021. "Power of Data in Quantum Machine Learning." *Nature Communications* 12(1):2631. doi.org/10.1038/S41467-021-22539-9.

Huggins, W. J., B. A. O'Gorman, N. C. Rubin, D. R. Reichmn, R. Babbush, and J. Lee. 2022. "Unbiasing Fermionic Quantum Monte Carlo with a Quantum Computer." *Nature* 603:416–420. doi.org/10.1038/s41586-021-04351-z.

Hussain, R., G. Allodi, A. Chiesa, E. Garlatti, D. Mitcov, A. Konstantatos, K. S. Pedersen, R. De Renzi, S. Piligkos, and S. Carretta. 2018. "Coherent Manipulation of a Molecular Ln-Based Nuclear Qudit Coupled to an Electron Qubit." *Journal of the American Chemical Society* 140(31):9814–9818. doi.org/10.1021/Jacs.8b05934.

Ishikawa, N., M. Sugita, and W. Wernsdorfer. 2005. "Nuclear Spin Driven Quantum Tunneling of Magnetization in a New Lanthanide Single-Molecule Magnet: Bis(Phthalocyaninato)Holmium Anion." *Journal of the American Chemical Society* 127(11):3650–3651. doi.org/10.1021/Ja0428661.

Jackson, C. E., C. Lin, S. H. Johnson, J. Tol, and J. M. Zadrozny. 2019. "Nuclear-Spin-Pattern Control of Electron-Spin Dynamics in a Series of V(IV) Complexes." *Chemical Science* 10(36):8447–8454. doi.org/10.1039/C9SC02899D.

Jackson, C. E., I. P. Moseley, R. Martinez, S. Sung, and J. M. Zadrozny. 2021. "A Reaction-Coordinate Perspective of Magnetic Relaxation." *Chemical Society Reviews* 50(12):6684–6699. doi.org/10.1039/D1CS00001B.

Jenkins, M. D., Y. Duan, B. Diosdado, J. J. García-Ripoll, A. Gaita-Ariño, C. Giménez-Saiz, P. J. Alonso, E. Coronado, and F. Luis. 2017. "Coherent Manipulation of Three-Qubit States in a Molecular Single-Ion Magnet." *Physical Review B* 95(6):064423. doi.org/10.1103/Physrevb.95.064423.

Johnson, M. W., M. H. S. Amin, S. Gildert, T. Lanting, F. Hamze, N. Dickson, R. Harris, A. J. Berkley, J. Johansson, P. Bunyk, E. M. Chapple, C. Enderud, J. P. Hilton, K. Karimi, E. Ladizinsky, N. Ladizinsky, T. Oh, I. Perminov, C. Rich, M. C. Thom, E. Tolkacheva, C. J. S. Truncik, S. Uchaikin, J. Wang, B. Wilson, and G. Rose. 2011. "Quantum Annealing with Manufactured Spins." *Nature* 473(7346):194–198. doi.org/10.1038/nature10012.

Jones, R. O. 2015. "Density Functional Theory: Its Origins, Rise to Prominence, and Future." *Reviews of Modern Physics* 87(3):897–923. doi.org/10.1103/RevModPhys.87.897.

Kanai, S., F. J. Heremans, H. Seo, G. Wolfowicz, C. P. Anderson, S. E. Sullivan, M. Onizhuk, G. Galli, D. D. Awschalom, and H. Ohno. 2022. "Generalized Scaling of Spin Qubit Coherence in Over 12,000 Host Materials." *Proceedings of the National Academy of Sciences USA* 119(15):E2121808119. doi.org/10.1073/Pnas.2121808119.

Kandala, A., K. Temme, A. D. Córcoles, A. Mezzacapo, J. M. Chow, and J. M. Gambetta. 2019. "Error Mitigation Extends the Computational Reach of a Noisy Quantum Processor." *Nature* 567(7749):491–495. doi.org/10.1038/S41586-019-1040-7.

Kayyalha, M., D. Xiao, R. Zhang, J. Shin, J. Jiang, F. Wang, Y.-F. Zhao, R. Xiao, L. Zhang, K. M. Fijalkowski, P. Mandal, M. Winnerlein, C. Gould, Q. Li, L. W. Molenkamp, M. H. W. Chan, N. Samarth, and C.-Z. Chang. 2020. "Absence of Evidence for Chiral Majorana Modes in Quantum Anomalous Hall-Superconductor Devices." *Science* 367(6473):64–67. doi.org/10.1126/Science.Aax6361.

Kennedy, E., N. Reynolds, L. R. Dacosta, F. Hellman, C. Ophus, and M. C. Scott. 2020. "Tilted Fluctuation Electron Microscopy." *Applied Physics Letters* 117(9):091903. doi.org/10.1063/5.0015532.

Kilgour, M., and L. Simine. 2022. "Inside the Black Box: A Physical Basis for the Effectiveness of Deep Generative Models of Amorphous Materials." *Journal of Computational Physics* 452:110885. doi.org/10.1016/J.Jcp.2021.110885.

Kim, E., A. Safavi-Naini, D. A. Hite, K. S. Mckay, D. P. Pappas, P. F. Weck, and H. R. Sadeghpour. 2017. "Electric-Field Noise From Carbon-Adatom Diffusion on a Au(110) Surface: First-Principles Calculations and Experiments." *Physical Review A* 95(3):033407. doi.org/10.1103/Physreva.95.033407.

Kitaev, A. Y. 2006. "Fault-Tolerant Quantum Computation by Anyons." *Annals of Physics* 303(1):2–30. doi.org/10.1016/S0003-4916(02)00018-0.

Koch, W., and M. C. Holthausen. 2001. *A Chemist's Guide to Density Functional Theory*. John Wiley and Sons. www.onlinelibrary.wiley.com/doi/book/10.1002/3527600043.

Krantz, P., M. Kjaergaard, F. Yan, T. P. Orlando, S. Gustavsson, and W. D. Oliver. 2019. "A Quantum Engineer's Guide to Superconducting Qubits." *Applied Physics Reviews* 6(2):021318. doi.org/10.1063/1.5089550.

Kundu, K., J. R. K. White, S. A. Moehring, J. M. Yu, J. W. Ziller, F. Furche, W. J. Evans, and S. Hill. 2022. "A 9.2-GHz Clock Transition in a Lu(II) Molecular Spin Qubit Arising from a 3,467-MHz Hyperfine Interaction." *Nature Chemistry* 14(4):392–397. doi.org/10.1038/S41557-022-00894-4.

Lee, J., D. W. Berry, C. Gidney, W. J. Huggins, J. R. McClean, N. Wiebe, and R. Babbush. 2021. "Even More Efficient Quantum Computations of Chemistry Through Tensor Hypercontraction." *PRX Quantum* 2(3):030305. doi.org/10.1103/Prxquantum.2.030305.

Lee, S., J. Lee, H. Zhai, Y. Tong, A. M. Dalzell, A. Kumar, P. Helms, J. Gray, Z. Cui, W. Liu, M. Kastoryano, R. Babbush, J. Preskill, D. R. Reichman, E. T. Campbell, E. F. Valeev, L. Lin, and G. K. L. Chan. 2023. "Evaluating the Evidence for Exponential Quantum Advantage in Ground-State Quantum Chemistry." *Nature Communications* 14(1). https://doi.org/10.1038/s41467-023-37587-6.

Li, Z., J. Li, N. S. Dattani, C. J. Umrigar, and G. K.-L. Chan. 2019. "The Electronic Complexity of the Ground-State of the FeMo Cofactor of Nitrogenase as Relevant to Quantum Simulations." *Journal of Chemical Physics* 150(2):024302. doi.org/10.1063/1.5063376.

Lilienfeld, O. A. V., K. R. Müller, and A. Tkatchenko. 2020. "Exploring Chemical Compound Space with Quantum-Based Machine Learning." *Nature Reviews Chemistry* 4(7):347–358. https://doi.org/10.1038/s41570-020-0189-9.

Liu, H., G. H. Low, D. S. Steiger, T. Häner, M. Reher, and M. Troyer. 2022. "Prospects of Quantum Computing for Molecular Sciences." *Materials Theory* 6(11). doi.org/10.1186/s41313-021-00039-z.

Lordi, V., and J. M. Nichol. 2021. "Advances and Opportunities in Materials Science for Scalable Quantum Computing." *MRS Bulletin* 46(7):589–595. doi.org/10.1557/S43577-021-00133-0.

Lovchinsky, I., A. O. Sushkov, E. Urbach, N. P. De Leon, S. Choi, K. De Greve, R. Evans, R. Gertner, E. Bersin, C. Müller, L. Mcguinness, F. Jelezko, R. L. Walsworth, H. Park, and M. D. Lukin. 2016. "Nuclear Magnetic Resonance Detection and Spectroscopy of Single Proteins Using Quantum Logic." *Science* 351(6275):836–841. doi.org/10.1126/Science.Aad8022.

Lubinski, T., S. Johri, P. Varosy, J. Coleman, L. Zhao, J. Necaise, C. H. Baldwin, K. Mayer, and T. Proctor. 2023. "Application-Oriented Performance Benchmarks for Quantum Computing." *IEEE Transactions on Quantum Engineering* 4:1–32. https://doi.org/10.1109/tqe.2023.3253761.

Luis, F., P. J. Alonso, O. Roubeau, V. Velasco, D. Zueco, D. Aguilà, J. I. Martínez, L. A. Barrios, and G. Aromí. 2020. "A Dissymmetric [Gd_2] Coordination Molecular Dimer Hosting Six Addressable Spin Qubits." *Communications Chemistry* 3(1):176. doi.org/10.1038/S42004-020-00422-W.

Luis, F., A. Repollés, M. J. Martínez-Pérez, D. Aguilà, O. Roubeau, D. Zueco, P. J. Alonso, M. Evangelisti, A. Camón, J. Sesé, L. A. Barrios, and G. Aromí. 2011. "Molecular Prototypes for Spin-Based CNOT and SWAP Quantum Gates." *Physical Review Letters* 107(11):117203. doi.org/10.1103/Physrevlett.107.117203.

Lukin, D. M., M. A. Guidry, and J. Vučković. 2020. "Integrated Quantum Photonics with Silicon Carbide: Challenges and Prospects." *PRX Quantum* 1(2):020102. doi.org/10.1103/Prxquantum.1.020102.

Ma, H., M. Govoni, and G. Galli. 2020. "Quantum Simulations of Materials on Near-Term Quantum Computers." *npj Computational Materials* 6(1):85. doi.org/10.1038/S41524-020-00353-Z.

Ma, H., N. Sheng, M. Govoni, and G. Galli. 2020. "First-Principles Studies of Strongly Correlated States in Defect Spin Qubits in Diamond." *Physical Chemistry Chemical Physics* 22(44):25522–25527. doi.org/10.1039/D0CP04585C.

Macaluso, E., M. Rubín, D. Aguilà, A. Chiesa, L. A. Barrios, J. I. Martínez, P. J. Alonso, O. Roubeau, F. Luis, G. Aromí, and S. Carretta. 2020. "A Heterometallic [LnLn′Ln] Lanthanide Complex as a Qubit with Embedded Quantum Error Correction." *Chemical Science* 11(38):10337–10343. doi.org/10.1039/D0SC03107K.

Martinis, J. M., K. B. Cooper, R. Mcdermott, M. Steffen, M. Ansmann, K. D. Osborn, K. Cicak, S. Oh, D. P. Pappas, R. W. Simmonds, and C. C. Yu. 2005. "Decoherence in Josephson Qubits from Dielectric Loss." *Physical Review Letters* 95(21):210503. doi.org/10.1103/Physrevlett.95.210503.

McArdle, S., S. Endo, A. Aspuru-Guzik, S. C. Benjamin, and X. Yuan. 2020. "Quantum Computational Chemistry." *Reviews of Modern Physics* 92(1):015003. https://doi.org/10.1103/RevModPhys.92.015003.

McClean, J. R., J. Romero, R. Babbush, and A. Aspuru-Guzik. 2016. "The Theory of Variational Hybrid Quantum-Classical Algorithms." *New Journal of Physics* 18(2):023023. doi.org/10.1088/1367-2630/18/2/023023.

Mclnnes, E. J. L. 2022. "Molecular Spins Clock in." *Nature Chemistry* 14(4):361–362. doi.org/10.1038/S41557-022-00919-Y.

Miao, K. C., J. P. Blanton, C. P. Anderson, A. Bourassa, A. L. Crook, G. Wolfowicz, H. Abe, T. Ohshima, and D. D. Awschalom. 2020. "Universal Coherence Protection in a Solid-State Spin Qubit." *Science* 369(6510):1493–1497. doi.org/10.1126/Science.Abc5186.

Morello, A., J. J. Pla, F. A. Zwanenburg, K. W. Chan, K. Y. Tan, H. Huebl, M. Möttönen, C. D. Nugroho, C. Yang, J. A. Van Donkelaar, A. D. C. Alves, D. N. Jamieson, C. C. Escott, L. C. L. Hollenberg, R. G. Clark, and A. S. Dzurak. 2010. "Single-Shot Readout of an Electron Spin in Silicon." *Nature* 467(7316):687–691. doi.org/10.1038/Nature09392.

Moreno-Pineda, E., M. Damjanović, O. Fuhr, W. Wernsdorfer, and M. Ruben. 2017. "Nuclear Spin Isomers: Engineering a $Et_4N[DyPc_2]$ Spin Qudit." *Angewandte Chemie International Edition* 56(33):9915–9919. doi.org/10.1002/Anie.201706181.

Moreno-Pineda, E., C. Godfrin, F. Balestro, W. Wernsdorfer, and M. Ruben. 2018. "Molecular Spin Qudits for Quantum Algorithms." *Chemical Society Reviews* 47(2):501–513. doi.org/10.1039/C5CS00933B.

Motta, M., D. M. Ceperley, G. K.-L. Chan, J. A. Gomez, E. Gull, S. Guo, C. A. Jiménez-Hoyos, T. N. Lan, J. Li, F. Ma, A. J. Millis, N. V. Prokof'ev, U. Ray, G. E. Scuseria, S. Sorella, E. M. Stoudenmire, Q. Sun, I. S. Tupitsyn, S. R. White, D. Zgid, and S. Zhang (Simons Collaboration on the Many-Electron Problem). 2017. "Towards the Solution of the Many-Electron Problem in Real Materials: Equation of State of the Hydrogen Chain with State-of-the-Art Many-Body Methods." *Physical Review X* 7(3):031059. doi.org/10.1103/Physrevx.7.031059.

Nayak, C. 2021. "Full Stack Ahead: Pioneering Quantum Hardware Allows for Controlling up to Thousands of Qubits at Cryogenic Temperatures." Microsoft Research. January 27, 2021. www.microsoft.com/en-us/research/blog/full-stack-ahead-pioneering-quantum-hardware-allows-for-controlling-up-to-thousands-of-qubits-at-cryogenic-temperatures.

Nayak, C., S. H. Simon, A. Stern, M. Freedman, and S. Das Sarma. 2008. "Non-Abelian Anyons and Topological Quantum Computation." *Reviews of Modern Physics* 80(3):1083–1159. doi.org/10.1103/Revmodphys.80.1083.

Noel, C., M. Berlin-Udi, C. Matthiesen, J. Yu, Y. Zhou, V. Lordi, and H. Häffner. 2019. "Electric-Field Noise from Thermally Activated Fluctuators in a Surface Ion Trap." *Physical Review A* 99(6):063427. doi.org/10.1103/Physreva.99.063427.

O'Brien, T. E., G. Anselmetti, F. Gkritsis, V. E. Elfving, S. Polla, W. J. Huggins, O. Oumarou, K. Kechedzhi, D. Abanin, R. Acharya, I. Aleiner, R. Allen, T. I. Andersen, K. Anderson, M. Ansmann, F. Arute, K. Arya, A. Asfaw, J. Atalaya, D. Bacon, J. C. Bardin, A. Bengtsson, S. Boixo, G. Bortoli, A. Bourassa, J. Bovaird, L. Brill, M. Broughton, B. Buckley, D. A. Buell, T. Burger, B. Burkett, N. Bushnell, J. Campero, Y. Chen, Z. Chen, B. Chiaro, D. Chik, J. Cogan, R. Collins, P. Conner, W. Courtney, A. L. Crook, B. Curtin, D. M. Debroy, S. Demura, I. Drozdov, A. Dunsworth, C. Erickson, L. Faoro, E. Farhi, R. Fatemi, V. S. Ferreira, L. Flores Burgos, E. Forati, A. G. Fowler, B. Foxen, W. Giang, C. Gidney, D. Gilboa, M. Giustina, R. Gosula, A. Grajales Dau, J. A. Gross, S. Habegger, M. C. Hamilton, M. Hansen, M. P. Harrigan, S. D. Harrington, P. Heu, J. Hilton, M. R. Hoffmann, S. Hong, T. Huang, A. Huff, L. B. Ioffe, S. V. Isakov, J. Iveland, E. Jeffrey, Z. Jiang, C. Jones, P. Juhas, D. Kafri, J. Kelly, T. Khattar, M. Khezri, M. Kieferová, S. Kim, P. V. Klimov, A. R. Klots, R. Kothari, A. N. Korotkov, F. Kostritsa, J. M. Kreikebaum, D. Landhuis, P. Laptev, K. Lau, L. Laws, J. Lee, K. Lee, B. J. Lester, A. T. Lill, W. Liu, W. P. Livingston, A. Locharla, E. Lucero, F. D. Malone, S. Mandra, O. Martin, S. Martin, J. R. McClean, T. McCourt, M. McEwen, A. Megrant, X. Mi, A. Mieszala, K. C. Miao, M. Mohseni, S. Montazeri, A. Morvan, R. Movassagh, W. Mruczkiewicz, O. Naaman, M. Neeley, C. Neill, A. Nersisyan, H. Neven, M. Newman, J. H. Ng, A. Nguyen, M. Nguyen, M. Y. Niu, S. Omonije, A. Opremcak, A. Petukhov, R. Potter, L. P. Pryadko, C. Quintana, C. Rocque, P. Roushan, N. Saei, D. Sank, K. Sankaragomathi, K. J. Satzinger, H. F. Schurkus, C. Schuster, M. J. Shearn, A. Shorter, N. Shutty, V. Shvarts, J. Skruzny, V. Smelyanskiy, W. C. Smith, R. Somma, G. Sterling, D. Strain, M. Szalay,

D. Thor, A. Torres, G. Vidal, B. Villalonga, C. Vollgraff Heidweiller, T. White, B. W. K. Woo, C. Xing, Z. J. Yao, P. Yeh, J. Yoo, G. Young, A. Zalcman, Y. Zhang, N. Zhu, N. Zobrist, C. Gogolin, R. Babbush, and N. C. Rubin. 2022. "Purification-Based Quantum Error Mitigation of Pair-Correlated Electron Simulations." *ArXiv*. doi.org/10.48550/arxiv.2210.10799.

O'Malley, P. J. J., R. Babbush, I. D. Kivlichan, J. Romero, J. R. McClean, R. Barends, J. Kelly, P. Roushan, A. Tranter, N. Ding, B. Campbell, Y. Chen, Z. Chen, B. Chiaro, A. Dunsworth, A. G. Fowler, E. Jeffrey, E. Lucero, A. Megrant, J. Y. Mutus, M. Neeley, C. Neill, C. Quintana, D. Sank, A. Vainsencher, J. Wenner, T. C. White, P. V. Coveney, P. J. Love, H. Neven, A. Aspuru-Guzik, and J. M. Martinis. 2016. "Scalable Quantum Simulation of Molecular Energies." *Physical Review X* 6(3):031007. doi.org/10.1103/Physrevx.6.031007.

Paz-Silva, G. A., and D. A. Lidar. 2013. "Optimally Combining Dynamical Decoupling and Quantum Error Correction." *Scientific Reports* 3(1):1530. doi.org/10.1038/srep01530.

Peng, B., and K. Kowalski. 2022. "Mapping Renormalized Coupled Cluster Methods to Quantum Computers through a Compact Unitary Representation of Nonunitary Operators." *Physical Review Research* 4(4):043172. doi.org/10.1103/PhysRevResearch.4.043172.

Peruzzo, A., J. McClean, P. Shadbolt, M.-H. Yung, X.-Q. Zhou, P. J. Love, A. Aspuru-Guzik, and J. L. O'Brien. 2014. "A Variational Eigenvalue Solver on a Photonic Quantum Processor." *Nature Communications* 5(1):4213. doi.org/10.1038/Ncomms5213.

Philbin, J. P., and P. Narang. 2021. "Computational Materials Insights into Solid-State Multiqubit Systems." *PRX Quantum* 2(3):030102. doi.org/10.1103/Prxquantum.2.030102.

Phillips, W. A. 1987. "Two-Level States in Glasses." *Reports on Progress in Physics* 50(12):1657. doi.org/10.1088/0034-4885/50/12/003.

Pruitt, S. R., C. Bertoni, K. R. Brorsen, and M. S. Gordon. 2014. "Efficient and Accurate Fragmentation Methods." *Accounts of Chemical Research* 47(9):2786–2794. doi.org/10.1021/Ar500097m.

Rabl, P., S. J. Kolkowitz, F. H. L. Koppens, J. G. E. Harris, P. Zoller, and M. D. Lukin. 2010. "A Quantum Spin Transducer Based on Nanoelectromechanical Resonator Arrays." *Nature Physics* 6(8):602–608. doi.org/10.1038/Nphys1679.

Ray, K. G., B. M. Rubenstein, W. Gu, and V. Lordi. 2019. "Van der Waals-Corrected Density Functional Study of Electric Field Noise Heating in Ion Traps Caused by Electrode Surface Adsorbates." *New Journal of Physics* 21(5):053043. doi.org/10.1088/1367-2630/Ab1875.

Reiher, M., N. Wiebe, K. M. Svore, D. Wecker, and M. Troyer. 2017. "Elucidating Reaction Mechanisms on Quantum Computers." *Proceedings of the National Academy of Sciences USA* 114(29):7555–7560. doi.org/10.1073/Pnas.1619152114.

Rose, B. C., D. Huang, Z.-H. Zhang, P. Stevenson, A. M. Tyryshkin, S. Sangtawesin, S. Srinivasan, L. Loudin, M. L. Markham, A. M. Edmonds, D. J. Twitchen, S. A. Lyon, and N. P. De Leon. 2018. "Observation of an Environmentally Insensitive Solid-State Spin Defect in Diamond." *Science* 361(6397):60–63. doi.org/10.1126/Science.Aao0290.

Rosen A. S., V. Fung, P. Huck, C. T. O'Donnell, M. K. Horton, D. G. Truhlar, K. A. Persson, J. M. Notestein, and R. Q Snurr. 2022. "High-Throughput Predictions of Metal–Organic Framework Electronic Properties: Theoretical Challenges, Graph Neural Networks, and Data Exploration." *npj Compuational Materials* 8(1):112. doi.org/10.1038/s41524-022-00796-6.

Rosen A. S., S. M. Iyer, D. Ray, Z. Yao, A. Aspuru-Guzik, L. Gagliardi, J. M. Notestein, and R. Q. Snurr. 2021. "Machine Learning the Quantum-Chemical Properties of Metal–Organic Frameworks for Accelerated Materials Discovery." *Matter* 4(5):1578–1597. doi.org/10.1016/j.matt.2021.02.015.

Ryabinkin, I. G., S. N. Genin, and A. F. Izmaylov. 2019. "Constrained Variational Quantum Eigensolver: Quantum Computer Search Engine in the Fock Space." *Journal of Chemical Theory and Computation* 15(1):249–255. doi.org/10.1021/Acs.Jctc.8b00943.

Salfi, J., J. A. Mol, R. Rahman, G. Klimeck, M. Y. Simmons, L. C. L. Hollenberg, and S. Rogge. 2016. "Quantum Simulation of the Hubbard Model with Dopant Atoms in Silicon." *Nature Communications* 7(1):11342. doi.org/10.1038/ncomms11342.

Sandhoefer, B., and G. K.-L. Chan. 2016. "Density Matrix Embedding Theory for Interacting Electron-Phonon Systems." *Physical Review B* 94(8):085115. doi.org/10.1103/Physrevb.94.085115.

Sangtawesin, S., B. L. Dwyer, S. Srinivasan, J. J. Allred, L. V. H. Rodgers, K. De Greve, A. Stacey, N. Dontschuk, K. M. O'Donnell, D. Hu, D. A. Evans, C. Jaye, D. A. Fischer, M. L. Markham, D. J. Twitchen, H. Park, M. D. Lukin, and N. P. De Leon. 2019. "Origins of Diamond Surface Noise Probed by Correlating Single-Spin Measurements with Surface Spectroscopy." *Physical Review X* 9(3):031052. doi.org/10.1103/Physrevx.9.031052.

Schuch, N., and F. Verstraete. 2009. "Computational Complexity of Interacting Electrons and Fundamental Limitations of Density Functional Theory." *Nature Physics* 5(10):732–735. https://doi.org/10.1038/nphys1370.

Scientific Library. n.d. "List of Quantum Chemistry and Solid State Physics Software." Accessed April 12, 2023. https://www.scientificlib.com/en/Chemistry/ListQuantumChemistrySolidStatePhysicsSoftware.html.

Sedlacek, J. A., J. Stuart, D. H. Slichter, C. D. Bruzewicz, R. Mcconnell, J. M. Sage, and J. Chiaverini. 2018. "Evidence for Multiple Mechanisms Underlying Surface Electric-Field Noise in Ion Traps." *Physical Review A* 98(6):063430. doi.org/10.1103/Physreva.98.063430.

Seifrid, M., R. Pollice, A. Aguilar-Granda, Z. M. Chan, K. Hotta, C. T. Ser, J. Vestfrid, T. C. Wu, and A. Aspuru-Guzik. 2022. "Autonomous Chemical Experiments: Challenges and Perspectives on Establishing a Self-Driving Lab." *Accounts of Chemical Research* 55(17):2454–2466. doi.org/10.1021/Acs.Accounts.2c00220.

Seo, H., H. Ma, M. Govoni, and G. Galli. 2017. "Designing Defect-Based Qubit Candidates in Wide-Gap Binary Semiconductors for Solid-State Quantum Technologies." *Physical Review Materials* 1(7):075002. doi.org/10.1103/Physrevmaterials.1.075002.

Shavitt, I., and R. J. Bartlett. 2009. *Many-Body Methods in Chemistry and Physics: MBPT and Coupled-Cluster Theory, Cambridge Molecular Science*. Cambridge: Cambridge University Press.

Sheng, N., C. Vorwerk, M. Govoni, and G. Galli. 2022. "Green's Function Formulation of Quantum Defect Embedding Theory." *Journal of Chemical Theory and Computation* 18(6):3512–3522. doi.org/10.1021/Acs.Jctc.2c00240.

Sheridan, E., T. F. Harrelson, E. Sivonxay, K. A. Persson, M. Altoé, I. Siddiqi, D. F. Ogletree, D. I. Santiago, and S. M. Griffin. 2021. "Microscopic Theory of Magnetic Disorder-Induced Decoherence in Superconducting Nb Films." *ArXiv preprint*. arXiv:2111.11684.

Shiddiq, M., D. Komijani, Y. Duan, A. Gaita-Ariño, E. Coronado, and S. Hill. 2016. "Enhancing Coherence in Molecular Spin Qubits via Atomic Clock Transitions." *Nature* 531(7594):348–351. doi.org/10.1038/Nature16984.

Siddiqi, I. 2021. "Engineering High-Coherence Superconducting Qubits." *Nature Reviews Materials* 6(10):875–891. doi.org/10.1038/S41578-021-00370-4.

Singh, M. K., G. Wolfowicz, J. Wen, S. E. Sullivan, A. Prakash, A. M. Dibos, D. D. Awschalom, F. J. Heremans, and S. Guha. 2022. "Development of a Scalable Quantum Memory Platform—Materials Science of Erbium-Doped TiO_2 Thin Films on Silicon." *ArXiv preprint*. arXiv:2202.05376.

Sørensen, M. A., H. Weihe, M. G. Vinum, J. S. Mortensen, L. H. Doerrer, and J. Bendix. 2017. "Imposing High-Symmetry and Tuneable Geometry on Lanthanide Centres with Chelating Pt and Pd Metalloligands." *Chemical Science* 8(5):3566–3575. doi.org/10.1039/C7SC00135E.

Su, Y., D. W. Berry, N. Wiebe, N. Rubin, and R. Babbush. 2021. "Fault-Tolerant Quantum Simulations of Chemistry in First Quantization." *PRX Quantum* 2(4):040332. doi.org/10.1103/Prxquantum.2.040332.

Sugisaki, K., H. Wakimoto, K. Toyota, K. Sato, D. Shiomi, and T. Takui. 2022. "Quantum Algorithm for Numerical Energy Gradient Calculations at the Full Configuration Interaction Level of Theory." *Journal of Physical Chemistry Letters* 13(48):11105–11111. doi.org/10.1021/acs.jpclett.2c02737.

Sun, Q., and G. K.-L. Chan. 2016. "Quantum Embedding Theories." *Accounts of Chemical Research* 49(12):2705–2712. doi.org/10.1021/Acs.Accounts.6b00356.

Szabo, A., and N. S. Ostlund. 1989. *Modern Quantum Chemistry: Introduction to Advanced Electronic Structure Theory*. New York: McGraw-Hill.

Thiel, W. 2014. "Semiempirical Quantum–Chemical Methods." *WIREs Computational Molecular Science* 4(2):145–157. doi.org/10.1002/wcms.1161.

Thiele, S., F. Balestro, R. Ballou, S. Klyatskaya, M. Ruben, and W. Wernsdorfer. 2014. "Electrically Driven Nuclear Spin Resonance in Single-Molecule Magnets." *Science* 344(6188):1135–1138. doi.org/10.1126/Science.1249802.

Tuckett, D. K., S. D. Bartlett, and S. T. Flammia. 2018. "Ultrahigh Error Threshold for Surface Codes with Biased Noise." *Physical Review Letters* 120(5):050505. doi.org/10.1103/Physrevlett.120.050505.

Van Meter, R., and K. M. Itoh. 2005. "Fast Quantum Modular Exponentiation." *Physical Review A* 71(5):052320. doi.org/10.1103/PhysRevA.71.052320.

von Kugelgen, S., M. D. Krzyaniak, M. Gu, D. Puggioni, J. M. Rondinelli, M. R. Wasielewski, and D. E. Freedman. 2021. "Spectral Addressability in a Modular Two Qubit System." *Journal of the American Chemical Society* 143(21):8069–8077. https://doi.org/10.1021/jacs.1c02417.

Vorwerk, C., N. Sheng, M. Govoni, B. Huang, and G. Galli. 2022. "Quantum Embedding Theories to Simulate Condensed Systems on Quantum Computers." *Nature Computational Science* 2(7):424–432. doi.org/10.1038/S43588-022-00279-0.

Wan, N. H., T.-J. Lu, K. C. Chen, M. P. Walsh, M. E. Trusheim, L. De Santis, E. A. Bersin, I. B. Harris, S. L. Mouradian, I. R. Christen, E. S. Bielejec, and D. Englund. 2020. "Large-Scale Integration of Artificial Atoms in Hybrid Photonic Circuits." *Nature* 583(7815):226–231. doi.org/10.1038/S41586-020-2441-3.

Wang, C., C. Axline, Y. Y. Gao, T. Brecht, Y. Chu, L. Frunzio, M. H. Devoret, and R. J. Schoelkopf. 2015. "Surface Participation and Dielectric Loss in Superconducting Qubits." *Applied Physics Letters* 107(16):162601. doi.org/10.1063/1.4934486.

Wang, D. S., M. Haas, and P. Narang. 2021. "Quantum Interfaces to the Nanoscale." *ACS Nano* 15(5):7879–7888. doi.org/10.1021/acsnano.1c01255.

Weber, J. R., W. F. Koehl, J. B. Varley, A. Janotti, B. B. Buckley, C. G. Van De Walle, and D. D. Awschalom. 2010. "Quantum Computing with Defects." *Proceedings of the National Academy of Sciences USA* 107(19):8513–8518. doi.org/10.1073/Pnas.1003052107.

Wecker, D., M. B. Hastings, and M. Troyer. 2015. "Progress Towards Practical Quantum Variational Algorithms." *Physical Review A* 92(4):042303. doi.org/10.1103/Physreva.92.042303.

Weissman, S. A., and N. G. Anderson. 2015. "Design of Experiments (DoE) and Process Optimization. A Review of Recent Publications." *Organic Process Research & Development* 19(11):1605–1633. doi.org/10.1021/Op500169m.

Williams, K. T., Y. Yao, J. Li, L. Chen, H. Shi, M. Motta, C. Niu, U. Ray, S. Guo, R. J. Anderson, J. Li, L. N. Tran, C.-N. Yeh, B. Mussard, S. Sharma, F. Bruneval, M. Van Schilfgaarde, G. H. Booth, G. K.-L. Chan, S. Zhang, E. Gull, D. Zgid, A. Millis, C. J. Umrigar, and L. K. Wagner (Simons Collaboration on the Many-Electron Problem). 2020. "Direct Comparison of Many-Body Methods for Realistic Electronic Hamiltonians." *Physical Review X* 10(1):011041. doi.org/10.1103/Physrevx.10.011041.

Wineland, D. J., C. Monroe, W. M. Itano, D. Leibfried, B. E. King, and D. M. Meekhof. 1998. "Experimental Issues in Coherent Quantum-State Manipulation of Trapped Atomic Ions." *Journal of Research of the National Institute of Standards and Technology* 103(3):259–328. doi.org/10.6028/Jres.103.019.

Wootters, W. K., and W. H. Zurek. 1982. "A Single Quantum Cannot Be Cloned." *Nature* 299(5886):802–803. doi.org/10.1038/299802a0.

Xiong Y., C. Bourgois, N. Sheremetyeva, W. Chen, D. Dahliah, H. Song, S.M. Griffin, A. Sipahigil, and G. Hautier. 2023. "High-Throughput Identification of Spin-Photon Interfaces in Silicon." *ArXiv*. arxiv.org/pdf/2303.01594.pdf.

Yamabayashi, T., M. Atzori, L. Tesi, G. Cosquer, F. Santanni, M.-E. Boulon, E. Morra, S. Benci, R. Torre, M. Chiesa, L. Sorace, R. Sessoli, and M. Yamashita. 2018. "Scaling Up Electronic Spin Qubits into a Three-Dimensional Metal–Organic Framework." *Journal of the American Chemical Society* 140(38):12090–12101. doi.org/10.1021/Jacs.8b06733.

Yamamoto, K., D. Z. Manrique, I. T. Khan, H. Sawada, and D. M. Ramo. 2022. "Quantum Hardware Calculations of Periodic Systems with Partition-Measurement Symmetry Verification: Simplified Models of Hydrogen Chain and Iron Crystals." *Physical Review Research* 4(3):033110. doi.org/10.1103/physrevresearch.4.033110.

Ye, M., H. Seo, and G. Galli. 2019. "Spin Coherence in Two-Dimensional Materials." *npj Computational Materials* 5(1):44. doi.org/10.1038/S41524-019-0182-3.

Yu, C.-J., M. J. Graham, J. M. Zadrozny, J. Niklas, M. D. Krzyaniak, M. R. Wasielewski, O. G. Poluektov, and D. E. Freedman. 2016. "Long Coherence Times in Nuclear Spin-Free Vanadyl Qubits." *Journal of the American Chemical Society* 138(44):14678–14685. doi.org/10.1021/Jacs.6b08467.

Yu, C.-J., S. von Kugelgen, M. D. Krzyaniak, W. Ji, W. R. Dichtel, M. R. Wasielewski, and D. E. Freedman. 2020. "Spin and Phonon Design in Modular Arrays of Molecular Qubits." *Chemistry of Materials* 32(23):10200–10206. doi.org/10.1021/Acs.Chemmater.0c03718.

Yu, C.-J., S. von Kugelgen, D. W. Laorenza, and D. E. Freedman. 2021. "A Molecular Approach to Quantum Sensing." *ACS Central Science* 7(5):712–723. doi.org/10.1021/Acscentsci.0c00737.

Yu, P., J. Chen, M. Gomanko, G. Badawy, E. P. A. M. Bakkers, K. Zuo, V. Mourik, and S. M. Frolov. 2021. "Non-Majorana States Yield Nearly Quantized Conductance in Proximatized Nanowires." *Nature Physics* 17(4):482–488. doi.org/10.1038/S41567-020-01107-W.

Zadrozny, J. M., A. T. Gallagher, T. D. Harris, and D. E. Freedman. 2017. "A Porous Array of Clock Qubits." *Journal of the American Chemical Society* 139(20):7089–7094. doi.org/10.1021/jacs.7b03123.

Zhang, X., C. Wolf, Y. Wang, H. Aubin, T. Bilgeri, P. Willke, A. J. Heinrich, and T. Choi. 2022. "Electron Spin Resonance of Single Iron Phthalocyanine Molecules and Role of Their Non-Localized Spins in Magnetic Interactions." *Nature Chemistry* 14(1):59–65. doi.org/10.1038/s41557-021-00827-7.

Zhao, L., J. Goings, K. Wright, J. Nguyen, J. Kim, S. Johri, K. Shin, W. Kyoung, J. I. Fuks, J.-K. K. Rhee, Y. M. Rhee, et al. 2022. "Orbital-Optimized Pair-Correlated Electron Simulations on Trapped-Ion Quantum Computers." *ArXiv preprint*. doi.org/10.48550/arXiv.2212.02482.

5

Building a Diverse, Quantum-Capable Workforce and Fostering Economic Development at the Intersection of QIS and Chemistry

Key Takeaways

- Many jobs in QIS industries do not require an extensive educational background in QIS; rather, other specialized skills or education paths, including in the chemical sciences, are necessary.
- The opportunities for developing the skills needed to pursue careers in QIS are largely concentrated at the graduate level, where the percentage of people from historically marginalized communities is low.
- While efforts exist at K–12 levels, most are focused at the high school level, with a particular focus on physics and computer science education.
- Most traditional undergraduate chemistry paths have limited exposure to QIS. This exposure typically occurs under the auspices of other departments (e.g., physics, materials science) and in later stages of education (i.e., upper-level undergraduate or graduate-level courses).
- The chemical sciences discipline is a more diverse field in many aspects than traditional QIS fields. Through intentional recruitment efforts to remove inequitable barriers to entry, diversity can be tapped and the QIS workforce strengthened.
- While efforts are under way, data on quantum-related careers are still limited.
- Traditional industry players, including those involved in the chemical sciences, are increasingly partnering with quantum computing companies.
- Continued and future research and development at the intersection of QIS and chemistry holds strong potential to transform science and technology in the private sector and foster economic development.
- A significant barrier limiting the size and breadth of the chemistry research community working in QIS is that principles of QIS and quantum behaviors are not incorporated significantly in traditional chemistry education and are incorporated only at a very late stage (i.e., upper-level baccalaureate degree programs).
- Continued progress and innovation will require an expanded and strengthened quantum-capable workforce.

The previous chapters have explored a multitude of ongoing and future research directions at the intersection of QIS and chemistry that hold the potential to benefit and advance each respective field. As the convergence of QIS and chemistry advances, the potential impact of innovation will not be limited to academia and government-funded research. New discoveries and novel applications hold additional promise for transformational changes in science and technology in the private sector. As these industries grow and the application space of QIS and chemistry–related research expands, a diverse, quantum-capable workforce could support them. In this chapter, the committee provides commentary on the potential transformational impacts that research in QIS and chemistry could have in science and technology through the lens of education and workforce development. During its investigation, the committee received expert information from professionals in scientific policy, economic development, small business creation, and diversity and inclusion (see agenda from Information-Gathering Meeting 3, Appendix B). The committee provides its assessment of the needs and challenges related to a diverse, quantum-capable workforce that would support this industry, including the barriers to entry that limit the size and breadth of the chemistry research community working in QIS.

5.1 GETTING THE SCIENCE RIGHT FOR QIS AND CHEMISTRY STUDENTS

As highlighted at the beginning of this report, the United States has made significant progress in developing a few key quantum technologies, such as atomic clocks, electric field sensors, and superconducting qubits. This acceleration can be attributed to the support of the National Quantum Initiative Act (NQIA) and the National Defense Authorization Act for Fiscal Year 2020. The ultimate vision of the NQIA is to have the United States lead the world in delivering novel quantum technologies to the commercial market, with an innovation pipeline bolstered by a quantum-capable workforce. In a public presentation given to the committee, Dr. Charles Tahan, Assistant Director at the Office of Science and Technology Policy, reviewed the National Quantum Initiative (NQI) Plan, which outlines the actions to be taken for the United States to achieve this vision. At the top of the list shown in Figure 5-1, the United States should prioritize taking a "science-first approach."

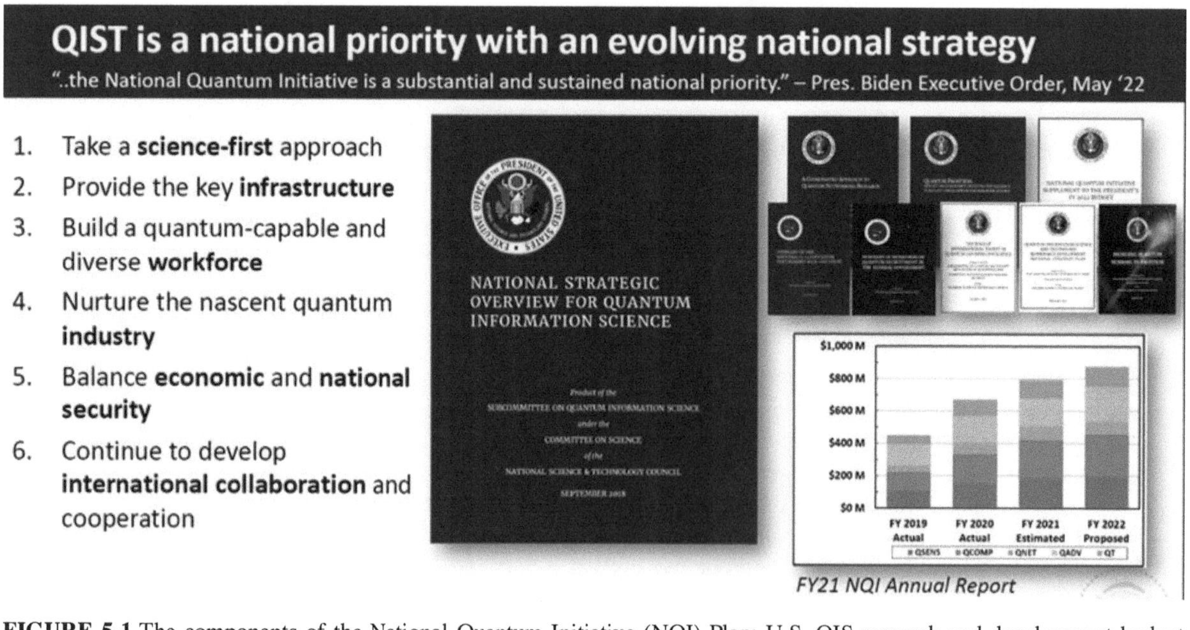

FIGURE 5-1 The components of the National Quantum Initiative (NQI) Plan: U.S. QIS research and development budget allocations by NQI Program Component Area topic for Fiscal Year (FY) 2019 and FY 2020 actual expenditures, FY 2021 estimated expenditures, and FY 2022 proposed budgets.
SOURCE: Tahan 2022.

The interpretation of the first step of the National Strategic Overview for Quantum Information Science is both mixed and urgent for the chemistry community. Educators may interpret "getting the science right" as developing curricula or pedagogies to teach important QIS concepts to undergraduate and graduate chemistry students. Another perspective is to expose students to science, technology, engineering, and mathematics (STEM)–related experiences broadly, and they may eventually apply this knowledge to QIS research and technology. However, this approach could limit students' opportunities to experience QIS earlier in their education and career, thus decreasing the chances for those students to enter the QIS workforce later. In sum, while scientists agree that conducting laboratory research in QIS is an important activity that advances the field, opinions differ on how and to what extent these research findings should be taught in the classroom or even presented to a broader audience. Unifying these different interpretations is the belief that clear scientific explanations of QIS phenomena to chemists through education are critical for facilitating the innovation of new quantum applications.

The postsecondary university curriculum for QIS-related concepts has been taught primarily in the departments of physics or engineering at most academic institutions. Some universities are offering minors and certificate programs specifically for QIS; however, for most of these initiatives, chemistry has been excluded from the course descriptions or prerequisites. There could be a great benefit if the research activities supported by the NQI involved more undergraduate chemistry students as well as students outside the normal pathway of physics and engineering. This topic will be elaborated further in Section 5.2, with explanations detailing how chemistry education could include more QIS concepts and vice versa.

Although six separate points are made in the national strategy, they are all deeply connected and tied to the intent of creating a quantum innovation ecosystem supported by a sustainable, capable, and diverse workforce (Raymer and Monroe 2019). Therefore, a natural question to be addressed is the following: what barriers to entry, if any, are impeding the growth of this emerging workforce? Furthermore, what opportunities are offered to students of different educational and socioeconomic backgrounds to experience hands-on QIS laboratory demonstrations or research? How are students with chemistry backgrounds exposed to QIS? As detailed in Section 5.3, the answers to these questions are diverse, and the proposed mechanisms for lowering these barriers to entry are also wide-ranging. Overall, providing valuable research experience as well as enhancing QIS education are two key actions of "getting the science right" for chemists as we move toward a future composed of chemically related applications in QIS. Finally, Section 5.4 expands on ways to build an inclusive and diverse quantum-capable workforce from the various sectors in chemistry and QIS. Section 5.5 examines the economic development challenges and opportunities of translating both basic and mature research into applications and the commercial space.

Undoubtedly, multiple aspects of NQI are related to research, education, and commercialization. The initiative emphasizes opportunities for enhanced activity and federal support in research to accelerate the transition to commercialization. One avenue that has been well executed is the creation of multidisciplinary centers that enable new QIS discoveries to occur in collaborative environments. This initiative focuses on getting the science correct, putting the science to use in applications, and translating the science to industry. While "science first" may seem obvious at first glance, the details for its implementation and its impact downstream on innovation are less obvious, especially for the chemistry community. The following sections attempt to address a few of the challenges and opportunities related to barriers to entry, chemical and QIS education development, and workforce and economic development. The following recommendation discusses the specific actors and activities needed at a broad level to strengthen the chemistry–QIS workforce.

5.2 QIS AND CHEMISTRY EDUCATION DEVELOPMENT

Ensuring a continued knowledge and expertise base that is familiar with QIS requires preparing students and workers at various educational and skill set levels. Education development at the intersection of QIS and chemistry involves a multidisciplinary approach underpinned by key STEM subjects and concepts: physics, chemistry, computer science, and mathematics. This section provides a key recommendation for improving education development for QIS and chemistry at different levels and expanding education outreach to the workforce. Previous consensus study reports by the National Academies of Sciences, Engineering, and Medicine have addressed science and engineering education in the United States. Some of these reports have addressed the lack of diversity, equity,

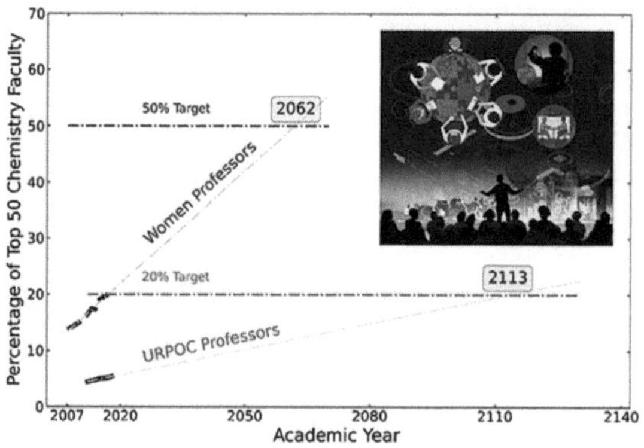

FIGURE 5-2 The projected year the percentages of women and underrepresented people of color (URPOC) professors match the percentages they represent in the general population.
SOURCE: Hernandez 2023.

and inclusion in academic research. For example, one of the previous reports regarding graduate STEM education recommended that faculty and administrators involved in graduate education should develop, adopt, and regularly evaluate a suite of strategies to accelerate increasing diversity and improving equity and inclusion. The report urged further that comprehensive recruitment, holistic review in admissions, and interventions to prevent attrition in the late stages of progress toward a degree should be carried out at universities (NASEM 2018). Another report by the National Academies regarding minority institution involvement in scientific research suggested that due to the large institutional resources required to compete for large grants and contracts, public and private funding agencies should reconsider the practicality of current competitive funding models for underresourced minority-serving institutions (NASEM 2019). Many of the previous reports were consistent with their message; however, they were not specific to chemistry or chemistry in QIS. Furthermore, an analysis of discipline-specific research that was published recently shows the state and projection of the gender and racial diversity in chemistry (Figure 5-2), and offers ways to measure and increase faculty diversity (Hernandez 2023). This chapter focuses on the educational and economic aspects, as well as barriers to entry for chemists attempting to enter the field of QIS.

5.2.1 Preparing Curricular Resources Related to Chemistry Concepts Guided by QIS Principles for K–12 and Undergraduate Educators

K–12 educators have many responsibilities, from academic instruction, to oversight, to fostering student social development and interaction. And to task them with the additional expectation to create specialized curricula and concepts (e.g., quantum chemistry, quantum mechanics, quantum algorithms) is a tall order. In some cases, just having the right materials to teach particular concepts can be a barrier. For example, a previous National Academies report devoted to elementary education suggested that state and district leaders should ensure that every school has the curriculum materials and instructional resources (as they become available) needed for engaging in science and engineering teaching that works toward equity and justice (NASEM 2022). Initiatives around the United States are creating resources for education, which will alleviate this extra burden. For example, Quantum Logical Electrons & Nuclei has examples of Python code that can be used to visually and programmatically teach QIS fundamentals without a large emphasis on theory. At the same time, these efforts will help sustain workforce development for emerging fields such as QIS. To address this challenge, the National Science Foundation (NSF) funded a program (Quantum for All) to introduce QIS to high school physics teachers, led by the University of Texas at Arlington (Quantum for All n.d.). This program hosts workshops to help train teachers on how to teach QIS in their classrooms. Although Quantum for All focuses on the physics curriculum, it could serve as a model to help train high school chemistry educators to teach aspects of QIS applied to chemistry. Furthermore, in 2020, NSF hosted a virtual workshop, Key Concepts for Future QIS Learners, and invited physics and computer science researchers and educators to come together to define a core set of key concepts for future QIS learners. The goal for these concepts is to serve as a starting point for further curricular and educator development activities (Center for Integrated Quantum Materials n.d.). Expanding on this effort, NSF funded the National Q–12 Education Partnership, which established a QIS K–12 Key Concepts Chemistry Focus Group. This initiative created a framework outlining expectations, learning goals, and crosscutting themes that could be used by curriculum developers and teachers seeking to develop chemistry lessons and activities for teaching QIS K–12 key concepts (National Q–12 Education Partnership n.d.). Additionally, the framework attempts to connect current Next Generation Science Standards to the proposed QIS concepts (Next Generation Science Standards n.d.). Showing these connections increases the likelihood that state education systems will formally adopt these concepts into classrooms throughout the United States (National Science Board 2018). Seventy percent of all American students have taken general chemistry in high school, making it the second most attended class in the country following biology. Therefore, introducing QIS in high school chemistry classes will reach the second largest group of science students across all education levels.

QuSTEAM (Quantum Science, Technology, Engineering, Arts and Mathematics), a nonprofit organization supported by the NSF Convergence Accelerator, is (as of 2022) providing education opportunities for two- and four-year undergraduate institutions by supplying instructors with customizable curriculum modules (QuSTEAM n.d.). Each module incorporates components necessary to receive a minor in quantum information science and engineering (QISE). Four different courses are offered as curricula samples to the organization's members: (1) Introduction to Quantum Information Science, (2) Classical and Quantum Logic, (3) Mathematical Methods for QISE, and (4) Quantum vs. Classical Lab. The European Union has a similar organization, known as Quantum Technology Education (QTEdu). QTEdu (n.d.) is funded by the European Union's Horizon 2020, and the goal of this initiative is to prepare the European Union's quantum workforce through comparable mechanisms as QuSTEAM. In both organizations, QuSTEAM and QTEdu, the courses and curricula largely focus on quantum computers, quantum algorithms and coding, and mathematics. In other words, the emphasis of these topics is on physics and computer science and engineering. This preference can also be observed in the organizations' committee structure, for which much of the team is from either a physics or engineering department. In "Building a Quantum Engineering Undergraduate Program," Asfaw and colleagues (2022) outline a detailed roadmap curated for engineering departments to develop QISE multidisciplinary programs. Appendix D lists several programs throughout the United States that offer a minor degree, master's degree, or certificate in QISE; a more updated map showing international master's degree programs in quantum technologies can also be found on the QURECA (Quantum Resources & Careers) website (QURECA n.d.).

Although chemistry is often associated as a subject deeply integrated into the multidisciplinary area of QISE, it is again absent from either the course descriptions or prerequisite lists of the QISE programs. Students in general chemistry courses are often introduced to quantum mechanics concepts (e.g., electronic structures, wave-particle duality, and electromagnetic radiation) early in the curriculum. Therefore, these chemistry courses are a natural place to introduce and elaborate on QIS concepts at the undergraduate level. Rather than creating a separate QIS elective, QIS could be integrated into the existing chemistry curriculum through small modifications, which ensures exposure to a broader demographic (e.g., premedical students, prepharmacy students, engineering majors, biology majors, and others) (NCSES 2023).

5.2.2 Engaging in Outreach Activities to Increase Exposure to QIS Chemical Technical Concepts at Varying Levels of Education

Building a workforce to engage actively in QIS and chemistry research and innovation will require the field to attract a broad labor force—in particular, students at varying levels of education. The Quantum Information Science and Technology Workforce Development National Strategic Plan (see Raymer and Monroe 2019; Subcommittee on Quantum Information Science of the National Science and Technology Council 2022) outlined four key strategies to advance the quantum information science and technology (QIST) workforce:

1. Develop and maintain an understanding of workforce needs in the QIST ecosystem, with both short-term and long-term perspectives;
2. Introduce broader audiences to QIST through public outreach and educational materials;
3. Address QIST-specific gaps in professional education and training opportunities; and
4. Make careers in QIST and related fields more accessible and equitable.

As highlighted earlier, most traditional chemistry education paths have limited exposure to QIS. And in circumstances where QIS programs are available, they are usually placed under the auspices of other departments (e.g., physics, computer science, electrical engineering). In addition, while efforts at K–12 levels also exist, these activities are focused at the high school level and skewed toward physics and computer science education. This lack of QIS introduction to chemistry students presents an opportunity to increase outreach activities, which aligns with the points made in the QIST Workforce Development National Strategic Plan. Deploying impactful campaigns requires an understanding of where and how to reach students; an efficient option may be to promote and highlight the importance of chemistry within the departments that already cover QIS topics.

The National Quantum Information Science Research Centers have made inroads in reaching talent at the undergraduate, graduate, and postdoctoral levels, and beyond, by hosting workshops and by offering internships, apprenticeships, fellowships, postdoctoral positions, and visiting-faculty appointments. However, the opportunities for developing the skills needed to pursue careers in QIS are largely concentrated at the graduate level, where the percentage of people from historically marginalized communities is low. In order to train a developing workforce in QIS, a multidisciplinary approach is needed in which engineers, physicists, chemists, and biologists cooperate to define the critical concepts and challenges within each of the disciplines and, more importantly, across disciplines. For example, training grants for the development of educational approaches to QIS that function across colleges within institutions (e.g., a college of science and a college of engineering) could lead toward the development of an effective pedagogy for teaching students, postdocs, and researchers. These courses could be introduced as early as the first or second year, similar to when engineering and physics fields introduce electromagnetics and semiconductor fundamental courses. QIS theory, equation derivation, modeling/simulation, quantum mechanics, and manipulation of quantum properties are appropriate topics that could be introduced and studied at this level. In sum, early introduction to the fundamentals of QIS would lead to deeper understanding in upper-level courses, increase interest and improve recruitment, and spur innovation.

To address this issue, the Department of Energy (DOE) offers Community College Internships (CCIs) (U.S. Department of Energy Office of Science n.d.) to encourage those students to enter technical careers related to the agency's mission space. CCIs offer mentorship and guidance at one of the 16 national laboratories which some include QIS research. However, a similar trend to that of the curriculum developments is observed here. According to the laboratory placement roster, the majority of internships offered in the past year (Fall 2022 and Summer 2022) were awarded to students with a major in computer science or engineering. Fewer than five percent of the awardees majored in chemistry. This observation reiterates the theme that chemistry at various education levels appears to be largely left out of the QIS research and workforce development enterprise.

Another avenue for reaching a diverse base of students is through lectures or a demonstration series to introduce QIS and chemistry. The topics could focus on technical concepts and ideas, a new consortium, an experiment, or activities at various QIS institutes. For example, these events could include monthly lectures focused on QIS and chemistry topics and accompanied by supplemental readings. Then, a comprehensive guide could be supplied to each student to explain the process of applying for internships (e.g., how to find opportunities, how to approach a principal investigator) and to give an overview of opportunities available to them both locally and nationally. Box 5-1 includes best practices that should be adopted to increase outreach efforts.

Remote learning is also a popular strategy for outreach. Computational chemistry allows traditionally underrepresented groups to build careers in the sciences through this path. The degree of theory and computational development in the QIS field enables an opportunity to remove or decrease the geographic restrictions of working and living near national laboratories and large research universities. The possibility of remote work extends the benefits that new technology affords the greater community by allowing the workforce to remain in their communities. Not only will this impact the demographics of the new workforce, but the option of working remotely will also extend the reach and influence of the programs supported by the government and academia by allowing people to work from the communities in which they are embedded.

BOX 5-1
Best Practices for Increasing Diversity through Outreach

The following are best practices that should be adopted to increase outreach efforts of introducing QIS concepts and activities to the broader scientific and engineering communities:

- Advertise QIS and chemistry research fellowships or scholarships to students at pre-graduate levels (e.g., students in general chemistry or organic chemistry courses) across the country at academic institutions;
- Partner with science communication experts to communicate QIS and chemistry technical concepts accessible to non-quantum experts at various education levels; and
- Highlight the significant relationship between QIS and chemistry at professional society conferences, in certified science curricula (including the related fields of physics, engineering, etc.), with standards, and through other means.

5.3 BARRIERS TO ENTRY INTO QIS AND CHEMISTRY

Workforce demands for QIS in chemistry are rapidly expanding. At the same time, industrial corporations recognize that the field lacks diversity, both in the current workforce and in new entrants. Thus, a unique opportunity exists in this emerging field to create a more diverse, equitable, and inclusive workforce. As illustrated in Section 5.1, another recruitment obstacle facing the QIS workforce is the lack of exposure chemists have to QIS activities, like learning about QIS concepts through multidisciplinary approaches, grant opportunities, mentorship, and even job and internship postings. Understanding the motivation behind "why" an individual would want to join the field will further the industry's understanding of how to reach a broader base. Examples of such motivations include salary, career track, company brand, and work culture. This limitation makes it difficult for the chemist to consider working in QIS as a viable career option.

5.3.1 Fostering Cross-Disciplinary and Cross-Sector Collaborations Using Projects Related to the Intersection of Chemistry and QIS

QIS is a highly interdisciplinary subject that integrates chemistry, physics, mathematics, and computer science. Because QIS consists of wide-ranging subjects, it presents a challenge for students to feel they have mastered a topic or are truly proficient in that area, making it difficult for them to "fit in" within the academic ecosystem. In an information-gathering meeting, Dr. Keeper Sharkey, founder and chief executive officer of a low-profit limited liability company (L3C) called ODE, shared her personal experience with this type of barrier to the committee. Sharkey (2022) pointed out that the interests that led her to a career in QIS did not fit into a traditional major or even into a single department. She indicated that she felt isolated as a woman entering the field and recognized the need for educational materials at the grade school and early college levels as well as the graduate levels. Motivated by her own experience, Sharkey established and leveraged the framework of an L3C to advance educational goals at the intersection of QIS and chemistry.

For example, theory and computation, which are often part of the physical chemistry discipline, are invaluable for providing models and predictions to understand experimental observables like chemical processes and properties. Computational tools such as interpreted and compiled programming languages (e.g., Unix environments) and high-performance computing technology are used to develop these theories and models. Being able to master these skills also requires, at a minimum, a basic understanding of computer science. Thus, one of ODE's goals is to create the space and opportunity for students to learn about the quantum nature of electrons and atoms in molecules using computational tools and mathematics.

An institute that combines actual research with broad outreach is one way of engaging students at the graduate and postdoctoral levels with multidisciplinary approaches. The Molecular Sciences Software Institute (MolSSI n.d.) has created a network of computational groups and software developers as well as a fellowship program that has connected several academic research institutions with industry and national laboratories. For students and companies, networking with a highly recognized institution can be an invaluable resource for connecting employers with the technically skilled employees needed for their field. A similar institute or close collaboration across academia, government, and industry could be the missing link between students and entities needing the specialized skills required to advance the QIS field. Although its sole focus is quantum computers, Quantum Futures (based in the United Kingdom) is another organization that adopts different techniques to acquire and connect talent to their optimal nodes inside the ecosystem (Quantum Futures n.d.). This program introduces university students and postdocs to available opportunities in the quantum computing industry.

5.3.2 Creating Internal Programmatic Strategies to Remove Implicit and Unconscious Bias During Review Processes

Implicit or unconscious bias can be defined as attitudes, stereotypes, and other hidden biases that influence perception, judgment, and action. Project Implicit, a nonprofit organization led by researchers studying cognition bias, educates the public about the effects of these types of bias on different demographics (e.g., race, ethnicity, religion, gender, career, and skin tone; see Project Implicit n.d.). Starting in 1998, this organization has collected data across surveys and consolidated its research findings into a database of associations about race, gender, and sexual orientation. Empirical evidence reveals that diversity—heterogeneity in race, ethnicity, cultural background, gender, sexual orientation, and other attributes—has material benefits for organizations, communities, and nations. However, because diversity can also incite detrimental forms of conflict and resentment, its benefits are not always realized, and it may impede further developments in emerging fields like QIS and chemistry.

However, if biases are not accounted for the organization or group may experience adverse effects. For example, the minimization of diverse backgrounds and insufficient mentoring can lead to underperformance by underrepresented groups. For example, Rodolfo Denton at Open Chemistry Collaborative in Diversity Equity (OXIDE) outlined areas where women were "under-encouraged" to publish in the field (Cimpian and Leslie 2015). Bias against diverse backgrounds and thoughts is not unique to QIS for chemistry. Other inequities in the recruitment process such as differentiated performance based on stereotype threats or inequitable access to the "hidden rules" for expectations of successful applicants also exist. These inequities can enter the climate and support structure once researchers enter the system. A recent study by the National Academies highlights these inequities in depth, as well as their impacts in the fields of science, technology, mathematics, and medicine (NASEM 2023).

Furthermore, these barriers are more pervasive in a nascent technology, such as quantum applications, with limited participants. This limits the pool of available applicants willing to pursue the field.

The committee also heard from Rigoberto Hernandez, who is a professor at Johns Hopkins University and the director of OXIDE, in an information-gathering meeting (see Appendix B). He discussed with the committee that middle managers in academia could be held more accountable because much of the culture of thought and diversity of teams is shaped at this organizational level. Hernandez called for an intentional plan and policy to foster the diversification of applicants and promotion through the ranks.

Another perspective related to these discussions was that studies have shown that it is a risky proposition for a candidate to enter the QIS field as a faculty member. For example, the National Diversity Equity Workshops in Chemical Sciences highlight the following barriers among many: implicit or unconscious bias, lack of universal design, stereotype threat, minimization of diverse backgrounds/color blindness, and insufficient mentoring or solo status (Stallings, Iyer, and Hernandez 2018). Intentional efforts to diversify the QIS workforce should not only lower the barriers to entry and success but also lower the risk for members of underrepresented groups to join the efforts. These efforts could include loan forgiveness programs (which would otherwise have kept low-income students from pursuing degrees beyond the bachelor's), postdoctoral pathway programs similar to the National Institutes of Health Pathway to Independence Award (National Institutes of Health n.d.), and mechanisms to soften the blow of possible tenure denials in tenure-stream pathways.

Similarly, a lack of universal design describes the absence of a process or practice to make environments welcoming, accessible, and usable by everyone. The Center for Universal Design at North Carolina State University (NCHPAD n.d.) is championing the need to design products and environments to be usable by all people, to the greatest extent possible, without the need for adaptation or specialized design. Stereotype threats describe performance and innovation risks in specific fields like QIS that come from the activation and confirmation of specific negative stereotypes. Continued and future research and development (R&D) at the intersection of QIS and chemistry, especially that supported by Small Business Innovation Research (SBIR) or Small Business Technology Transfer (STTR) programs, holds strong potential to transform science and technology in the private sector and foster economic development.

Moving beyond the academic workforce, research and innovation also play a critical role in driving progress at the intersection of QIS and chemistry. The early stages of R&D in QIS are ripe for innovation but will need more diversity. Many of these developments are currently supported by SBIR or STTR programs. The applications for SBIR and STTR undergo a peer-review process, which is one of the gateways for introducing a more diverse class of entrepreneurs. Box 5-2 elaborates further on key obstacles faced in QIS R&D that are slowing technological growth.

BOX 5-2
Areas in QIS Research and Development (R&D) That Need More Small Business Innovation Research (SBIR) and Small Business Technology Transfer (STTR) Support

Whether applied in the laboratory or in industry, the instrumentation to measure and confirm quantum and entangled effects rapidly needs to be built; usually, this is done with customized components. Universal techniques such as nuclear magnetic resonance (NMR) and Fourier-transform infrared (FTIR) that are essential for most chemistry studies are often maintained and operated by trained technicians and staff scientists. Unlike NMR and FTIR, most techniques needed to study quantum processes do not have this level of support. The closest universal instrument in QIS is transient electron paramagnetic resonance (EPR)/electron spin resonance instrumentation. However, as mentioned at the end of Chapter 3, the high cost of a turnkey transient EPR instrument lies in the fact that many of the components need to be designed and assembled in house. Additionally, this situation exists because only a few companies produce the technology, even though the advanced microwave sources are being improved continually and made cheaper by communications. Moving forward, a decrease in price and increase in production ability for transient EPR could be a major moving force for QIS in chemistry. Currently, most QIS instrumentation is built in specialized optics laboratories at universities. While this leads to fruitful collaborations, it is also a significant barrier to progress. Moreover, the average chemist cannot make a molecule and test if it is a good qubit or reacts differently to entangled photons than classical photons. Achieving a step of more universal instrumentation needs to be preceded by a focused commercial creation of specialized QIS components. Having modular parts will allow integration with existing instruments produced by larger companies that are serving industry and academia alike. Figure 5-2-1 shows one type of technique could be made more accessible to the QIS and chemistry research community. Hence, the economic value of such sub-components goes far beyond QIS in chemistry, and these sub-components are translatable to the entire QIS and quantum computing fields both as network components and analysis devices. Outside of the competition of large companies, smaller or startup companies have an opportunity to create and scale up these specialized components that are translating chemistry in practical QIS applications. These efforts tend to draw on the support from SBIR and STTR investments.

BOX 5-2 Continued

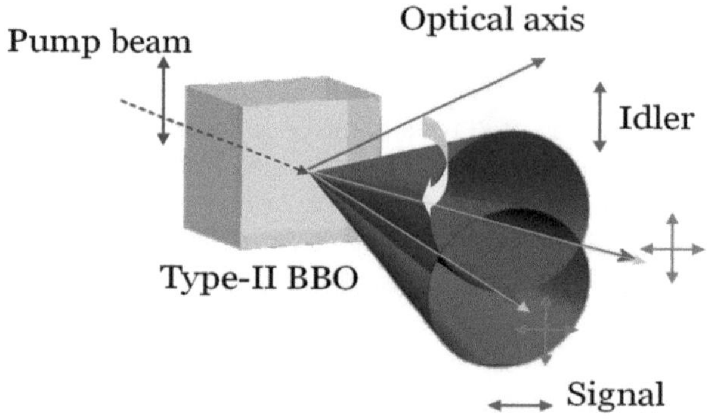

FIGURE 5-2-1 Generation of entangled photons by the process of spontaneous parametric down-conversion.

Other example opportunities for SBIR/STTR investments are in R&D works aimed at creating and measuring entangled photons. Entangled photon sources can be bought from a few companies for quantum networking purposes, but they usually have narrow frequency ranges (telecom at 1550 nm), low powers, and limited energy–time correlation bandwidths as needed for spectroscopy. While useful for QIS in chemistry demonstration kits for laboratories, these techniques do not have the properties needed for academia or industry. Most sources remain custom built. Creating turnkey entangled photon sources with broader wavelength ranges, shorter correlation times, and higher fluxes would be a precursor to the spectrometers and characterization techniques needed to explore quantum and entangled effects in chemistry. At the same time, these sources are needed for multiplexed quantum key distribution and other QIS technologies that are already experiencing large investment. A finding of this report is that many of the techniques lacking for QIS in chemistry are also needed for QIS and characterization of other quantum-based systems in other fields. Where large overlap is possible, as is common for spectroscopy and communication fields, increased economic viability is present and should be a target beyond just narrowly focusing on, per se, molecular chemistry. Investment in chemistry in QIS instrumentation is integral and congruent for the more funded QIS and quantum computing fields in terms of startups and large industry but currently is not a recognized goal of these programs, which mainly focus only on the end product. Characterization and spectroscopy started in chemistry in QIS could be instrumental for almost every industry, quantum or not, and should be considered in the same funding portfolios where possible.

5.3.3 Providing Support for Incubator Space Dedicated to Those Pursuing QIS and Chemistry Innovation R&D

Some industry practitioners are finding the direct implementation and commercialization of QIS-informed chemistry approaches to be challenging. A significant factor for this challenge is the current lack of proven market potential for products and tools that arise from QIS-informed chemistry R&D, especially physical technologies (i.e., those that are not software or computational services). Few, if any, successful products of this type from this field have resulted in significant returns on investment. As such, comparatively fewer investments have been made toward the development and commercialization of products in the QIS–chemistry space than in other fields.

Owing to the nascent market for QIS–chemistry products, some researchers and inventors in initial phases of development and commercialization have limited access to capabilities, tools, and resources necessary for creating products, including access to a physical space or location (including inside and outside of an academic institution) devoted to entrepreneurial work. Unlike university laboratories where academic research can be conducted on site in a straightforward process, setting up commercialization spaces that also require laboratory setup is complicated by a different set of governance arrangements. For example, setting up a separate space may include complex licensing agreements involving both the university and startups, as well as outside entities including scientists and engineers with a broad scope of skills, and negotiation of rental space. Many universities do not offer incubation spaces for researchers in this field; thus, a space for "tough tech" (e.g., for R&D where there is no clear market) is limited or missing.

A molecular foundry and processing center could serve as an alternative, but as mentioned above, this would require novel governance arrangements. Several factors should be considered in commercializing technologies from the intersection of QIS and chemistry, including the need for an appropriate innovation ecosystem that includes a physical innovation space and talent.

5.4 DEVELOPMENT OF A DIVERSE, QUANTUM-CAPABLE WORKFORCE

Up to this point, many of the discussions have centered on including more chemistry content into QIS curricula and activities and supporting diversity in a nascent field. Here, the committee provides a recommendation aimed at increasing the talent in the existing QIS and chemistry workforce by recruiting nontraditional students, retraining the existing workforce, and appealing to talent from adjacent fields.

5.4.1 Recruitment Opportunities

The extensive network of two-year degree-granting institutions makes higher education opportunities accessible to nontraditional students, people from historically excluded demographics, and low-income families. According to the National Center for Education Statistics, as of January 2021, there were 1,587 public and 2,344 private degree-granting institutions in the United States. More than 50 percent of the public and about 20 percent of the private institutions were two-year degree-granting institutions (National Center for Education Statistics 2021). In 2020, the percentage of non-white students at two-year degree-granting institutions was greater than that of non-white students at four-year degree-granting institutions (National Center for Education Statistics 2022). The cultural diversity in the workforce could thus be increased by reaching out to and helping students transfer successfully from two-year programs to four-year programs, or by recruiting graduates of two-year institutions directly to industry. Public–private partnerships could then train graduates directly for the jobs made available through the development of QIS.

Furthermore, access to, and participation with, advanced STEM materials is often not available to these various populations, particularly those in underserved communities. As reported by Jones (2018), "Almost one in five African American high-school students attends a school that does not offer any advanced placement courses. But even in schools that do, under-represented students are more likely to be placed on courses that are less academically demanding than are white, middle-class children."

One way to facilitate the transition from other STEM disciplines into chemistry–QIS is to adopt a model of transparent pathways for obtaining research experiences (e.g., guides to explain how to apply to internships, descriptions of local and national opportunities to do research, and details about mentorship programs). Jensen and colleagues (2021) showed in their results from surveying representatives across different demographics that pay is the dominant factor affecting or limiting the respondents' willingness to participate in internships within their fields (Figure 5-3). Since equitable pay is a major hurdle for reaching different demographics, including two-year college students, paid internships would be a key consideration when planning to reach out to these populations. In addition, while internships offer a fully immersive research experience, often the relationship between interns and groups/mentors naturally dwindles following the internship program. Historically, many students who have been exposed to research wish to continue their work during their subsequent academic terms. Therefore, building in optional follow-on funding for interns to remain as part-time researchers for an extended time period may increase the likelihood that students will remain in the QIS field. This strategy will also help

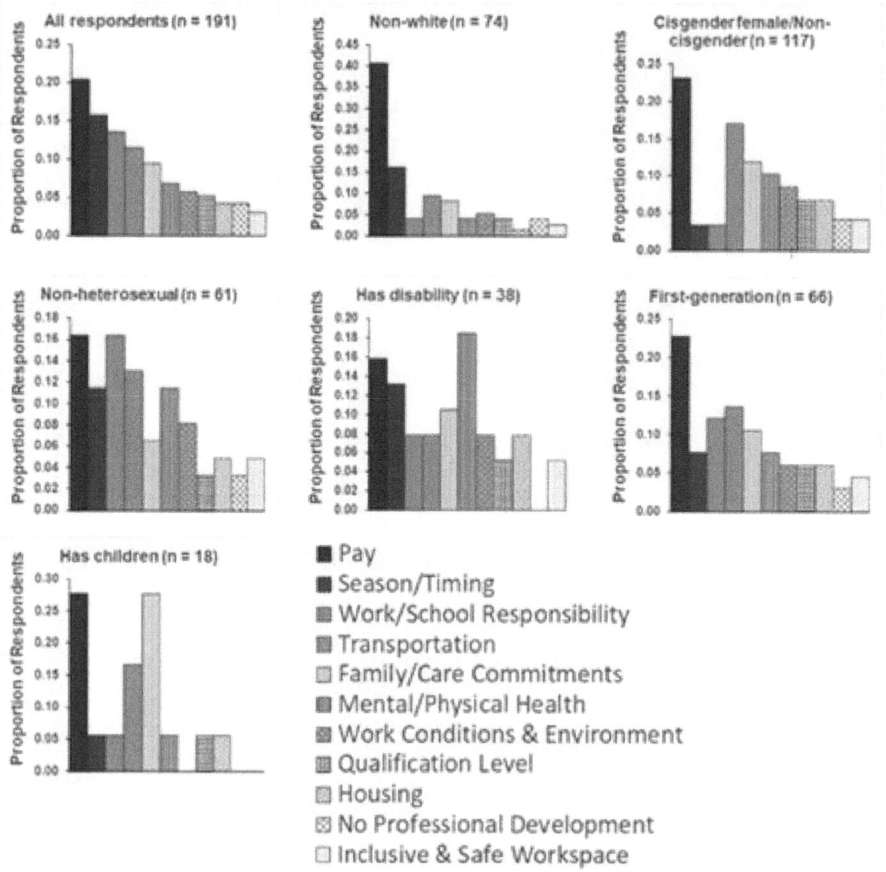

FIGURE 5-3 Proportion of the subset of respondents (n = 191 out of N = 879 total responses) that indicated there were other factors that affected or limited their willingness to participate in field internships that Jensen and colleagues (2021) did not ask about. Results are presented for all respondents in this subset and for different underrepresented groups. Categories with less than five responses are not shown, such as location (4), application process (3), citizenship status (3), cultural or religious commitments (3), lack of employee benefits (2), lack of advertising (2), pets (2), language barrier (1), type of work (1), and relationships (1).
SOURCE: Jensen et al. 2021.

remove the income inequality barrier associated with these longer-term research opportunities (e.g., enough time to produce publishable results) while helping the interns to build a network and develop relationships in the research community.

In sum, as economic activity at the intersection of QIS and the chemical sciences expands, demand for a quantum-capable chemical sciences workforce will increase. Efforts could be made to ensure that the fields of QIS and chemistry are accessible to people from diverse backgrounds, which would help to ensure that the future workforce is diverse and inclusive. These needs underscore the importance of placing more emphasis on reaching out to nontraditional QIS and chemistry students, particularly from two-year colleges.

5.4.2 Retraining the Current QIS and Chemistry Workforce

In cutting-edge research fields such as QIS, where few education paths (e.g., QIS minors, majors, or certificate programs) are available, greater opportunities may exist for on-the-job training for the potential employee.

Instead of requiring a four-year degree, the employment prerequisite can focus more on the applicant's skill set. Another strategy to increase talent in this space is to retrain the existing workforce to be more versed in QIS and chemistry's technical concepts and trade. The third tactic is to recruit talent from adjacent STEM fields, such as the semiconductor, pharmaceutical, and other industries. These approaches allow employers to recruit and train diverse quantum scientists, engineers, and technicians to support QIS in chemistry.

A report on the skills necessary to work in the QIS industry produced an analysis that surveyed QIS companies and their perspective employees (Hughes et al. 2022). While jobs such as error-correction scientist, experimental physicist, theoretical physicist, and computational chemist have a real need for detailed quantum technical expertise and experience, many of the other QIS-related jobs do not necessarily require quantum skills. Here, "quantum skill" is defined as one that is specific to the quantum industry and includes quantum knowledge. Hughes and colleagues (2022) concluded that quantum skills tend to cluster by type and into specific job roles, as illustrated in Figure 5-4. In general, the more specific the job title, the more specific the quantum skill required for the position will be.

Students in non-quantum, STEM-related undergraduate programs possess the skill sets needed for the career paths that do not require in-depth quantum knowledge, such as maintenance technician. Next Generation Quantum Science and Engineering (Q-NEXT), a multidisciplinary research center supported by DOE, is working with this demographic and training the next quantum workforce through several different pathways. Q-NEXT's program consists of retraining certificate programs to build foundational skills for quantum careers, innovative cooperative training programs with industry, and quantum-focused institutional degree programs with the center's university partners (Q-NEXT n.d.). Similarly, the Quantum Systems Accelerator offers new education and workforce development programs to provide immediate retraining and to feed the pipeline (Quantum Systems Accelerator n.d.). It is important that young chemists advance their STEM skills in general in order to be competitive for opportunities in QIS. Excellence in their own areas of study and research may pave a path for future inclusion in QIS even without a great deal of quantum knowledge initially.

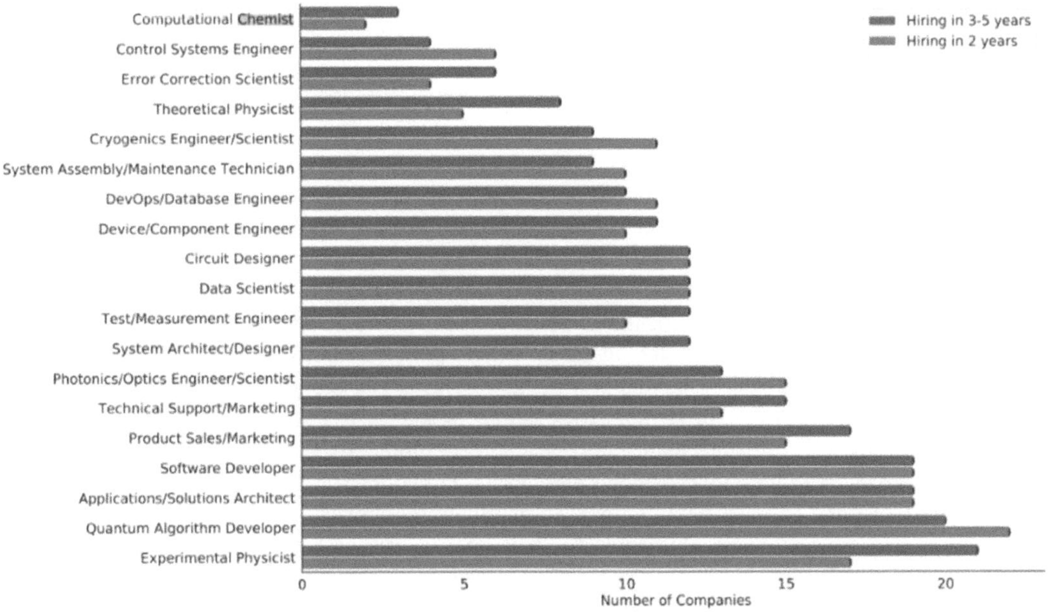

FIGURE 5-4 The number of companies looking to hire people with certain skill sets in the next two years (orange) and three to five years (blue).
SOURCE: Hughes et al. 2022.

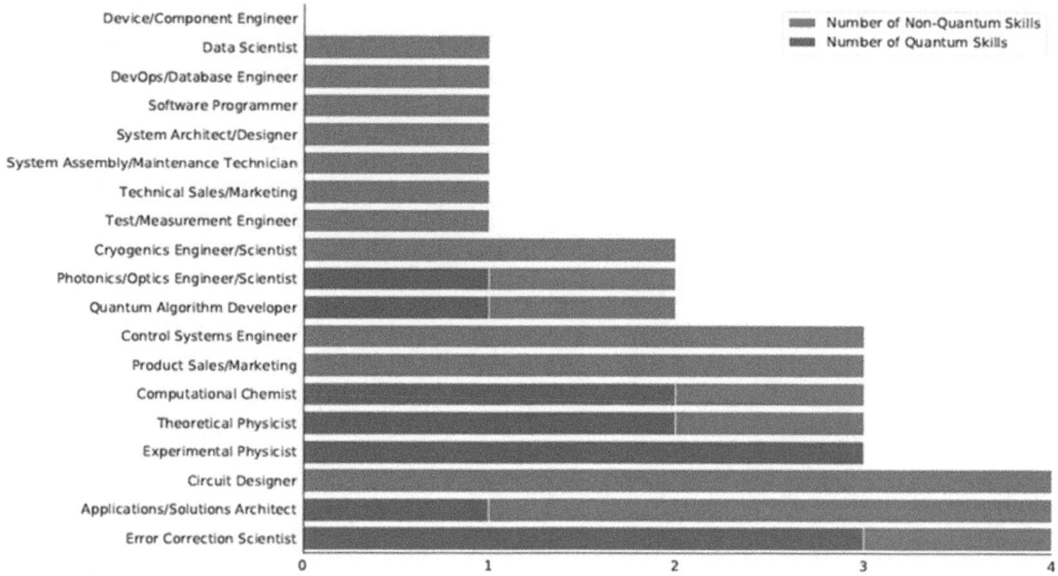

FIGURE 5-5 The number of important quantum (blue) and non-quantum (orange) skills needed for each job type.
SOURCE: Hughes et al. 2022.

5.4.3 Examining Job Requirements Beyond Doctoral Prerequisites

Career opportunities in QIS require a range of skills, most of which may not be related specifically to quantum chemistry. Effective communication between students who are potential applicants and those with available openings to address industry needs is paramount to support the education and training of a new workforce. Students need to know how to tailor their education, skill development, and choice of mentors to career opportunities in the QIS field. Likewise, corporations, academic institutions, national laboratories, and government agencies must create job descriptions or labor categories that align with the skills needed for a nascent field. The Quantum Economic Development Consortium (QED-C) was established by the National Institute of Standards and Technology in 2018 as part of the NQIA (QED-C n.d.a). An early product of the QED-C was a published paper titled "Assessing the Needs of the Quantum Industry" (Hughes et al. 2022).

Hughes and colleagues (2022) include survey results from top companies in the field, which indicate the need for positions to fill vacancies that require a variety of degree levels. Degree levels can range from technician/certificate level to four-year degrees and postgraduate degrees. As shown in Figure 5-5, computational chemists are sought as potential hires in the next two years by several companies. Furthermore, the QED-C maintains a near real-time job board from member companies, academic organizations, and national laboratories to address their needs as the field grows (QED-C). Finally, new curricula and internships are needed to foster workforce development and align with potential career tracts. For example, Indian River State College (IRSC) received NSF funding to develop curricula and certificate programs to foster diversity and development of technicians as well as high-end QIS practitioners (Seldes 2021). IRSC is working with the QED-C as well as other two-year degree-granting institutions to build the relevant curriculum. Also, internships are an important bridge between academia and industry for undergraduates and recent graduates to develop skills and acquire mentors. The Chicago Quantum Exchange partners with companies, national laboratories, and others to provide opportunities for its students and trainees, including internships and collaborative research opportunities, that expose them to a broad array of experiences in industrial, academic, and government facilities (Chicago Quantum Exchange n.d.).

The quantum industry can also recruit personnel from the current non-quantum workforce, thereby opening an avenue for chemists to provide new ideas and approaches to this industry. Box 5-3 highlights how the semiconductor industry has a strong overlap with the quantum space in terms of skill sets required of the workforce and is thus a potential area from which to recruit QIS and chemistry candidates.

BOX 5-3
The Semiconductor Industry Holds Promise for Chemistry and QIS Workforce Development

Economic development of semiconductor chips is indeed a major focus to limit offshore manufacturing and bring this important production back to the United States. In addition to increasing the workforce in the U.S. semiconducting industry, developments of new chips can have far-reaching implications in medicine, energy, defense, and national security. The national effort to bolster this area is reflected by the federal agency's large investments (Table 5-3-1). Chemists and chemical engineers are involved heavily in chip technology development. Even though scientists may not be trained in QIS, the semiconductors that they develop have important quantum properties that drive the application; hence, their skills are transferable. The recent CHIPS Act could play a role in the education and experience of young scientists in fields not necessarily quantum related but with skills that could be used in quantum applications. Additionally, the development of QIS is connected with the semiconductor industry's research to overcome the computing power limitations of Moore's law. As illustrated in some examples in Appendix C, academic and government researchers have made advances in quantum computing by working in tandem or in partnership with the semiconductor industry.

TABLE 5-3-1 Authorization Budget Over Five Years Across Federal Agencies

Key Programs	Five-Year Authorization	Increase over Baseline
National Science Foundation (NSF)	**$81 billion**	**$36 billion**
• NSF Tech Directorate	$20 billion	$20 billion
• NSF Core Activities	$61 billion	$16 billion
Department of Commerce (DOC)	**$11 billion**	**$11 billion**
• Regional Technology Hubs	$10 billion	$10 billion
• RECOMPETE Pilot	$1 billion	$1 billion
National Institute of Standards and Technology (NIST)	**$10 billion**	**$5 billion**
• NIST Research	$6.9 billion	$2.8 billion
• Manufacturing USA	$829 million	$744 million
• Manufacturing Extension Partnership	$2.3 billion	$1.5 billion
Department of Energy (DOE)*	**$67.9 billion**	**$30.5 billion**
• DOE Office of Science	$50.3 billion	$12.9 billion
• Additional DOE Science and Innovation	$17.6 billion	$17.6 billion
Total	**$169.9 billion**	**$82.5 billion**

*Across all the DOE sections, there is:

- A total of $14.7 billion for infrastructure, equipment, and instrumentation across 17 DOE National Laboratories.
- A total of$16.5 billion in new or above baseline authorizations for research in the 10 technology areas identified in USICA across the Office of Science and DOE's applied R&D offices in advanced energy and industrial efficiency technologies, artificial intelligence and machine learning, advanced manufacturing, cybersecurity, biotechnology, high performance computing, advanced materials, and quantum information science.

SOURCE: Chips and Science Act of 2022 Division B Summary – Research and Innovation National Science Foundation, Department of Commerce, National Institute of Standards and Technology, National Aeronautics and Space Administration, and Department of Energy 2022e.

BOX 5-3 Continued

Similar to the semiconductor industry, quantum computing requires sophisticated fabrication capabilities, specialized materials, and advanced technologies. Again, many of these skills do not require detailed knowledge of quantum-related phenomena themselves but instead a detailed knowledge of the science, technology, engineering, and mathematics–related skill. It is also suggested that future activities in QIS seek these cross-disciplinary approaches toward solving hard QIS problems. The National Science Foundation and the Department of Energy are poised to take leading roles in the development of research and technology related to this effort.

Figures 5-4 and 5-5 show that the quantum and non-quantum skills needed by industry are outpacing available applications. Hence, the potential to overhire in QIS is minimal at present. Two references can be reviewed to get a snapshot on the actual number of workers needed for the QIS industry. Quantum Futures (n.d.) produces a quarterly industry guide, which includes a survey of QIS-related companies and their needs for the future. The QED-C job board posts job vacancies and descriptions for current needs from member corporations, academic institutions, national laboratories, and government agencies (QED-C n.d.b).

5.4.4 Increasing the Number of Permanent Positions for Those with Varying Education and Experience Levels

Academic institutions are less amenable to hiring a greater number of scientific staff due to their educational missions. However, it may be possible to expand an institution's research and teaching capabilities if more permanent staff aided in maintaining the continuity of the personnel and supporting the education and training of budding scientists. Within chemistry alone, the traditional sub-disciplines are organic, inorganic, analytical, physical, and biological chemistry. Being proficient in these sub-disciplines can prepare students for careers in QIS. For example, many analytical techniques, such as electron paramagnetic resonance and nuclear magnetic resonance, are becoming used routinely to characterize and evaluate materials with quantum properties (see Chapter 3 for an in-depth discussion). Understanding the theory of such methods fits within the traditional physical chemistry major and can prepare students to further develop, combine, or improve analytical instruments. The dearth of specialized equipment and expert users greatly limits the opportunity for new students to become proficient users.

Another barrier to hiring scientists into long-term positions is the dearth of guaranteed continuous funding opportunities. The soft-money nature of most programs is a disincentive and makes hiring long-term staff nearly impossible. Postdocs, with their transitory nature, are well served through the current funding mechanisms, but the timeline from high school to postdoctoral scholar is poorly suited for a quickly developing field.

Examples of Career Structures in Academia and National Laboratories

Academic research personnel:

- Professors (principal investigators)
- Postdoctoral Scholars
- Graduate students
- Staff scientists (less common)

National laboratory personnel:

- Scientists (principal investigators)
- Postdoctoral Scholar
- Senior and Assistant Staff Scientists

BOX 5-4
Best Practices to Increase the Job Applicant Pool

Establish a centralized posting location (e.g., Broad Agency Announcement) with information about QIS research grants and Small Business Innovation Research contracts for both the National Science Foundation and the Department of Energy.

A single website that aggregates all of the open positions—in which a user can filter the positions based on various features such as geographic location, remote-work possibility, and skills required—will be beneficial to prospective applicants. This type of consolidated job announcement would be a useful tool for finding and applying to open positions, while also increasing the likelihood of matching the talent to the position.

Increase the diversity of advertisement platforms and information distribution outlets (e.g., leveraging email listservs of professional societies, nonprofits, consortia, and others) for national laboratory, industry, and academic positions in QIS.

The creation of an industry-informed resource will provide insight into career paths and drive engagement with new communities. An example that can be emulated by QIS is the field of computer science. Like QIS, computer science is a diverse and rapidly changing field. The website computerscience.org provides a thorough taxonomy of degree programs, positions, potential salaries, and scholarships in the field of computer science. QIS needs a similar resource where potential entrants into the field of QIS are well informed of potential paths for a career (Swed 2022).

To normalize and bring equality to the salaries of scientists at various levels of seniority, human resource departments are attempting to standardize the positions and pay bands for people with the same title regardless of demographic information. This also allows the institution to monitor the demographics of employees with various titles. In academia, the typical researchers are graduate and undergraduate students and postdoctoral scholars. Staff scientists are uncommon, except for some specialized positions such as instrument facility management. In government laboratories, the typical researchers are staff scientists at senior and assistant levels. Generally, senior staff scientists can serve as principal investigators, and assistant staff scientists are more specialized doctoral-level researchers. Postdoctoral scholar positions are also available at government laboratories, and as in academia, they are temporary positions. The positions at government laboratories are almost exclusively doctoral-level positions. More bachelor's and master's level positions would benefit and help improve diversity among the laboratories. Box 5-4 shows best practices that could be adopted to increasing the talent pool.

5.5 IMPACTS OF QIS AND CHEMISTRY ON ECONOMIC DEVELOPMENT

Chemistry-based approaches in QIS and the application of QIS technologies in traditional chemistry industries hold strong potential to transform areas of science and technology, and by extension might have significant impacts on the domestic and global economies. While quantum chemistry historically has been limited to the confines of academia and government-funded national laboratories, a sufficient body of evidence now exists to show that R&D at the intersection of QIS and chemistry is gaining traction in emerging and traditional industries.

5.5.1 Development of New Chemical Systems with QIS Methods Including the Potential for Lowering the Cost of Business

The development of new molecular architectures could have a significant economic impact beyond the current QIS investments. This includes the use of molecules as optical elements in quantum communication networks, such as repeaters, transducers, memory elements, and so forth. However, the development of QIS in chemistry also can produce new technologies not feasible within the existing QIS fields.

For example, molecules that interact strongly with entangled light would allow for the simplification of various nonlinear microscopies that are pervasive in bioimaging, such as two-photon fluorescence and stimulated Raman scattering. These classical light form techniques are already moving from the laboratory to specialized hospital clinics, and they seem to be on track to move eventually to the doctor's office (just as fluorescence microscopy did). Nonlinear microscopies are also branching beyond medicine into semiconductor and photonic device analysis, where the same three-dimensional imaging capabilities allow for quick analysis of architectures and more subtle features like electric fields not detectable with linear optical or X-ray techniques. Entangled photon sources can improve these technologies and their distribution, by removing the ultrafast laser system and replacing it with a continuous wave laser. This not only decreases the cost of the microscopes from more than $100,000 to less than $10,000 (at the laboratory scale, without large manufacturing added in) but also enables a portable and even photonic on-chip architecture to allow access to more sensing environments. Such sources could remove technical hurdles like alignment and stability, and could fit in mobile devices with the use of small packaging through modern photonics. However, despite this potential, the ideal molecular tags to make entangled nonlinear imaging competitive with classical techniques have yet to be fully discovered, and this is a task for which quantum-based computing will come into play.

Similarly, as outlined in this report, entanglement is theorized to play a role in controlling reactions, spin states, and other chemical properties (beyond optical). If entanglement is present in fundamental Hamiltonians (a task ideally suited to be solved using quantum computing), then it can be utilized for a new control of chemical products and reaction chains. How or if entanglement can influence the molecular processes important to the industry, or create new ones, is a task for which QIS methods will play a domineering role.

5.5.2 Collaborations between Quantum Sectors and the Chemical Industry

The committee found that many current collaborations between industry and academia (national laboratories) are still in their initial stages of discovery and are searching for particular areas of interest that would be of great benefit to both the industrial and university partners. Many of these companies, including Google, Microsoft, Amazon, and IBM, for example, are quite interested in quantum computing for chemistry and openly publish papers on the topic as well as open-source packages devoted to advancing the field of quantum computing for chemistry (see "OpenFermion" [Google Quantum AI n.d.] and "Open Source Quantum Development" [Qiskit n.d.]). This work is usually conducted in collaboration with quantum computing startups that focus on applications and algorithms including QC Ware (n.d.), Cambridge Quantum Computing (Quantinuum n.d.), Zapata Computing (n.d.), Q-CTRL (n.d.), 1Qbit (n.d.), Rigetti Computing (n.d.), PsiQuantum (n.d.), HQS Quantum Simulations (n.d.), Riverlane (n.d.), and others.

Below, the committee surveyed a few different examples of large companies that work in the traditional chemistry domain and collaborate with quantum computing companies on research. The committee describes the research problems addressed by these collaborations to demonstrate what type of research is being conducted in such partnerships.

Boehringer Ingelheim (BI) is a pharmaceutical company that has done work in quantum computing. A paper by Malone and colleagues (2022) represents a collaboration between BI and quantum computing software/services startup QC Ware. Here, they focus on methods of development of problem decomposition techniques that allow one potentially to "fit" larger quantum chemistry problems into the constraints of smaller quantum computers. BI has also had several publications with Google. For example, a paper by O'Brien and colleagues (2022) focuses on developing quantum algorithms for computing molecular forces more efficiently. This is valuable for BI because it potentially would enable using quantum computers to design better force fields in order to improve molecular dynamics simulations, which are an important component of modeling for drug discovery. Finally, in another paper with Google, BI conducts a study of the resources that would be required to simulate the drug anti-target cytochrome P450 using a fault-tolerant quantum computer (Goings et al. 2022). Goings and colleagues (2022) conclude that with the error rates targeted by groups such as Google, several million physical qubits would be required to surpass the state of the art for modeling P450.

Covestro is a materials science company (formerly the materials science arm of Bayer) based in Germany. It has worked with two smaller companies (QC Ware and Qu & Co) in the quantum computing space. Their collaboration seeks to improve algorithms for using near-term quantum computers within the framework of variational quantum eigensolver (VQE) algorithms. In particular, Covestro developed a way of parameterizing quantum circuits for VQE that exploits symmetries (e.g., particle number is a good quantum number) known to exist in chemical systems. This potentially reduces circuit depth and improves prospects for error mitigation since one can detect errors by noticing if, for example, the system leaves the correct particle-number manifold. Elfving and colleagues (2021) focus on simulating quantum chemistry on quantum computers in a seniority-zero subspace, which is a subspace of the complete active space where all configurations involve paired electrons. This is advantageous for quantum computing because the pairs of electrons can be treated bosonically, thus alleviating the requirement that fermionic antisymmetry is imposed (reducing circuit size) while halving the number of qubits. The downside of this approach is that the seniority-zero subspace is not a sufficient representation for many strongly correlated chemical systems. Finally, Parrish, Anselmetti, and Gogolin (2021) developed ways of computing gradients of parameterized quantum circuits. In the context of VQE, this allows one to train the parameterized circuits more efficiently, thus alleviating the classical cost of these quantum–classical algorithms (and potentially requiring fewer queries to the quantum computer).

Daimler (the parent company of Mercedes Benz) has a division devoted to battery research that has taken an interest in quantum computing and worked with both IBM and Google. For example, Daimler worked with Google to show a method of perturbatively extending active space calculations performed on quantum computers into a larger extended space (Takeshita et al. 2020). The technique requires more circuit repetitions (a classical resource) but does not require any additional qubits or quantum gates. Daimler also collaborated with IBM to implement a toy (proof-of-principle) demonstration of using an actual noisy intermediate-scale quantum computer to extract properties such as dipole moments from active spaces of compounds related to lithium-sulfate batteries (Rice et al. 2021).

Furthermore, Dow is a large chemical company based in the United States. Dow has recently worked with the ion trap quantum hardware startup IonQ and the quantum software/services startup 1QBit to develop problem decomposition methods for representing quantum chemical problems on quantum hardware (Kawashima et al. 2021). Like the energy decomposition techniques that Covestro was working on, this is a technique for fitting a larger molecule into fewer resources to implement on the quantum computer. Dow then demonstrated these techniques by realizing them on IonQ's quantum computer—again a toy experiment with just a few qubits involved.

BASF is a German company and the world's largest chemical producer. It has established a group to work in quantum computing. An example of work from BASF is a paper by Kühn and colleagues (2019), which involved a collaboration with HQS Quantum Simulations (a Germany-based quantum computing software/services startup). The goal of this project was to do a preliminary estimation of the resources (e.g., number of gates) required to implement certain approaches to VQE (especially certain forms of unitary coupled cluster) to solve a handful of classically intractable molecules. This work is somewhat outdated now, however, as most approaches to VQE being explored today are more resource efficient. But the work nonetheless reveals BASF's interest in quantum computing.

Bosch is a German multinational engineering and technology company with a large materials science division that has an interest in quantum computing. Bosch has worked with HQS and Google on several projects related to simulating lattice models of electrons and chemistry. With HQS, they developed new parameterizations of circuits for VQE that claim to reduce the circuit depth required to reach target accuracies (Vogt et al. 2021). Furthuremore, Arute and colleagues (2020) implemented an experiment on quantum hardware with Google (involving more than 20 qubits) that simulates the dynamics of fermions in the Hubbard model.

IBM has worked with JSR, the Japanese Synthetic Rubber manufacturer. Together, they implemented a small-scale experiment on IBM's superconducting qubit quantum hardware simulating the photochemistry of the sulfonium cation using a VQE-related approach (Motta et al. 2022). Again, this is an experiment with only a few qubits. IBM has also worked with the aerospace company Boeing, which is presumably interested in developing new materials for aviation using quantum computers. In particular, IBM worked with Boeing to develop methods of simulating surface chemistry on quantum computers. Their approach concerns how to best represent the Hamiltonians of these systems to minimize the number of qubits required while being able to perform Brillouin zone integration to mitigate finite-size effects (Gujarati et al. 2022).

IBM has also worked with Mitsubishi Chemical in order to implement a proof-of-principle demonstration of simulating electronic transitions in phenylsulfonyl-carbazole thermally activated delayed fluorescence emitters on a quantum computer (Gao et al. 2021). The application that Mitsubishi is interested in here seems to be organic light-emitting diodes. Like many of these "proof-of-principle" experiments, only a few qubits are involved in the experiment, and it is unclear (perhaps even unlikely) that the approach can scale to classically intractable instances without quantum error correction.

Astex Pharmaceuticals has recently worked with the quantum computing startup Riverlane to perform yet another proof-of-principle experiment on a few qubits (Izsak et al. 2022). They also further developed methods of representing the active space on a quantum computer to minimize the number of qubits required. Like almost all of these experiments, the active space is solved with some form of VQE. In this case, the experimental platform used was Rigetti Computing's superconducting qubit device.

As a final example in a non-exhaustive list of recent industry papers, Roche Pharmaceuticals has worked with the startup Cambridge Quantum Computing (CQC). CQC recently merged with Honeywell's quantum computing arm to form Quantinuum. In a paper by Kirsopp and colleagues (2022), Roche and CQC run prototype demonstrations (again, a few qubits) of modeling protein–ligand binding on several different quantum computers. In particular, they compare Honeywell's ion trap systems to IBM's superconducting qubit hardware.

The motivation for these companies to work in quantum computing varies. Most seem to believe that quantum computing is a technology that they should eventually master in order for their research units to stay competitive. As a result, they are interested in learning about the technology now and conducting research to understand when it might be expected to help with problems related to their interests. Some are interested in performing calculations on near-term quantum computers, although most have not managed to use more than just a few qubits. Although sometimes producing publishable research, this is currently just an exercise in exploring the technology. No company has yet deployed quantum computations to solve problems in chemistry that could not be solved with a regular laptop computer. Whether quantum computing will ever impact industrially relevant chemistry problems without quantum error correction remains speculative. However, if an error-corrected quantum computer is built, then there are some known examples of industrial problems that could be solved with a quantum advantage over classical computers. See, for example, the study on the cost of simulating the cytochrome P450 enzyme that was conducted jointly between Google and BI (Goings et al. 2022). Still, the devices that could deliver such computations appear to be years away from realization.

While the list of publications from domain companies working on quantum computing for chemistry is by no means comprehensive, a trend can be observed from these efforts. There are certainly important issues related to further partnerships with industry moving forward that will require detailed analysis and considerations of intellectual property before both sides of the collaboration may want to proceed. Furthermore, an interesting trend is that while the United States has more developed quantum computing companies and capabilities than Europe, a disproportionately high proportion of the domain companies exploring quantum computing solutions appear to be in Europe and Japan. In March 2023, the United Kingdom released its National Quantum Strategy, for which the U.K. government will double its current investment and commit £2.5 billion to develop quantum technologies in the United Kingdom by 2033 (U.K. Department for Science, Innovation, and Technology 2023). The United Kingdom's efforts in supporting the developments have similarities to the U.S. NQI, for which the emphasis is placed on fostering talent, strengthening fundamental research, expanding infrastructure, and supporting quantum businesses. In addition to a current collaboration with the United States (United Kingdom of Great Britain and Northern Ireland and the United States of America 2021), the United Kingdom is aiming to expand its international partnership through bilateral arrangements with five other leading quantum nations. From these observations, the committee concluded that increasing R&D efforts, public–private partnerships, workforce and chemistry–QIS education development, and other forms of scientific collaboration at the interface of chemistry and QIS will support economic development in the United States and strengthen the U.S. quantum advantage on the global stage.

5.6 SUMMARY OF RECOMMENDATIONS

RECOMMENDATION 5-1. Achieving the goal of a diverse and inclusive workforce will require participation from various members across the quantum information science (QIS) and chemistry enterprise. The Department of Energy, the National Science Foundation, and other U.S. federal agencies should support efforts to create a more diverse and inclusive chemical QIS workforce. Private and public stakeholders such as educators at various levels, nonprofit organizations, human resource personnel, and professional societies should also foster talent development and recruitment and increase public awareness related to QIS and chemistry activities. These efforts should aim to strengthen QIS in K–12,

two-year degree-granting institutions, and beyond. The efforts should also lower barriers to entry for all scientists in QIS and develop the necessary skills in participants at multiple levels of education. Agencies and relevant stakeholders should prioritize actions to address the following topics:

1. **QIS and chemistry education development;**
2. **Barriers to entry at the intersection of QIS and chemistry; and**
3. **Development of a diverse, quantum-capable workforce.**

Recommendation 5-2. Efforts to enhance curriculum resources and opportunities for students to gain exposure to concepts and skills at the intersection of quantum information science (QIS) and chemistry should be made. These efforts will support more learners in traditional educational and academic environments interested in pursuing research and careers at the intersection of QIS and chemistry.

- **Education development initiatives and curriculum developers should prepare curricular resources that include chemistry concepts guided by QIS principles for K–12 and undergraduate levels.**
- **Educators, human resource personnel, program managers, and communication teams should engage in outreach activities to increase exposure to QIS chemical technical concepts at varying levels of education.**

Recommendation 5-3. Efforts should be made to lower the current barriers to entry that limit members of the chemistry research community from entering quantum information science (QIS)–related research and careers. Efforts should also be made to lower barriers to entry for nontraditional participants to provide equitable pathways to careers at the intersection of QIS and chemistry and to expand access to broader, more diverse groups of talent.

- **Industry consortiums, education organizations, federal agencies, and other relevant entities should foster cross-disciplinary and cross-sector collaborations that explore projects related to the intersection of chemistry and QIS.**
- **Program managers and administrators should create internal programmatic strategies to remove implicit and unconscious bias during the review process of grant applications and other peer-reviewed applications (e.g., Small Business Innovation Research [SBIR]).**
- **Academic institutions, SBIR programs, and other relevant stakeholders should provide support to establish incubator spaces dedicated to those pursuing QIS and chemistry innovation research and development (e.g., academic institutions, SBIR/Small Business Technology Transfer programs).**

Recommendation 5-4. Increasing broader participation and diversity remains a challenge in recruiting and retaining talent in the field of quantum information science (QIS) and chemistry. Dedicated and focused efforts should be made to foster a diverse, quantum-capable chemical sciences workforce.

- **Program coordinators, researchers, educators, and other relevant personnel should recruit students from two-year colleges who typically do not engage in QIS research and who are likely to transfer to four-year academic institutions.**
- **Federal agencies and professional development coordinators should provide retraining opportunities for the academic, industrial, and national laboratory workforce of potential QIS participants with requisite professional skills that are useful for employment in a QIS field.**
- **Human resource personnel and hiring managers should provide detailed descriptions of the technical skill sets beyond doctoral prerequisites needed for jobs at the intersection of QIS and chemistry.**
- **National laboratories, industry, federal agencies, and academic institutions should increase support for hiring more permanent, professional, and diverse (in terms of demographics) technical staff at varying education and experience levels.**

REFERENCES

1Qbit. n.d. "Research Papers | 1QBit." Accessed April 12, 2023. https://1qbit.com/our-thinking/research-papers/.

Arute, F., K. Arya, R. Babbush, D. Bacon, J. C. Bardin, R. Barends, A. Bengtsson, S. Boixo, M. Broughton, and B. B. Buckley. 2020. "Observation of Separated Dynamics of Charge and Spin in the Fermi-Hubbard Model." *ArXiv*. doi.org/10.48550/arXiv.2010.07965.

Asfaw, A., A. Blais, K. R. Brown, J. Candelaria, C. Cantwell, L. D. Carr, J. Combes, D. M. Debroy, J. M. Donohue, and S. E. Economou. 2022. "Building a Quantum Engineering Undergraduate Program." *IEEE Transactions on Education* 65(2):220–242. doi.org/10.1109/TE.2022.3144943.

Center for Integrated Quantum Materials. n.d. "CIQM Course Modules." Accessed April 8, 2023. ciqm.harvard.edu/course-modules.html.

Chicago Quantum Exchange. n.d. "Education and Training: Internships." Accessed December 30, 2022. chicagoquantum.org/education-and-training/internships.

Chips and Science Act of 2022 Division B Summary – Research and Innovation National Science Foundation, Department of Commerce, National Institute of Standards and Technology, National Aeronautics and Space Administration, and Department of Energy 2022.

Cimpian, A., and S.-J. Leslie. 2015. "Response to Comment on 'Expectations of Brilliance Underlie Gender Distributions Across Academic Disciplines.'" *Science* 349(6246):391. doi.org/10.1126/science.aaa9892.

Elfving, V. E., M. Millaruelo, J. A. Gámez, and C. Gogolin. 2021. "Simulating Quantum Chemistry in the Seniority-Zero Space on Qubit-Based Quantum Computers." *Physical Review A* 103(3):032605. doi.org/10.1103/PhysRevA.103.032605.

Gao, Q., G. O. Jones, M. Motta, M. Sugawara, H. C. Watanabe, T. Kobayashi, E. Watanabe, Y.-Y. Ohnishi, H. Nakamura, and N. Yamamoto. 2021. "Applications of Quantum Computing for Investigations of Electronic Transitions in Phenylsulfonyl-Carbazole TADF Emitters." *npj Computational Materials* 7(1):70. doi.org/10.1038/S41524-021-00540-6.

Goings, J. J., A. White, J. Lee, C. S. Tautermann, M. Degroote, C. Gidney, T. Shiozaki, R. Babbush, and N. C. Rubin. 2022. "Reliably Assessing the Electronic Structure of Cytochrome P450 on Today's Classical Computers and Tomorrow's Quantum Computers." *Proceedings of the National Academy of Sciences USA* 119(38):e2203533119. doi.org/10.1073/pnas.2203533119.

Google Quantum AI. n.d. "OpenFermion." Accessed April 8, 2023. quantumai.google/openfermion.

Gujarati, T. P., M. Motta, T. Nguyen Friedhoff, J. E. Rice, N. Nguyen, P. Kl. Barkoutsos, R. J. Thompson, T. Smith, M. Kagele, and M. Brei. 2022. "Quantum Computation of Reactions on Surfaces Using Local Embedding." *ArXiv*. doi.org/10.48550/arXiv.2203.07536.

Hernandez, R. 2023. "Discipline-Based Diversity Research in Chemistry." *Accounts of Chemical Research* 56(7):787–797. doi.org/10.1021/acs.accounts.2c00797.

HQS Quantum Simulations. n.d. "Solutions." Accessed April 12, 2023. https://quantumsimulations.de/solutions.

Hughes, C., D. Finke, D.-A. German, C. Merzbacher, P. M. Vora, and H. J. Lewandowski. 2022. "Assessing the Needs of the Quantum Industry." *IEEE Transactions on Education* 65(4):592–601. doi.org/10.1109/TE.2022.3153841.

Izsák, R., C. Riplinger, N. S. Blunt, B. de Souza, N. Holzmann, O. Crawford, J. Camps, F. Neese, and P. Schopf. 2022. "Quantum Computing in Pharma: A Multilayer Embedding Approach for near Future Applications." *Journal of Computational Chemistry* 44(3):406–421. https://doi.org/10.1002/jcc.26958.

Jensen, A. J., S. P. Bombaci, L. C. Gigliotti, S. N. Harris, C. J. Marneweck, M. S. Muthersbaugh, B. A. Newman, S. L. Rodriguez, E. A. Saldo, K. E. Shute, K. L. Titus, A. L. Williams, S. Wing Yu, and D. S. Jachowski. 2021. "Attracting Diverse Students to Field Experiences Requires Adequate Pay, Flexibility, and Inclusion." *Bioscience* 71(7):757–770. doi.org/10.1093/biosci/biab039.

Jones, N. 2018. "Boosting the Number of Students from Underrepresented Groups in Physics." *Nature* 562:S12–S14. doi.org/10.1038/d41586-018-06834-y.

Kawashima, Y., E. Lloyd, M. P. Coons, Y. Nam, S. Matsuura, A. J. Garza, S. Johri, L. Huntington, V. Senicourt, and A. O. Maksymov. 2021. "Optimizing Electronic Structure Simulations on a Trapped-Ion Quantum Computer Using Problem Decomposition." *Communication Physics* 4(1):1–9. doi.org/10.1038/s42005-021-00751-9.

Kirsopp, J. J., C. Di Paola, D. Zsolt Manrique, M. Krompiec, G. Greene-Diniz, W. Guba, A. Meyder, D. Wolf, M. Strahm, and D. Muñoz Ramo. 2022. "Quantum Computational Quantification of Protein–Ligand Interactions." *International Journal of Quantum Chemistry* 122(22):E26975. doi.org/10.1002/qua.26975.

Kühn, M., S. Zanker, P. Deglmann, M. Marthaler, and H. Wei. 2019. "Accuracy and Resource Estimations for Quantum Chemistry on a Near-Term Quantum Computer." *Journal of Chemical Theory and Computation* 15(9):4764–4780. doi.org/10.1021/Acs.Jctc.9b00236.

Malone, F. D., R. M. Parrish, A. R. Welden, T. Fox, M. Degroote, E. Kyoseva, N. Moll, R. Santagati, and M. Streif. 2022. "Towards the Simulation of Large Scale Protein–Ligand Interactions on NISQ-Era Quantum Computers." *Chemical Science* 13(11):3094–3108. doi.org/10.1039/d1sc05691c.

MolSSI (Molecular Sciences Software Institute). n.d. "MolSSI's Goals." Accessed December 30, 2022. molssi.org.

Motta, M., G. O. Jones, J. E. Rice, T. P. Gujarati, R. Sakuma, I. Liepuoniute, J. M. Garcia, and Y.-Y. Ohnishi. 2022. "Quantum Chemistry Simulation of Ground- and Excited-State Properties of the Sulfonium Cation on a Superconducting Quantum Processor." *Chemical Science* 14(11):2915–2927. doi.org/10.1039/d2sc06019a.

NASEM (National Academies of Sciences, Engineering, and Medicine). 2018. *Graduate STEM Education for the 21st Century*. Washington, DC: The National Academies Press. doi.org/10.17226/25038.

NASEM. 2019. *Minority Serving Institutions: America's Underutilized Resource for Strengthening the STEM Workforce*. Washington, DC: The National Academies Press. doi.org/10.17226/25257.

NASEM. 2022. *Science and Engineering in Preschool Through Elementary Grades: The Brilliance of Children and the Strengths of Educators*. Washington, DC: The National Academies Press. doi.org/10.17226/26215.

NASEM. 2023. *Advancing Antiracism, Diversity, Equity, and Inclusion in STEMM Organizations: Beyond Broadening Participation*. Washington, DC: The National Academies Press. doi.org/10.17226/26803.

National Center for Education Statistics. 2021. "Degree-Granting Postsecondary Institutions, by Control and Level of Institution: Selected Years, 1949-50 Through 2020-21." Last modified January 2021. nces.ed.gov/programs/digest/d20/tables/dt20_317.10.asp.

National Center for Education Statistics. 2022. "Total Fall Enrollment in Degree-Granting Postsecondary Institutions, by Level and Control of Institution and Race/Ethnicity or Nonresident Status of Student: Selected Years, 1976 Through 2021." Last modified December 2022. nces.ed.gov/programs/digest/d22/tables/dt22_306.20.asp.

National Institutes of Health. n.d. "Pathway to Independence Awards (K99/R00)." Accessed April 9, 2023. https://www.nigms.nih.gov/training/careerdev/Pages/PathwayIndependence.aspx.

National Q–12 Education Partnership. n.d. "QIS Key Concepts for K-12 Chemistry." Accessed April 8, 2023. https://q12education.org/wp-content/uploads/2023/01/K-12-Quantum-Education-Framework-Chemistry-final.pdf.

National Science Board. 2018. "High School Coursetaking in Mathematics and Science." *Science and Engineering Indicators 2018*. https://nsf.gov/statistics/2018/nsb20181/report/sections/elementary-and-secondary-mathematics-and-science-education/high-school-coursetaking-in-mathematics-and-science.

NCHPAD (National Center on Health, Physical Activity and Disability). n.d. "Center for Universal Design - North Carolina State University - Raleigh, North Carolina, USA: NCHPAD - Building Inclusive Communities." Accessed April 8, 2023. https://www.nchpad.org/Directories/Organizations/2558/Center~for~Universal~Design~-~North~Carolina~State~University.

NCSES (National Center for Science and Engineering Statistics). 2023. *Diversity and STEM: Women, Minorities, and Persons with Disabilities 2023. Special Report NSF 23-315*. Alexandria, VA: National Science Foundation. https://ncses.nsf.gov/wmpd.

"Next Generation Science Standards." n.d. Accessed April 8, 2023. https://www.nextgenscience.org/.

O'Brien, T. E., M. Streif, N. C. Rubin, R. Santagati, Y. Su, W. J. Huggins, J. J. Goings, N. Moll, E. Kyoseva, and M. Degroote. 2022. "Efficient Quantum Computation of Molecular Forces and Other Energy Gradients." *Physical Review Research* 4(4):043210. doi.org/10.1103/physrevresearch.4.043210.

Parrish, R. M., G.-L. R. Anselmetti, and C. Gogolin. 2021. "Analytical Ground- and Excited-State Gradients for Molecular Electronic Structure Theory from Hybrid Quantum/Classical Methods." *ArXiv*. doi.org/10.48550/arxiv.2110.05040.

Project Implicit. n.d. "About Us: Project Implicit." Accessed April 8, 2023. https://implicit.harvard.edu/implicit/aboutus.html.

PsiQuantum. n.d. "Quantum Computing | Technical Resources." Accessed April 12, 2023. https://psiquantum.com/resources.

Q-CTRL. n.d. "Advanced Topics | Q-CTRL." Accessed April 12, 2023. https://q-ctrl.com/learning-center/advanced-topics.

Q-NEXT (Next Generation Quantum Science and Engineering). n.d. "About Q-Next." Accessed December 30, 2022. q-next.org.

QC Ware. n.d. "QC Ware Research." Accessed April 12, 2023. https://www.qcware.com/research.

QED-C (The Quantum Economic Development Consortium). n.d.a. "Home: QED." Accessed February 28, 2023. https://quantumconsortium.org/.

QED-C. n.d.b. "Quantum Jobs." Accessed April 12, 2023. https://quantumconsortium.org/quantum-jobs/.

Qiskit. n.d. "Open-Source Quantum Development." qiskit.org/.

QTEdu (Quantum Technology Education). n.d. "About page: Why QTEdu Csa." Accessed April 8, 2023. https://qtedu.eu/.

Quantinuum. n.d. "A Center for Gravity for Quantum Computing." Accessed April 12, 2023. https://www.quantinuum.com/resources.

Quantum for All. n.d. "Background: Quantum for All." Accessed April 9, 2023. https://quantumforall.org/background/.

Quantum Futures. n.d. "Seeking Talent." Accessed December 30, 2022. quantum-futures.com/ecosystem/.

Quantum Systems Accelerator. n.d. "Our Science." Accessed April 4, 2023. quantumsystemsaccelerator.org.

QURECA (Quantum Resources & Careers). n.d. "Masters in Quantum Technologies." Accessed April 8, 2023. qureca.com/masters-in-quantum-technologies/.

QuSTEAM. n.d. "QuSTEAM Curriculum." Accessed December 30, 2022. qusteam.org/courses.

Raymer, M. G., and C. Monroe. 2019. "The US National Quantum Initiative." *Quantum Science and Technology* 4(2):020504. doi.org/10.1088/2058-9565/ab0441.

Rice, J. E., T. P. Gujarati, M. Motta, T. Y. Takeshita, E. Lee, J. A. Latone, and J. M. Garcia. 2021. "Quantum Computation of Dominant Products in Lithium–Sulfur Batteries." *Journal of Chemical Physics* 154(13):134115. doi.org/10.1063/5.0044068.

Rigetti Computing. n.d. "Quantum Computing Research." Accessed April 12, 2023. https://www.rigetti.com/research.

Riverlane. n.d. "Riverlane Research-Quantum Error Correction, Systems Engineering & Industry Applications." Accessed April 12, 2023. https://www.riverlane.com/research.

Seldes, S. 2021. "Curriculum in Advanced Optics and Quantum Research-Enabled Technologies." Indian River State College, May 26. https://irsc.edu/news/articles/irsc-awarded-nsf-grant-052721.html.

Sharkey, K. 2022. "Theoretical Chemistry to Enhance Quantum Information Science: Education, Funding, Research." Information-Gathering Meeting 3, Virtual Meeting, July 15, 2022.

Stallings, D., S. Iyer, and R. Hernandez. 2018. "National Diversity Equity Workshops: Advancing Diversity in Academia." *ACS Symposium Series, Vol. 1277, National Diversity Equity Workshops in Chemical Sciences (2011–2017),* 1–19. doi.org/10.1021/bk-2018-1277.ch001.

Subcommittee on Quantum Information Science of the National Science and Technology Council. 2022. *Quantum Information Science and Technology Workforce Development National Strategic Plan*. quantum.gov/wp-content/uploads/2022/02/qist-natl-workforce-plan.pdf.

Swed, K. 2022. "Top Careers in Computer Science." Last modified December 22, 2022. computerscience.org/careers/.

Tahan, C. 2022. "Pursuing Quantum Information Science Together." Information-Gathering Meeting 3, Virtual Meeting, July 15, 2022.

Takeshita, T., N. C. Rubin, Z. Jiang, E. Lee, R. Babbush, and J. R. McClean. 2020. "Increasing the Representation Accuracy of Quantum Simulations of Chemistry Without Extra Quantum Resources." *Physical Review X* 10(1):011004. doi.org/10.1103/physrevx.10.011004.

U.K. Department for Science, Innovation, and Technology. 2023. *National Quantum Strategy*. London: Open Government License. https://assets.publishing.service.gov.uk/government/uploads/system/uploads/attachment_data/file/1142942/national_quantum_strategy.pdf.

United Kingdom of Great Britain and Northern Ireland and the United States of America. 2021. "Joint Statement of the United Kingdom of Great Britain and Northern Ireland and the United States of America on Cooperation in Quantum Information Sciences and Technologies." https://www.gov.uk/government/publications/uk-us-joint-statement-on-cooperation-in-quantum-information-sciences-and-technologies/joint-statement-of-the-united-kingdom-of-great-britain-and-northern-ireland-and-the-united-states-of-america-on-cooperation-in-quantum-information-sci.

U.S. Department of Energy Office of Science. n.d. "Community College Internships (CCI)." Accessed December 30, 2022. science.osti.gov/wdts/cci.

Vogt, N., S. Zanker, J. M. Reiner, M. Marthaler, T. Eckl, and A. Marusczyk. 2021. "Preparing Ground States with a Broken Symmetry with Variational Quantum Algorithms." *Quantum Science and Technology* 6(3):035003. https://doi.org/10.1088/2058-9565/abe568.

Zapata Computing. n.d. "The Quantum State." Accessed April 12, 2023. https://www.zapatacomputing.com/blog/?cat=169.

Appendix A

Committee Members' Biographical Sketches

Theodore G. Goodson, III is the Richard Barry Bernstein Collegiate Professor of Chemistry at the University of Michigan. He is also a professor in the College of Engineering and Macromolecular Science as well as in applied physics at the University of Michigan. His research is in the area of physical chemistry of novel materials for optical and electronic applications. His laboratory investigates the properties of nanomaterials with time-resolved, nonlinear, and quantum optical methods. These studies have probed optical effects in organic chromophores and polymers as well as inorganic metal cluster systems. He served on a previous National Research Council committee concerning laboratory safety in the chemical sciences. A native of Indianapolis, Indiana, Goodson studied chemistry at Wabash College for his undergraduate degree and at the University of Nebraska-Lincoln for his Ph.D.

David D. Awschalom is the Liew Family Professor and vice dean of the Pritzker School for Molecular Engineering at the University of Chicago, a senior scientist at Argonne National Laboratory, and director of the Chicago Quantum Exchange. He is also the inaugural director of Q-NEXT, one of the U.S. Department of Energy Quantum Information Science Research Centers. Before arriving in Chicago, he was the director of the California NanoSystems Institute and professor of physics, electrical engineering, and computer engineering at the University of California, Santa Barbara. He served as a research staff and manager at the IBM Watson Research Center. He works in spintronics and quantum information engineering, studying the quantum states of electrons, nuclei, and photons in semiconductors and molecules for quantum information processing. Awschalom received the American Physical Society Oliver Buckley Prize and Julius Edgar Lilienfeld Prize, the European Physical Society Europhysics Prize, the Materials Research Society David Turnbull Award and Outstanding Investigator Prize, the American Association for the Advancement of Science Newcomb Cleveland Prize, the International Magnetism Prize from the International Union of Pure and Applied Physics, and an IBM Outstanding Innovation Award. He is a member of the American Academy of Arts and Sciences, the National Academy of Sciences, the National Academy of Engineering, and the European Academy of Sciences. Professor Awschalom received his B.Sc. in physics from the University of Illinois at Urbana-Champaign and his Ph.D. in experimental physics from Cornell University.

Ryan J. Babbush is a researcher leading the quantum algorithms team at Google Quantum AI (Google's flagship quantum computing effort). He was the first intern of this group back in 2013 when it was four people; it has now grown to more than 100 scientists. He joined the effort full time in 2015 after finishing a Ph.D. focusing on quantum algorithms for chemistry. In 2021, he received the TR35 award from *MIT Tech Review* for his leading work in this area. Babbush received his Ph.D. in chemical physics from Harvard University.

Lawrence W. Cheuk is an assistant professor in the Department of Physics at Princeton University. His research expertise is in using atoms and laser-cooled molecules at ultracold temperatures for quantum simulation of many-body systems and quantum information processing. Cheuk is a member of the American Physics Society (APS). His relevant honors include being a finalist for the APS Division of Atomic and Molecular and Optical Physics thesis award and being awarded a Max-Planck-Harvard Quantum Optics Center postdoctoral fellowship. Cheuk obtained his Ph.D. in physics in 2017 from the Massachusetts Institute of Technology, where he worked on ultracold fermionic atomic gases.

Scott K. Cushing is an assistant professor at the California Institute of Technology, with a multidisciplinary background spanning chemistry, materials science, and physics. His research focuses on the creation of new scientific instrumentation that can translate quantum phenomena to practical devices and applications, including attosecond X-ray, ultrafast transmission electron microscopy/electron energy loss spectroscopy, and entangled photon studies. The Cushing Lab is pioneering entangled photon microscopy and spectroscopy methods that can measure chemical systems without perturbation, while also obtaining new information about classical and quantum correlations. The Cushing Lab has developed new entangled photon sources for chemical spectroscopy and is working toward completely on-chip, portable entangled photon spectroscopy to explore molecular systems such as qubits, quantum optical components, and quantum sensors. Cushing is also exploring if excited-state entanglement can lead to new reaction pathways. He has been awarded Department of Energy, Air Force Office of Scientific Research, Rose Hill, and American Chemical Society–related Early Career awards, among others. Cushing also leads a four-institution, seven-person collaboration exploring nanographene-molecular qubit systems. Cushing received his Ph.D. in physics from West Virginia University.

Natia L. Frank is an associate professor at the University of Nevada, Reno. She began her independent career in 2000 as an assistant professor at the University of Washington-Seattle in the study of multifunctional magnetic materials for spintronics and biosensing. In 2005, she was recruited as a Canada Research Chair Tier II in Multifunctional Materials Chemistry at the University of Victoria where she developed optically switchable spin-based qubits for quantum science. Her primary expertise is at the interface of organic chemistry, inorganic chemistry, spin-based materials, and photochemistry/electron transfer theory, which allows her to be well situated to address current challenges in molecular quantum information science: the design of molecular qubits with long decoherence times, multiqubit arrays, and qubits/qudits that can respond to external stimuli for quantum computing and sensing. Frank currently serves on two funded Department of Energy, Energy Frontiers Research Center advisory boards in quantum science and on the American Chemical Society-Petroleum Research Fund Advisory Board, and has served on numerous National Science Foundation funding panels in quantum relevant areas. Frank received her bachelor's degree with honors from Bard College in chemistry, math, and music; an M.Sc. in inorganic chemistry at the University of Wisconsin-Madison; and a Ph.D. at the University of California, San Diego in organic chemistry.

Danna E. Freedman is the Keyes Professor of Chemistry at the Massachusetts Institute of Technology. Freedman began her independent career as assistant professor at Northwestern University, where she was promoted to associate professor with tenure and subsequently to full professor. The Freedman Group applies the atomistic control inherent to synthetic chemistry to address fundamental questions in physics. Within this paradigm, the group is creating the next generation of materials for quantum information and harnessing high pressure to synthesize new emergent materials. Notable accomplishments include realizing millisecond coherence times in molecular qubits and developing molecular analogs of nitrogen-vacancy centers—enabling optical readout of spin information in molecules. Freedman's research has been recognized by a number of awards including the American Chemical Society Award in Pure Chemistry, the Presidential Early Career Award for Scientists and Engineers, the Camille Dreyfus Teacher-Scholar Award, and a National Science Foundation CAREER award. Freedman received her A.B. from Harvard University and her Ph.D. from the University of California, Berkeley, where she studied magnetic anisotropy in molecules.

Sinéad M. Griffin is a staff scientist in the Materials Sciences Division and Molecular Foundry at Lawrence Berkeley National Laboratory. Her work uses first-principles calculations and phenomenological models to describe and predict novel and enhanced phenomena in quantum materials. Applications of her work range from next-generation quantum sensors for dark matter detection and coherence enhancement in materials for quantum information science, to predicting new forms of topological and multiorder quantum matter. Her awards include the Swiss Physical Society's Award in General Physics and the Material and Processes Awards of the ETH Zürich. Dr. Griffin obtained a B.A. (mod) hons in theoretical physics at Trinity College Dublin, followed by an M.Sc. D.I.C. in quantum field theory from Imperial College London. Her doctorate work was carried out at the University of California, Santa Barbara, and ETH Zürich, where she received her Dr.Sc. in materials physics.

Stephen O. Hill currently holds the title of professor of physics at Florida State University (FSU) while also serving as director of the Electron Magnetic Resonance Facility at the U.S. National High Magnetic Field Laboratory (NHMFL). He previously held postdoctoral positions at Boston University and at the NHMFL, then took up faculty positions at Montana State University and the University of Florida before moving back to FSU in 2008. Hill has 30 years of experience performing microwave and far-infrared magneto-optical spectroscopy of materials in high magnetic fields, using a wide array of compact radiation sources and measurement techniques. Through this work, he has gained an international reputation in the spectroscopy of low-dimensional conducting, superconducting, and magnetic systems in high magnetic fields, including significant technique development. Hill's recent research has focused on fundamental studies of quantum phenomena in molecule-based magnets, as well as structure–property relationships in a variety of inorganic coordination compounds. Hill was elected fellow of the American Physical Society (APS) in 2014; he won the Silver Medal for Instrumentation from the International EPR Society in the same year and then served in the chair's line of the APS Topical Group on Magnetism from 2016 to 2020. Hill received both his bachelor's and Ph.D. degrees from the University of Oxford in the United Kingdom.

Hongbin Liu is a senior researcher at Microsoft Quantum. He currently leads Microsoft's effort to understand how quantum computers could benefit computational chemistry applications. As a chemist by training, he is driving innovations with modern and future computer architectures to help address a wide variety of chemistry problems, from catalysis to photovoltaics. His research interests range from developing novel classical electronic structure methods and improving quantum algorithms for chemistry simulations to building hybrid classical–quantum computational chemistry solutions. He has developed multiple quantum chemistry packages, including Gaussian and Chronus Quantum. Liu obtained his Ph.D. from the University of Washington in 2019, focusing on developing electronic structure methods to understand excited states and relativistic effects of molecules and materials.

Marilu Perez Garcia is a scientist at Ames Laboratory, a national laboratory operated by Iowa State University. She has helped organize conferences such as the Graduate Minority Assistantship Program Research Conference and was in the inaugural class of the Critical Materials Institute (CMI) Emerging Leaders Academy. She currently leads a project within CMI studying rare-earth element chemistry using a combination of computational methods: molecular mechanics, quantum mechanics, and machine learning. She also leads a study on understanding the quantum properties of organolanthanide molecules in a combined synthesis, magnetic analysis, and theoretical chemistry collaboration. Perez Garcia graduated from Idaho State University as an American Chemical Society Scholar and was a George Washington Carver Fellow at Iowa State University. She graduated with research and teaching excellence awards and, as a postdoc, served on the American Chemical Society Graduate Education Advisory Board. Perez Garcia received her Ph.D. from Iowa State University where she used solid-state nuclear magnetic resonance and isotopically labeled plant cell walls to characterize the molecular interactions between intact plant cell wall molecules and computational methods to characterize highly concentrated particles of volatile organic molecules as they exist in the atmosphere.

Brenda M. Rubenstein is currently the Joukowsky Family Assistant Professor of Chemistry at Brown University. Her research focuses on developing new electronic structure methods and alternative computing paradigms. Prior to arriving at Brown, she was a Lawrence Distinguished Postdoctoral Fellow at Lawrence Livermore National Laboratory. She is the recipient of a Camille and Henry Dreyfus Teacher-Scholar Award, the Cottrell Teacher-Scholar Award, a Sloan Research Fellowship, and the Department of Energy Computational Science Graduate Fellowship. She received her Sc.B. in chemical physics and applied mathematics at Brown University, her M.Phil. in computational chemistry while a Churchill Scholar at the University of Cambridge, and her Ph.D. in chemical physics at Columbia University.

Eric J. Schelter is a professor of chemistry at the University of Pennsylvania and the director of the National Science Foundation Center for Sustainable Separations of Metals. Schelter's research group studies synthetic inorganic chemistry, especially of the lanthanide and actinide elements, to address problems in chemical separations, electronic structure and bonding, photophysics and photocatalysis, bioinorganic chemistry, magnetism, and quantum materials. Schelter has been awarded a Department of Energy Early Career Research Program Award, the Research Corporation for Science Advancement Cottrell Scholar, an American Chemical Society Harry Gray Award for Creative Work in Inorganic Chemistry by a Young Investigator, the U.S. Environmental Protection Agency Green Chemistry Challenge Award, and the American Chemical Society Inorganic Chemistry Lectureship. He received his Ph.D. in inorganic chemistry from Texas A&M University and was a postdoc at Los Alamos National Laboratory.

Michael R. Wasielewski is currently the Clare Hamilton Hall Professor of Chemistry at Northwestern University; executive director of the Institute for Sustainability and Energy at Northwestern; and director of the Center for Molecular Quantum Transduction, a U.S. Department of Energy, Energy Frontier Research Center. His research has resulted in more than 730 publications and focuses on light-driven processes in molecules and materials, artificial photosynthesis, molecular electronics, quantum information science, ultrafast optical spectroscopy, and time-resolved electron paramagnetic resonance spectroscopy. His honors and awards include membership in the National Academy of Sciences and the American Academy of Arts & Sciences; the Bruker Prize in Electron Paramagnetic Spectroscopy (EPR); the Josef Michl American Chemical Society Award in Photochemistry; the International EPR Society Silver Medal in Chemistry; the Royal Society of Chemistry Physical Organic Chemistry Award; the Chemical Pioneer Award of the American Institute of Chemists; the Royal Society of Chemistry Environment Prize; the Humboldt Research Award; the Arthur C. Cope Scholar Award of the American Chemical Society; the Porter Medal for Photochemistry; and the James Flack Norris Award in Physical Organic Chemistry of the American Chemical Society. He received his B.S., M.S., and Ph.D. degrees from the University of Chicago.

Damian Watkins is currently chief of research and innovation for Aperio Global. He is a recognized expert in network protocol processing and analytic development for Department of Defense (DOD) agencies. His software development background includes more than 15 years of experience in Java and C++ development of data-driven analytics. He also provides consulting to the development of requirements of analytics and provides analysis of network protocols. He has written custom algorithms on live data to analyze network protocol behavior. He has also written Hadoop MapReduce–based analytics to support customer questions. He possesses a Cyber Security Analysis (CySA+) Certification as well as hands-on experience with malware analysis, signals intelligence (SIGINT), and cyber metadata generation and processing. He is currently developing artificial intelligence applications on SIGINT data using TensorFlow for intelligence/DOD customers. Watkins was awarded Patent No. 11037073, "Data Analysis System Using Artificial Intelligence." He was also a finalist for the National Aeronautics and Space Administration Science Mission Directorate Entrepreneurs Challenge in the area of Quantum Sensing for Dark Energy. Watkins received his doctorate in electrical and computer engineering from Morgan State University.

Appendix B

Agendas for Information-Gathering Meetings 1–3

Information-Gathering Meeting 1
March 11, 2022, 2:15–5:30 pm EST

2:15 Welcome and Introductions
Theodore Goodson, Committee Chair
The Richard Bernstein Collegiate Professor of Chemistry
University of Michigan

Session I. History and Outlook on Scalability

2:20 Opportunities for Chemistry in Quantum Information Science
David Awschalom
Liew Family Professor
University of Chicago
QIS Group Leader, Materials Science Division
Argonne National Laboratory

2:50 Q&A with Committee
Moderated by Brenda Rubenstein
Joukowsky Family Assistant Professor of Chemistry
Brown University

3:05 Molecular Systems for Quantum Information Science
Michael Wasielewski
Clare Hamilton Hall Professor of Chemistry
Northwestern University

3:35 Q&A with Committee
Moderated by Stephen Hill
Professor of Physics
Director of Electron Magnetic Resonance User Program
Department of Physics, Florida State University
National High Magnetic Field Laboratory

3:50 Break

Session II. Addressability, Scalability, and Molecular Architecture

4:00 Introduction
Theodore Goodson, Committee Chair
The Richard Bernstein Collegiate Professor of Chemistry
University of Michigan

4:05 Magnetic Molecules in Hybrid Architectures for QIS
Roberta Sessoli
Professor of General and Inorganic Chemistry
University of Florence

4:20 Q&A with Committee
Moderated by Danna Freedman
F. G. Keyes Professor of Chemistry
Massachusetts Institute of Technology

4:35 Quantum Simulations and Quantum Information: A Key Feedback Loop
Giulia Galli
Liew Family Professor of Electronic Structure and Simulations
University of Chicago

4:50 Q&A with Committee
Moderated by Ryan Babbush
Head of Quantum Algorithms
Staff Research Scientist
Google, LLC

5:05 Closing Remarks
Theodore Goodson, Committee Chair
The Richard Bernstein Collegiate Professor of Chemistry
University of Michigan

5:10 Adjourn Open Session

Information-Gathering Meeting 2
June 10, 2022, 2:15–6:00 pm EST

2:15 Welcome and Introductions
Theodore Goodson, Committee Chair
The Richard Bernstein Collegiate Professor of Chemistry
University of Michigan

Session I. Instrumentation and Tools to Study Chemical Approaches for QIS Applications

2:20 Probing Coherence in Nonlinear Molecular Spectroscopy with Quantum Light
Shaul Mukamel
Distinguished Professor of Chemistry and of Physics and Astronomy
University of California, Irvine

2:50 Q&A with Committee
Moderated by Lawrence Cheuk
Assistant Professor of Physics
Princeton University

3:05 Spectroscopy Methods for Molecular Quantum Spin Science
Stephen Hill
Professor of Physics
Director of Electron Magnetic Resonance Facility
Florida State University

3:35 Q&A with Committee
Moderated by Scott Cushing
Assistant Professor of Chemistry
California Institute of Technology

3:50 Break

Session II. Biological Applications and Computational Applications for Quantum Information Science

4:00 Introduction
Theodore Goodson, Committee Chair
The Richard Bernstein Collegiate Professor of Chemistry
University of Michigan

4:05 Chemical Design Principles for Controlling Quantum Dynamics in Complex Environments: Quantum *IN* Biology and Quantum *FOR* Biology
Greg Engel
Professor of Chemistry and Molecular Engineering
University of Chicago

4:20 Q&A with Committee
Moderated by Eric Schelter
Professor of Chemistry
University of Pennsylvania

4:35 Chemical Approaches to Quantum Information Science
Danna Freedman
F. G. Keyes Professor of Chemistry
Massachusetts Institute of Technology

4:50 Q&A with Committee
Moderated by Hongbin Liu
Senior Researcher
Microsoft Quantum

5:05 Quantum for Chemistry & Chemistry for Quantum: A Theory Perspective
Prineha Narang
Assistant Professor
John A. Paulson School of Engineering and Applied Sciences
Harvard University

5:20 Q&A with Committee
Moderated by Sinéad Griffin
Staff Scientist
Lawrence Berkeley National Laboratory
Materials Sciences Division and Molecular Foundry

5:35 Closing Remarks
Ted Goodson, Committee Chair
The Richard Bernstein Collegiate Professor of Chemistry
University of Michigan

5:40 Adjourn Open Session

Information-Gathering Meeting 3
July 15, 2022, 2:15–5:40 pm EST

2:15 Welcome and Introductions
Theodore Goodson, Committee Chair
The Richard Bernstein Collegiate Professor of Chemistry
University of Michigan

Session I. Quantum Workforce Development

2:20 Pursuing Quantum Information Science Together
Charles Tahan
Assistant Director for Quantum Information Science
Director of National Quantum Coordination Office
Office of Science and Technology and Policy

2:35 Q&A with Committee
Moderated by Theodore Goodson
The Richard Bernstein Collegiate Professor of Chemistry
University of Michigan

2:50 QIS-Chemistry and Education: Barriers to Entry from an Academic Perspective
Rigoberto Hernandez
Director of the Open Chemistry Collaborative in Diversity Equity
Gompf Family Professor of Chemistry
Department of Chemistry
John Hopkins University

3:05 Q&A with Committee
Moderated by Marilu Perez Garcia
Scientist II
Critical Materials Institute
Iowa State University
Ames Laboratory

3:20 Economic Development and Barriers to Entry at the Interface of Chemistry and QIS
Joseph Broz
Vice President Quantum Growth and Markets
IBM in Yorktown

3:35 Q&A with Committee
Moderated by David Awschalom
Liew Family Professor in Spintronics and Quantum Information
Pritzker School of Molecular Engineering
The University of Chicago

3:50 Break

Session II. Economic Development in QIS & Chemistry

4:00 Introduction
Theodore Goodson, Committee Chair
The Richard Bernstein Collegiate Professor of Chemistry
University of Michigan

4:05 Theoretical Chemistry to Enhance Quantum Information Science: Education, Funding, Research
Keeper Sharkey
Founder and CEO
ODE

4:20 Q&A with Committee
Moderated by Theodore Goodson
The Richard Bernstein Collegiate Professor of Chemistry
University of Michigan

4:35 Disrupting Computational Chemistry with Quantum Computing
Hongbin Liu
Quantum Solution Lead
Microsoft Quantum

4:50 Q&A with Committee
Moderated by Michael Wasielewski
Clare Hamilton Hall Professor of Chemistry
Northwestern University

5:05 Q for Quantum Chemistry
Arman Zaribafiyan
CEO & Founder
Good Chemistry

5:20 Q&A with Committee
Moderated by Damian Watkins
Chief of Research and Innovation
Aperio Global

5:35 Closing Remarks
Theodore Goodson, Committee Chair
The Richard Bernstein Collegiate Professor of Chemistry
University of Michigan

5:40 Adjourn Open Session

Appendix C

Multidisciplinary Centers Established under the National Quantum Initiative Act (NQIA) or National Defense Authorization Act (NDAA) and Related QIS Programs

TABLE C-1 Multidisciplinary Centers Established under the National Quantum Initiative Act (NQIA) or National Defense Authorization Act (NDAA)

	Acronym	Full Name of Center	Lead Laboratory *or University*	Funding over Five Years (millions)	Website
Department of Energy: The National Quantum Information Science Research Centers *(Authorized under NQIA)*	C^2QA	Co-design Center for Quantum Advantage	Brookhaven National Laboratory	$115[a]	bnl.gov/quantumcenter
	QSA	Quantum Systems Accelerator	Lawrence Berkeley National Laboratory	$115[a]	quantumsystemsaccelerator.org
	QSC	Quantum Science Center	Oak Ridge National Laboratory	$115[a]	qscience.org
	Q-NEXT	Next Generation Quantum Science and Engineering	Argonne National Laboratory	$115[a]	q-next.org
	SQMS	Superconducting Quantum Materials and Systems Center	Fermi National Accelerator Laboratory	$115[a]	sqms.fnal.gov
National Science Foundation: Quantum Leap Challenge Institutes *(Authorized under NQIA)*	Q-SEnSE	Quantum Systems through Entangled Science and Engineering	University of Colorado, Boulder	$25[a]	colorado.edu/research/qsense
	HQAN	Hybrid Quantum Architectures and Networks	University of Illinois —Illinois Quantum Information Science and Technology Center (IQUIST)	$25[a]	hqan.illinois.edu/
	CIQC	Challenge Institute for Quantum Computation	University of California, Berkeley	$24.9[a]	ciqc.berkeley.edu
	QuBBE	Quantum Sensing for Biophysics and Bioengineering	University of Chicago	$25[b]	qubbe.uchicago.edu
	RQS	Institute for Robust Quantum Simulation	University of Maryland, College Park	$25[b]	https://rqs.umd.edu/
Department of Defense: Quantum Information Science Research Centers *(Authorized under NDAA)*	LQC	LPS (Laboratory Physical Sciences) Qubit Collaboratory, U.S. Army Research Office	University of Maryland, College Park		qubitcollaboratory.org
	NRL	Naval Research Laboratory	U.S. Naval Research Laboratory		nrl.navy.mil/Our-Work/Areas-of-Research/Quantum-Research
	AFRL	Air Force Research Laboratory	Air Force Research Laboratory		afresearchlab.com/technology/quantum/

[a]Period of performance start date is 2020.
[b]Period of performance start date is 2021.

TABLE C-2 Large-Group Research and Development Efforts Related to QIS Topics by the National Science Foundation (NSF) and the Department of Energy (DOE)

Federal Agency	Acronym	Full Name of Center or Program	Lead University	Website
NSF	NSF Q-AMASE-i	Quantum Foundry for Accelerated Development of Quantum Materials	University of California, Santa Barbara	quantumfoundry.ucsb.edu
	NSF ERC-CQN	Engineering Research Center for Quantum Networks	University of Arizona	cqn-erc.org
	JILA	Physics Frontiers Center at JILA	University of Colorado, Boulder	jila-pfc.colorado.edu
	JQI	Joint Quantum Institute	University of Maryland, College Park	jqi.umd.edu
	IQIM	Institute for Quantum Information and Matter	California Institute of Technology	iqim.caltech.edu
	CUA	Center for Ultracold Atoms	Massachusetts Institute of Technology	cua.mit.edu
	CQuIC	Center for Quantum Information and Control	University of New Mexico	cquic.unm.edu/
	ITAMP	Institute for Theoretical Atomic and Molecular Physics	Harvard	pweb.cfa.harvard.edu
	CIQM	Center for Integrated Quantum Materials	Harvard	ciqm.harvard.edu
	EPIQC	Enabling Practical Scale Quantum Computing	The University of Chicago	epiqc.cs.uchicago.edu
	STAQ	Software-Tailored Architecture for Quantum co-design	Duke	staq.pratt.duke.edu
	QISE-NET	Quantum Information Science and Engineering Network	The University of Chicago	qisenet.uchicago.edu
DOE	ASCR	Advanced Scientific Computing Research QIS Program	N/A	https://science.osti.gov/ascr
	BER	Biological and Environmental Research	N/A	https://science.osti.gov/Initiatives/QIS/Program-Offices-QIS-Pages
	BES	Basic Energy Science QIS Program	N/A	science.osti.gov/bes/Research/qis
	FES	Fusion Energy Sciences	N/A	https://science.osti.gov/Initiatives/QIS/Program-Offices-QIS-Pages%20
	HEP	High Energy Physics QIS Program	N/A	science.osti.gov/hep/Research/Quantum-Information-Science-QIS

Appendix D

Programs Offering QISE (Quantum Information Science and Engineering)-Related Degrees or Certificates

QISE Minor

- Washington University in St. Louis; /bulletin.wustl.edu/undergrad/engineering/electrical-and-systems/minor-quantum-engineering/
- Colorado University; catalog.colorado.edu/undergraduate/colleges-schools/engineering-applied-science/programs-study/electrical-computer-energy-engineering/quantum-engineering-minor/
- California Institute of Technology; iqim.caltech.edu/quantum-science-and-engineering-minor/
- Saint Anslem College; www.anselm.edu/majors-minors/quantum-information-science

QISE Certificate

- Washington University in St. Louis; https://www.quantumx.washington.edu/training/aqet/
- University of Texas at Austin; catalog.utexas.edu/undergraduate/natural-sciences/minor-and-certificate-programs/
- Drexel University; https://drexel.edu/coas/academics/graduate-programs/physics/certificate-quantum-technology/

QISE Masters

- University of California, Los Angeles; https://qst.ucla.edu/program.html

All Three Programs Available

- Colorado School of Mines, Engineering; https://quantum.mines.edu/program/
- University of Arizona, Optical Science, Quantum Information; https://www.optics.arizona.edu/prospective-students/graduate-programs/ms-optical-sciences-quantum-information

Appendix E

Acronyms and Glossary

LIST OF ACRONYMS

2D	two dimensional
2DES	two-dimensional electronic spectroscopy
2DIR	two-dimensional infrared
3D	three dimensional
4D	four dimensional
AC	alternating current
AMO	atomic molecular and optical
BBO	β-barium borate
BI	Boehringer Ingelheim
CC	coupled cluster theory
CCI	Community College Internship
CCNOT	controlled-controlled NOT gate
CISS	chirality-induced spin selectivity
CNOT	controlled NOT gate
CQC	Cambridge Quantum Computing
CW	continuous wave
DEER	double electron-electron resonance
DFT	density functional theory
DMRG	density matrix renormalization group
DNP	dynamic nuclear polarization
DOD	Department of Defense
DOE	Department of Energy
DQ	double quantum

EDM	electric dipole moment
ELDOR	electron-electron double resonance
EMF	endohedral metallofullerene
ENDOR	electron-nuclear double resonance
EPR	electron paramagnetic resonance
ES	electronic spectroscopy
ESEEM	electron-spin-echo envelope modulation
ESR	electron spin resonance
ET	electron transfer
ETPA	entangled two-photon absorption
FAIR	Findability, Accessibility, Interoperability, and Reusability
FCI	full configuration interaction
FeMoCo	FeMo cofactor
FMO	Fenna–Matthews–Olson
FTQC	fault-tolerant quantum computing
HOM	Hong–Ou–Mandel
HR	hyper-Raman
HX	heavy-hole
INS	inelastic neutron scattering
IR	infrared
IRSC	Indian River State College
KDP	potassium dideuterium phosphate
L3C	low-profit limited liability company
LX	light-hole
MMCC	method of moments coupled cluster
MOF	metal–organic framework
MRFM	magnetic resonance force microscopy
μSR	muon spin relaxation
NDAA	National Defense Authorization Act
NHMFL	National High Magnetic Field Laboratory
NIR	near-infrared
NISQ	noisy intermediate-scale quantum
NIST	National Institute of Standards and Technology
NMR	nuclear magnetic resonance
NP	nondeterministic polynomial time
NQI	National Quantum Initiative
NQIA	National Quantum Initiative Act
NSF	National Science Foundation
NSMM	near-field scanning microwave microscopy
NSOM (SNOM)	near-field scanning optical microscopy
NV	nitrogen vacancy

OCC	optical cycling center
ODMR	optically detected magnetic resonance
OOP-ESEEM	out-of-phase electron-spin-echo envelope modulation
OXIDE	Open Chemistry Collaborative in Diversity Equity
PBS	polarizing beam splitter
PI	principal investigator
QED-C	Quantum Economic Development Consortium
QIS	quantum information science
QISE	quantum information science and engineering
QIST	quantum information science and technology
QMA-Hard	Quantum Merlin Arthur-Hard
QMC	quantum Monte Carlo
QML	quantum machine learning
Q-NEXT	Next Generation Quantum Science and Engineering
QPE	quantum phase estimation
QTEdu	Quantum Technology Education
QTM	quantum tunneling of magnetization
R&D	research and development
RF	radiofrequency
SBIR	Small Business Innovation Research
SF	singlet fission
SMM	single-molecule magnet
SNL	shot noise limit
SNR	signal-to-noise ratio
SPDC	spontaneous parametric down-conversion
SQP	spin qubit pair
STEM	science, technology, engineering, and mathematics
STM	scanning tunneling microscope
STTR	Small Business Technology Transfer
T_1	spin-lattice relaxation time
T_2	transverse spin relaxation time
TEM	transmission electron microscopy
TLS	two-level system
TPA	two-photon absorption
URPOC	underrepresented people of color
VQE	variational quantum eigensolver
ZFS	zero-field splitting
ZPL	zero-phonon line
ZQ	zero quantum

GLOSSARY

Bottom-up chemical synthesis - Creation of atomistically precise and reproducible chemical structures using molecular design, such as classic organic or inorganic methodologies, or atom-by-atom growth through methods such as chemical vapor deposition.

Chemical informatics - The generation, study, and application of chemical information to make data-driven decisions for the understanding, prediction, and design of new molecules and materials.

Chirality-induced spin selectivity - A strategy that is used in chiral molecular systems to manipulate the direction of the electron spin relative to the electron transfer displacement vector. Depending on the chirality of the molecule, the electron spin can align either parallel or antiparallel to the displacement vector.

Clock Transition - A transition between two states where the transition frequency is unchanged by environmental perturbations, such as changes in magnetic or electric field. Shiddiq, M., D. Komijani, Y. Duan, A. Gaita-Ariño, E. Coronado, and S. Hill. 2016. "Enhancing Coherence in Molecular Spin Qubits via Atomic Clock Transitions." *Nature* 531(7594):348–351. doi.org/10.1038/nature16984.

Coherence - Classical: Temporal—the time over which the wave phase remains stable; Quantum: A superposition of two states where the resulting state may be described as a linear combination of the two states. Coherence time is a measure of how long a qubit stores quantum information without succumbing to noise. Fox, M. 2006. *Quantum Optics: An Introduction*. New York: Oxford University Press.

Database - A usually large collection of data organized especially for rapid search and retrieval (as by a computer). Merriam Webster.

Decoherence - Decoherence can be viewed as the loss of information from a system into the environment, degradation of the quantum state, or collapse of the superposition state. Decoherence is often induced by noise. Nielsen, M. A., and I. L. Chuang. 2011. *Quantum Computation and Quantum Information*, 10th ed. New York: Cambridge University Press.

Defect centers - Spin-bearing defects hosted in extended solids, including, but not limited to, substitutional defects, interstitial defects, dopant atoms, vacancy-dopant pairs, or divacancy defects. Wolfowicz, G., F. J. Heremans, C. P. Anderson, S. Kanai, H. Seo, A. Gali, G. Galli, and D. D. Awschalom. 2021. "Quantum Guidelines for Solid-State Spin Defects." *Nature Reviews Materials* 6(10):906–925. doi.org/10.1038/s41578-021-00306-y.

Electronic structure algorithms - These are algorithms (on either quantum or classical computers) that aim to solve or approximate solutions to the electronic structure problem. The electronic structure problem is the problem of solving for the energy (and sometimes wave functions) of electrons interacting with each other and in the Coulomb field of nuclei.

Embedding - Combining high-accuracy calculations (e.g., quantum simulations) with less accurate but more universal calculations to enable the exploration of complex phenomena and systems. Examples of embedding include density matrix embedding theory, dynamical mean field theory, and the concept of a complete active space. Wouters, S., C. A. Jiménez-Hoyos, Q. Sun, and G. K.-L. Chan. 2016. "A Practical Guide to Density Matrix Embedding Theory in Quantum Chemistry." *Journal of Chemical Theory and Computation* 12(6):2706–2719. https://doi.org/10.1021/acs.jctc.6b00316.

High-throughput evaluations - Rapidly studying a large set of targets (e.g., molecules) through the use of automation.

Hybrid quantum architectures - This can refer to quantum technologies that make use of disparate quantum technologies (e.g., a system of nitrogen-vacancy center sensors that are coupled to superconducting microwave cavities), or it can refer to systems of processing information that make use of both quantum and conventional classical hardware (e.g., quantum variational algorithms, which use both a quantum computer and a classical computer). McClean, J. R., J. Romero, R. Babbush, and A. Aspuru-Guzik. 2016. "The Theory of Variational Hybrid Quantum-Classical Algorithms." *New Journal of Physics* 18(2):023023. doi.org/10.1088/1367-2630/18/2/023023.

Ligand field - The coordination environment around a metal ion is described by a combination of electrostatic interactions (i.e., crystal field theory) and ligand orbitals (i.e., molecular orbital theory). Figgis, B. N. and M. A. Hitchman. 2000. *Ligand Field Theory and Its Applications*. New York: Wiley-VCH.

Noise models - Models of how noise acts on a quantum system. These range from very idealized/simplistic and easy to simulate to extremely difficult to simulate and also more representative of the physics of a particular noisy device.

Optical cycling - Repeated photon scattering (i.e., relaxation to the starting state) upon continuous excitation between two states, either resonantly or off-resonantly.

Quantum computing - Computation that uses quantum phenomena, such as superposition and entanglement, to solve problems that may be challenging or intractable for classical computing algorithms. Nielsen, M. A., and I. L. Chuang. 2011. *Quantum Computation and Quantum Information, 10th ed.* New York: Cambridge University Press.

Quantum entanglement - Interaction between two (or more) particles or qubits where the resulting state cannot be described as a product of linear combinations of each individual particle. Entanglement is a type of nonclassical correlation that can be inefficient to simulate classically. Nielsen, M. A., and I. L. Chuang. 2011. *Quantum Computation and Quantum Information, 10th ed.* New York: Cambridge University Press.

Quantum error mitigation - Distinct from quantum error correction, quantum error mitigation is a large collection of techniques that are employed with the purpose of suppressing the effects of noise on a quantum experiment or computation. Usually, these techniques boil down to capabilities that are (at best) equivalent to being able to detect when an error occurs so that one can postselect on data corresponding to an error-free realization. But since devices with finite error rates per operation compound errors exponentially, these techniques are still intrinsically unscalable. Techniques that fundamentally change this scaling are error correction. Cai, Z., R. Babbush, S. C. Benjamin, S. Endo, W. J. Huggins, Y. Li, J. R. McClean, and T. E. O'Brien. 2022. "Quantum Error Mitigation." *ArXiv*. https://arxiv.org/abs/2210.00921.

Quantum information transduction - Coherent conversion of quantum information from one energy scale or storage medium to another (e.g., microwave-to-optical energy conversion). Lauk, N., N. Sinclair, S. Barzanjeh, J. P. Covey, M. Saffman, M. Spiropulu, and C. Simon. 2020. "Perspectives on Quantum Transduction." *Quantum Science and Technology* 5(2):020501. doi.org/10.1088/2058-9565/ab788a.

Quantum sensing - Use of a quantum system, quantum properties, or quantum phenomena to perform a measurement of a physical quantity. Degen, C. L., F. Reinhard, and P. Cappellaro. 2017. "Quantum Sensing." *Reviews of Modern Physics* 89:035002. doi.org/10.1103/RevModPhys.89.035002.

Sampling approaches - For quantum computing, sampling approaches refer to a method of estimating quantities from a wave function using repeated measurements of local observables. This method is different from a "quantum" approach such as quantum phase estimation.

Scalable quantum architecture - An architecture for quantum computing that can scale to much larger sizes and, ideally, scale to the point of realizing scientifically or commercially valuable computations. Here "architecture" refers to a complete specification for how an information processing system functions. For example, in an error-corrected quantum computer, it would refer not only to the qubit hardware, the control electronics, and the like but also to the way error correction is realized and the classical co-processing that is used to decode the error-correction procedures, etc.